Bioinorganic Electrochemistry

Bioinorganic Electrochemistry

Edited by

Ole Hammerich
University of Copenhagen, Denmark

and

Jens Ulstrup
Technical University of Denmark

Springer

A C.I.P. Catalogue record for this book is available from the Library of Congress.

ISBN 978-1-4020-6499-9 (HB)
ISBN 978-1-4020-6500-2 (e-book)

Published by Springer,
P.O. Box 17, 3300 AA Dordrecht, The Netherlands.

www.springer.com

Printed on acid-free paper

Contents

Contributing Authors

Tim Albrecht
Department of Chemistry and NanoDTU, Building 207, Technical University of Denmark, Kemitorvet, DK-2800 Kgs. Lyngby, Denmark

Fraser A. Armstrong
Department of Chemistry, Inorganic Chemistry Laboratory, University of Oxford, Oxford, OX1 3QR, UK.

Julea N. Butt
Centre for Metalloprotein Spectroscopy and Biology, School of Chemical Sciences and Pharmacy, University of East Anglia, Norwich, NR4 7TJ, UK.

Qijin Chi
Department of Chemistry and NanoDTU, Building 207, Technical University of Denmark, Kemitorvet, DK-2800 Kgs. Lyngby, Denmark

Alexander R. Dunn
Beckman Institute, MC 139-74, California Institute of Technology, Pasadena, California 91125, USA.

David G. Fernig
The School of Biological Sciences and The Centre for Nanoscale Science, Chemistry Department, University of Liverpool, Liverpool, L69 7ZB, UK.

Harry B. Gray
Beckman Institute, MC 139-74, California Institute of Technology, Pasadena, California 91125, USA.

Wolfgang Haiss
Centre for Nanoscale Science, Chemistry Department, University of Liverpool, Liverpool, L69 7ZD, UK.

Corinna R. Hess
Beckman Institute, MC 139-74, California Institute of Technology, Pasadena, California 91125, USA.

Michael G. Hill
Occidental College, Department of Chemistry, Los Angeles, CA 90041, USA.

Shana O. Kelley
Boston College, Eugene F. Merkert Chemistry Center, Chestnut Hill, MA 02467, USA.

Alexander M. Kuznetsov
The A.N. Frumkin Institute of Physical Chemistry and Electrochemistry of the Russian Academy of Sciences, Leninskij Prospect 31, 119071 Moscow, Russia

Richard J. Nichols
Centre for Nanoscale Science, Chemistry Department, University of Liverpool, Liverpool, L69 7ZD, UK.

David J. Schiffrin
Centre for Nanoscale Science, Chemistry Department, University of Liverpool, Liverpool, L69 7ZD, UK.

Jens Ulstrup
Department of Chemistry and NanoDTU, Building 207, Technical University of Denmark, Kemitorvet, DK-2800 Kgs. Lyngby, Denmark.

Mårten Wikström
Helsinki Bioenergetics Group, Institute of Biotechnology, University of Helsinki, PB 65, 00014 University of Helsinki, Finland.

Bilha Willner
Institute of Chemistry, The Hebrew University of Jerusalem, Jerusalem 91904, Israel.

Itamar Willner
Institute of Chemistry, The Hebrew University of Jerusalem, Jerusalem 91904, Israel.

Jay R. Winkler
Beckman Institute, MC 139-74, California Institute of Technology, Pasadena, California 91125, USA.

Harm Van Zalinge
Centre for Nanoscale Science, Chemistry Department, University of Liverpool, Liverpool, L69 7ZD, UK.

Jingdong Zhang
Department of Chemistry and NanoDTU, Building 207, Technical University of Denmark, Kemitorvet, DK-2800 Kgs. Lyngby, Denmark.

Preface

Interfacial electrochemistry of redox metalloproteins and DNA-based molecules is presently moving towards new levels of structural and functional resolution. Underlying fundamentals of electron and proton transfer are increasingly well understood although also with new challenges relating to the composite interfacial solid-electrolyte environment. The new inorganic bioelectrochemistry draws further from comprehensive studies of the interfacial environments for retaining biological charge transfer function of these highly sensitive macromolecular systems. Other biotechnology has been the use of mutant proteins, DNA-base variability and *de novo* synthetic metalloproteins.

Physical electrochemistry underwent a remarkable evolution from the 1970's, almost to be likened by a renaissance and prompted by close interaction between electrochemistry and surface physics. The introduction of single-crystal atomically planar electrode surfaces was a major breakthrough and laid the foundation for other new technology such as spectroscopic techniques and statistical mechanical and electronic structure theories. Only slightly later scanning tunneling (STM) and atomic force microscopy (AFM) signalled a lift of interfacial electrochemistry to unprecedented structural resolution. Atomic resolution of pure metal and semiconductor electrode surfaces and at least sub-molecular resolution of electrochemical adsorbates could be achieved, opening new worlds of microscopic structures and processes.

Similar boundary-traversing efforts are now visible in interfacial electrochemistry of proteins and DNA-based molecules. This has led to improved voltammetric sensitivity and structural mapping of the bioelectrochemical solid-liquid interface to single-molecule resolution. It is in fact remarkable that molecules as large and fragile as redox metalloproteins adsorbed on electrode surfaces can be mapped in their functional state by a subtle physical phenomenon, the quantum mechanical tunneling effect. These openings have offered new theoretical challenges for electron tunneling through biological macromolecules, the role of the metal centres, and the finite-size stochastic nature of the systems. Combination of protein and DNA biotechnology with electrochemistry has come to offer other perspectives in bioelectrochemical signal transfer between target molecules and external electrochemical circuitry based on strategic surface preparation and functional linker molecules.

The chapters in this volume offer overviews of electronic properties, electron transfer and electron-proton coupled charge transfer of biological molecules and macromolecules both in the natural aqueous solution environment and on metallic electrode surfaces, where the electrochemical potential controls biomolecular function. Redox metalloproteins and DNA-based molecules are primary targets, but amino acid and nucleobase building blocks are also addressed. Novel environments where proteins and DNA-based molecules are inserted in metallic nanoparticle hybrids or *in situ* STM configurations are other focus areas.

The chapters by Winkler *et al.* and by Wikström overview electron and electron-proton coupled charge transfer of both small electron transfer metalloproteins and large metalloenzyme complexes such as cyt *c* oxidase. A key point is the mechanistic detail now available. Percolation of electrons through the protein structure and proton hopping through conduction channels in the enzyme structures, triggered by electron transfer are important issues. Understanding of these patterns is a prerequisite for protein and enzyme voltammetry and hybrid systems for working enzyme-devices. Such perspectives are addressed by Willner and Willner, and by Butt and Armstrong. It is notable that large and fragile, sometimes multi-component enzymes can be controlled and retain close to full enzyme function at the electrochemical interface to the extent where molecular mechanisms of enzyme electrocatalysis can be mapped. This holds perspectives for multifarious technology for example in metalloenzyme biosensor function. Functional units which respond to optical, magnetic, and other signals can further be inserted between the electrode and the enzymes and biological redox chains constructed.

The chapter by Hill and Kelley addresses the interfacial electronic conductivity of DNA-based molecules controlled by the electrochemical potential. Binding of redox probes is a probe for electronic communication between the probe and the electrode through the DNA-molecular frame and therefore of the tunneling conductivity of the latter. This remains an intriguing issue as the redox-based electronic energies of the nucleobases are strongly off-resonance with the electrode Fermi energy and the redox level of the probe molecule.

Single-crystal, atomically planar electrode surfaces have paved the way for the scanning probe microscopies, STM and AFM in bioelectrochemistry. The chapter by Nichols *et al.* illuminates this powerful technology which has increased the structural resolution of the (bio)electrochemical electrode surfaces to sub-molecular levels. High-resolution images have been achieved for the biological building blocks, i.e. DNA nucleobases and amino acids, both of which form highly ordered monolayers on Au(111)- and Pt(111) -electrode surfaces. Dynamic surface phenomena such as adlayer phase transitions can also be followed. The image detail of individual molecules and patterns in their lateral organization holds clearly perspectives for understanding the interaction of biological liquids with solid surfaces.

The use of STM/AFM to biological macromolecules is discussed further in the chapters by Kuznetsov and Ulstrup, and by Zhang *et al.* Single-molecule resolution has been achieved for both redox metalloproteins and DNA-based molecules under conditions where the molecules are active in electron transfer or enzyme function. Not only structural mapping but given adequate theoretical support, electron transfer and redox enzyme function can be addressed at the single-molecule level. The Os-complexes and the redox metalloprotein, *P aeruginosa* azurin discussed in chapter 8 illuminate these perspectives which extend towards ultra-sensitive biological sensors and other "device" function. STM and in *situ* STM are theoretically demanding because the electrical current recorded does not translate directly into molecular topography. Long-range off-resonance conductivity is broadly understood in terms of electron exchange and energy gaps of appropriate atomic or molecular orbitals with exponential distance dependence of the tunneling current expected. This is sometimes observed but weak current attenuation emerges in other cases such as for single- and double-stranded oligonucleotides. This issue presently appears unsettled.

Electronic conductivity of (bio)molecules with low-lying redox states show a quite different pattern, namely two- (or multi-)step hopping through the redox state(s) induced by environmental configurational fluctuations.

Theoretical notions rest on electrochemical electron transfer but the novel environments have also disclosed new electron transfer phenomena. Switching or negative differential resistance, quite different from electrochemical electron transfer at semi-infinite electrode surfaces is immediately conspicuous. Coherent, multi-electron transfer in a single *in situ* STM event is another non-traditional electron transfer phenomenon.

(Bio)molecular electronics, enzyme electrochemistry, oligo-nucleotide organization, and high-resolution biological screening are exciting parts of new bioelectrochemistry. Networks of hybrid biomolecular structures are other novel targets. In a biotechnological perspective, fundamental bioelectrochemical innovation remains, however, essential. The objective of this volume is to illuminate these exciting new stages of bioinorganic electrochemistry.

Copenhagen and Lyngby, June 2007

Ole Hammerich, Jens Ulstrup

Chapter 1

ELECTRON TUNNELING THROUGH IRON AND COPPER PROTEINS

JAY R. WINKLER, ALEXANDER R. DUNN, CORINNA R. HESS, AND HARRY B. GRAY
Beckman Institute, MC 139-74, California Institute of Technology, Pasadena, California 91125, USA

1. INTRODUCTION

Iron and copper redox centers facilitate the transfer of electrons through proteins that are part of the respiratory and photosynthetic machinery of cells. Much work has been done with the goal of understanding the factors that control electron flow through these proteins.[1–18] The results of many of the key experiments have been interpreted in terms of semiclassical theory.

The rate of electron transfer (ET) from a donor (**D**) to an acceptor (**A**) held at fixed distance and orientation depends on of temperature (T), reaction driving force ($-\Delta G^{o}$) a nuclear reorganization parameter (λ), and an electronic coupling matrix element (H_{AB}).[4,12] The reorganization parameter reflects the changes in structure and solvation that result when an electron moves from **D** to **A**. A balance between nuclear reorganization and reaction driving force determines both the transition-state configuration and the height of the barrier associated with the ET process. At the optimum driving force ($-\Delta G^{o} = \lambda$), the reaction is activationless, and the rate ($k_{ET}{}^{o}$) is limited only by the strength of the **D**/**A** electronic coupling. When **D** and **A** are in van der Waals contact, the coupling strength is usually so large that the ET reaction is adiabatic, that is, it occurs every time the transition-state configuration is formed. In this adiabatic limit, the ET rate is independent of H_{AB} and the prefactor depends only on the frequency of motion along the reaction coordinate. An ET reaction is nonadiabatic when the **D**/**A**

O. Hammerich and J. Ulstrup (eds.), Bioinorganic Electrochemistry, 1–23.

interaction is weak and the transition state must be reached many times before an electron is transferred. The electronic coupling determines the frequency of crossing from reactants (**D** + **A**) to products (**D**$^+$ + **A**$^-$) in the region of the transition state.

The singular feature of electron transfer is that reactions can proceed at very high rates when **D** and **A** are separated by long distances. The electron tunnels through a potential barrier between **D** and **A**; for a square barrier, H_{AB} displays an exponential dependence on the distance (**R**) between the reactants.[19] The medium between redox centers potentially can control long-range ET. Owing to a 3.5–Å^{-1} distance-decay constant (β), the time required for electron exchange between hydrated ferrous and ferric ions is estimated to be 5×10^{16} years if the complexes are separated by 20 Å in a vacuum.[14] Superexchange coupling via hole and electron states of the intervening medium enhances the **D**/**A** electronic interaction and produces a more gradual decrease in rate with distance. Fill the void between hydrated ferrous and ferric ions with water (β = 1.59 Å^{-1})[20] and the time constant for 20–Å electron exchange decreases dramatically (400 years), but the reaction is still far too slow to support biological activity. If the distance decay factor for ET across a polypeptide is comparable to that found for electron tunneling across hydrocarbon bridges (β = 0.8–1.0 Å^{-1}),[14] then the time for a 20 Å electron exchange between complexed ferrous and ferric ions in the hydrophobic interior of a protein could be in the millisecond to microsecond range.

2. REORGANIZATION ENERGY

The protein fold plays a central role in lowering the reorganization energy of a biological ET reaction.[21,22] A large part of the λ-reduction results from sequestering a redox center from the aqueous solvent environment. Continuum models suggest that embedding a redox center inside a low dielectric cavity can lower the outer-sphere reorganization energy by as much as 50%.[23] Moreover, by constraining the coordination environment around metal centers, inner-sphere reorganization energy can be reduced as well.[22] Indeed, metals that are typically poor redox reagents because of large reorganization barriers can be extremely efficient when embedded in protein interiors. Copper is a case in point. The reorganization energy for electron self-exchange in $Cu(phen)_2^{2+/+}$ is ~2.4 eV; the value for Cu(II/I) in *Pseudomonas aeruginosa* azurin is 0.7 eV (*vide infra*). The 1.7 eV reduction in λ reflects the transition-state stabilization imposed by the azurin fold.[21]

Two approaches are commonly employed to estimate reorganization energies in ET reactions: the first involves driving-force variations; and the second is based on temperature dependences. Carefully designed studies of the variation of rates with driving force can lead to reliable estimates of λ,[24–29] but there are several pitfalls. To begin with, it can be difficult to modify natural redox sites sufficiently to define λ with confidence. At low driving forces, $\ln(k_{ET})$ varies linearly with $-\Delta G^{o}$, and the slope is independent of λ. Reorganization energies extracted from these data are fraught with uncertainty because they involve the analysis of small deviations from linearity. In principle, more reliable values of λ can come from measurements at driving forces close to the reorganization energy, where rates reach a maximum and the driving-force curve flattens. Complications can arise, however, in redox processes where ET is not rate limiting. Conformationally gated reactions are a prime example of this type of behavior.[30,31]

Reorganization energies also can be extracted from studies of the temperature dependence of ET rates.[32] These would seem to be more direct and reliable than driving-force studies. The drawback, however, is that accurate λ values can be extracted only when the temperature dependence of ΔG^{o} has been determined as well.[33] In many cases, particularly with deeply buried redox sites, reduction potentials are not known with great accuracy and temperature dependences are beyond available measurement methods.

2.1 Cytochrome *c*

Early work on Ru-ammine cytochrome *c*[34,35] (104 amino acids in the horse protein; 12.5 kDa; E^{o} = 0.28 V *vs.* NHE)[36,37] included the replacement of the native Fe center with Zn.[38] Long-range ET reactions in Zn-cyt *c* were initiated by excitation of the Zn-porphyrin (ZnPor) to its long-lived (>10 ms), strongly reducing (E^{o} ~ −0.8 V *vs.* NHE) triplet excited state.[38] A driving-force study of ET rates in $Ru(NH_3)_4L$(His33)-Zn-cyt *c* (L = NH_3, pyridine, isonicotinamide) was consistent with λ_{12} = 1.2 eV.[27] Since self-exchange reorganization energies for Ru-ammine complexes (λ_{11}) are in the vicinity of 1.6 eV,[4,39] the ET kinetics suggest that λ_{22} = 0.8 eV for Zn-cyt *c*.

Studies of high-driving-force ET in heme and nonheme proteins were made possible by Ru-diimine labeling protocols and the flash-quench method.[40] The driving-force dependence of ET in $Ru(diimine)_2$(im)(His33)-Fe-cyt *c* indicates λ_{12} = 0.8 eV.[11,29] This value is lower than that found for Ru-ammine-Zn-cyt *c*, because the diimine ligands coordinated to the Ru center are larger and more hydrophobic than ammines. Consequently, the self-exchange reorganization energy for $Ru(diimine)_2(im)(His)^{3+/2+}$ is substantially smaller (λ_{11} = 0.8 eV) than that of the ammine.[4,9] The combined results from ET measurements in the Ru-ammine and Ru-diimine proteins suggest that the reorganization energy for electron exchange between Fe(II)- and Fe(III)-cyt *c* is 0.8(1) eV.

2.2 Azurin

P. aeruginosa azurin has a Cu(II/I) reduction potential of 0.31 V *vs.* NHE.[41–43] Analysis of the driving-force dependence of Cu(I)→M(III) (M = Ru, Os) ET in M(diimine)$_2$(im)(His83)-azurin gives a reorganization energy of 0.8 eV.[44] In accord with this finding, the temperature independence (240–300 K) of Cu(I)→Ru(III) ET in Ru(bpy)$_2$(im)(His83)-azurin can be described by $\lambda_{12} = 0.7 \pm 0.1$ eV, although the observed two-fold increase in rate constant as the temperature is lowered to 170 K cannot be explained by changes in the exponential term of the semiclassical rate expression.[32] It is more likely that the Ru-Cu electronic coupling (*vide infra*) increases as the protein is cooled to 170 K.

Rates of Ru(III) and Os(III) reduction by Cu(I) have been measured in single crystals of *P. aeruginosa* M(diimine)$_2$(im)(His83)-azurin; in these cases, protein conformation and surface solvation are precisely defined by high-resolution X-ray structure determinations.[45] The time constants for electron tunneling in crystals are roughly the same as those measured in solution, indicating very similar protein structures in the two states. High-resolution structures of the oxidized (1.5 Å) and reduced (1.4 Å) forms of Ru(tpy)(bpy)(im)(His83)-azurin establish that very small changes in copper coordination accompany reduction.[45] Although Ru(bpy)$_2$(im)(His83)-azurin is less solvated in the crystal lattice, the reorganization energy for Cu(I)→Ru(III) electron transfer falls in the same range (0.6–0.8 eV) determined experimentally for the reaction in solution. It is striking that driving forces, reorganization energies, and rates of Cu(I)→M(III) (M = Ru, Os) ET are virtually unchanged when labeled azurins lose one-third of their solvent-accessible surface upon transfer from dilute solutions to crystal lattices with just 40% water.

3. ELECTRONIC COUPLING

Nonadiabatic ET reactions are characterized by weak electronic interaction between the reactants and products at the transition-state nuclear configuration ($H_{AB} \ll k_B T$). This coupling is directly related to the strength of the electronic interaction between the donor and acceptor.[46] For electron tunneling through a square potential barrier, H_{AB} drops off exponentially with increasing **D**/**A** separation.[19,47] The height of the tunneling barrier relative to the energies of the **D**/**A** states determines the distance-decay constant (β). Hopfield estimated $\beta \sim 1.4$ Å^{-1} on the basis of measurements of the temperature dependence of ET from a cytochrome to the oxidized special pair of chlorophylls in the photosynthetic reaction center of *Chromatium vinosum*.[19] His analysis predicted that the heme edge of the cytochrome would be 8 Å from the nearest edge of the special pair (structural studies revealed that the actual distance is 12.3 Å).[48]

3.1 Superexchange Coupling

Coupling **D** to **A** through electronic interactions with the intervening bridge is called superexchange. If oxidized states of the bridge mediate the coupling, the process is referred to as "hole transfer"; mediation by reduced bridge states is known as "electron transfer". In 1961 McConnell developed a nearest-neighbor superexchange coupling model to describe charge-transfer interactions between donors and acceptors separated by spacers comprised of identical repeat units.[49] The total coupling (H_{AB}) is given as the product of the coupling-decays for each bridge site (ε). For a bridge built from identical repeat units separated by *m* bonds, H_{AB} will be proportional to ε^m. In this model, the ET rate constant will exhibit an exponential dependence on the number of bonds separating **D** and **A**.

3.2 Tunneling Pathways in Proteins

The McConnell superexchange model is too simplistic for a protein intervening medium because of the complex array of bonded and nonbonded contacts between **D** and **A**.[5,6,50–64] An important advance was made by Beratan and Onuchic, who developed a generalized tunneling pathway (TP) superexchange coupling model that reduces the diverse interactions between the atoms in a folded polypeptide to a set of three types of contacts: covalent bonds, hydrogen bonds, and through space contacts.[10,50–56] Each type of contact is assigned a coupling decay value (ε_C, ε_H, and ε_S), which permits implementation of a search algorithm for finding optimal coupling pathways through proteins. The total coupling of a single pathway is given as a product of the couplings for the individual links (expression (1-1)).[10,50–56]

$$H_{AB} \propto \Pi\varepsilon_C \Pi\varepsilon_H \Pi\varepsilon_S \qquad (1\text{-}1)$$

A tunneling pathway can be described in terms of an effective covalent tunneling path comprised of n (nonintegral) covalent bonds, with a total length equal to σ_l (expressions (1-2 and 1-3)).

$$H_{AB} \propto (\varepsilon_C)^n \quad (1\text{-}2)$$

$$\sigma_l = n \times 1.4 \text{ Å/bond} \quad (1\text{-}3)$$

The relationship between σ_l and the direct **D-A** distance (**R**) reflects the coupling efficiency of a pathway. The variation of ET rates with **R** is expected to depend upon the coupling decay for a single covalent bond (ε_C); the magnitude of ε_C depends critically upon the energy of the tunneling electron relative to the energies of the bridge hole and electron states.

3.3 Rate/Distance Dependence

The **D-A** distance decay of protein ET rate constants depends on the capacity of the polypeptide matrix to mediate distant electronic couplings. If dominant coupling pathways mediate long-range electron transfer in proteins, then single-site mutations could have profound effects on enzyme function. In addition, if single pathways operate in biological ET reactions, then they have presumably been optimized through natural selection. These consequences of tunneling pathways impart a certain lack of robustness into the protein structure/function relationship. Concerns about this issue led Dutton and coworkers to propose that a folded polypeptide matrix behaves like a glassy solvent, imposing a uniform barrier (UB) to electron tunneling.[48] Analysis of a variety of ET rates, especially those from the photosynthetic reaction center, produced a universal distance-decay constant for protein electron transfer that was in remarkable agreement with Hopfield's estimate (1.4 $Å^{-1}$).[48] Disagreements over the appropriate **D/A** distance measure (edge-to-edge, center-to-center) fueled disputes about whether the large body of protein ET data supports a homogeneous barrier model, or whether a structure-dependent model is necessary. Recently, the UB model has been amended to include the packing density of the protein matrix.[65] Although this model ignores bond connectivity, it docs cmbody many of the same elements as the TP model by accounting for, in a rudimentary fashion, the protein structure dependence of long-range couplings.

The great strength of the TP model is that a straightforward analysis of a protein structure identifies residues that are important for mediating long-range coupling. Employing this model, Beratan, Betts, and Onuchic predicted in 1991 that proteins comprised largely of β-sheet structures would be more effective at mediating long-range couplings than those built from α helices.[54] This analysis can be taken a step further by comparing the coupling efficiencies of individual protein secondary structural elements (β-sheets, α-helices). The coupling efficiency can be determined from the variation of σ_l as a function of **R**. A linear σ_l/**R** relationship implies that k_{ET}^{o} will be an exponential function of **R**; the distance-decay constant is determined by the slope of the σ_l/**R** plot and the value of ε_C.[12]

A β-sheet is comprised of extended polypeptide chains interconnected by hydrogen bonds; the individual strands of β-sheets define nearly linear coupling pathways along the peptide backbone spanning 3.4 Å per residue. The tunneling length for a β-strand exhibits an excellent linear correlation with β-carbon separation ($\mathbf{R}_\beta$,); the best linear fit with zero intercept yields a slope of 1.37 $\sigma_l/\mathbf{R}_\beta$ (distance-decay constant = 1.0 $Å^{-1}$). Couplings across a β-sheet depend upon the ability of hydrogen bonds to mediate the **D/A** interaction. The standard parameterization of the TP model defines the coupling decay across a hydrogen bond in terms of the heteroatom separation. If the two heteroatoms are separated by twice the 1.4–Å covalent-bond distance, then the hydrogen-bond decay is assigned a value equal to that of a covalent bond. Longer heteroatom separations lead to weaker predicted couplings but, as yet, there is no experimental confirmation of this relationship.

In the coiled α-helix structure a linear distance of just 1.5 Å is spanned per residue. In the absence of mediation by hydrogen bonds, σ_l is a very steep function of $\mathbf{R}_\beta$, implying that an α-helix is a poor conductor of electronic coupling (2.7 $\sigma_l/\mathbf{R}_\beta$, distance-decay constant = 1.97 $Å^{-1}$). If the hydrogen-bond networks in α-helices mediate coupling, then the Beratan-Onuchic parameterization of hydrogen-bond couplings suggests a $\sigma_l/\mathbf{R}_\beta$ ratio of 1.72 (distance-decay constant = 1.26 $Å^{-1}$). Treating hydrogen bonds as covalent bonds further reduces this ratio (1.29 $\sigma_l/\mathbf{R}_\beta$, distance-decay constant = 0.94 $Å^{-1}$). Hydrogen-bond interactions, then, will determine whether α-helices are vastly inferior to or are slightly better than β-sheets in mediating long-range electronic couplings. It is important to note that the coiled helical structure leads to poorer $\sigma_l/\mathbf{R}_\beta$ correlations, especially for values of $\mathbf{R}_\beta$ under 10 Å. In this distance region, the TP model predicts little variation in coupling efficiencies for the different secondary structures.

4. TUNNELING TIMETABLES

4.1 Ru-proteins

Plots of coupling-limited tunneling times ($1/k_{ET}^{o}$) *vs.* distance (**R**) are called tunneling timetables.[12–14,66] To determine the efficiency of coupling across β-sheet structures, we examined distance dependences of ET in azurin.[67,68] The copper center in azurin is situated at one end of an eight-stranded β-barrel, ligated in a trigonal plane by two imidazoles (His46, His117) and a thiolate (Cys112); in addition, there are weak axial interactions (Met121 thioether sulfur, Gly45 carbonyl oxygen).[42,69] The azurin from *P. aeruginosa* has two additional His residues, one of which (His83) reacts readily with Ru-labeling reagents. A H83Q base mutant was prepared and individual mutant His residues were introduced at five locations on β-strands extending from the Cys112 and Met121 ligands (K122H, T124H, T126H, Q107H, M109H). Measurements of Cu(I)→ $Ru(bpy)_2(im)(HisX)^{3+}$ ET ($-\Delta G^o$ = 0.7 eV) provide a calibration for the distance dependence of electron transfer along β-strands, Fig. 1-1.

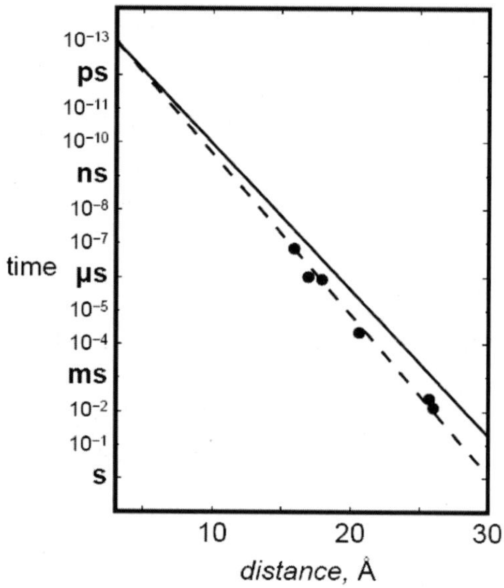

Figure 1-1. Tunneling timetable for $Ru(bpy)_2(im)(HisX)$-azurins [k_{ET}^o (X, Cu-Ru): 7.1×10^6 (122, 15.9); 2.2×10^4 (124, 20.6); 1.3×10^2 (126, 26.0); 8.5×10^5 (109, 17.9); 2.4×10^2 (107, 25.7); 1.0×10^6 s^{-1} (83, 16.9 Å)]. Solid line, β = 1.0 Å^{-1}; dashed line, β = 1.1 Å^{-1}. Adapted from ref. 13.

The driving-force-optimized electron tunneling timetable for azurin reveals a nearly perfect exponential distance dependence, with a decay constant (β) of 1.1 Å^{-1}, and an intercept at close contact (r_o = 3 Å) of 10^{13} s^{-1}. This decay constant is quite similar to that found for superexchange-mediated tunneling across saturated alkane bridges (β ≈ 1.0 Å^{-1}),[70,71] strongly indicating that a similar coupling mechanism is operative in the polypeptide.

The validity of the azurin tunneling timetable rests on the assumption that Ru-azurin structures are not very different in crystals and aqueous solutions. Our measurements of ET kinetics on crystalline samples of labeled azurins directly tested this assumption.[45] We found that the rate constants for oxidation of Cu(I) by Ru(III) and Os(III) in solutions and crystals are nearly identical for each donor-acceptor pair. It follows that the crystal structures of reduced and oxidized azurin are the relevant reactant and product states for solution ET.

It is important to distinguish between superexchange-mediated electron tunneling and multistep mechanisms that also can move charge over large molecular distances. In tunneling processes, quantum mechanical mixing of localized donor and acceptor states with oxidized (and/or reduced) bridge states couples the reactant and product states, producing an avoided crossing between the free-energy surfaces at the transition state. Importantly, oxidized (reduced) bridge states are not populated in tunneling reactions; electron transfer occurs in a single elementary reaction step. Because redox centers in metalloenzymes cannot come into close contact, electrons must tunnel between them. There is a practical upper limit to the separation distance between redox sites; if charges must be transferred farther than this range, then multiple tunneling steps are required. Long-range electron transfer can proceed by either mechanism, but each has distinct energetic and coupling requirements, and can respond quite differently to changes in reaction parameters.

The energy gap between the donor/acceptor redox levels and those of oxidized or reduced intermediate states is the primary criterion in determining when multistep tunneling becomes important (*vide infra*). In proteins with a single redox cofactor, the opportunities for multistep tunneling are limited. Extreme redox potentials are necessary to oxidize and reduce polypeptide backbones; thus multistep tunneling via backbone states will not contribute to observed ET kinetics under most solution conditions. The side chains of certain amino acids (*e.g.*, Tyr, Trp) have redox potentials that are more accessible than those of the peptide backbone.[72–76] Oxidized Trp and Tyr residues have been characterized spectroscopically in a large number of proteins, although there is a paucity of direct evidence for their involvement in multistep tunneling reactions.[77–79]

The $Ru(bpy)_2(im)(His)^{3+/2+}$ reduction potential (E° = 1.0 V *vs.* NHE)[40] may be high enough to oxidize Trp and Tyr residues in Ru-azurin. If the reactive Cu(I) center is replaced by redox-inert Zn(II) in the protein, however, we find that photogenerated holes in $Ru(bpy)_2(im)(HisX)^{3+}$ complexes remain localized on the Ru center. The energy gap between the Ru(III) hole state and oxidized bridge states must therefore be greater than 75 meV ($3k_BT$ at 295 K). The fact that oxidized bridge states lie at higher energy than the Ru(III) hole does not rule out multistep tunneling; endergonic steps can be compensated by favorable reactions later in a sequence. Endergonic reactions, however, become less effective as the temperature decreases, so that multistep tunneling with highly endergonic steps will exhibit a strong dependence on temperature. Our finding that the rate of Cu(I)→Ru(III) ET in $Ru(bpy)_2(im)$(HisX)-azurin is nearly independent of temperature between 240 and 300 K coupled with the observation that decreasing the temperature to 160 K produces a twofold *increase* in the ET rate demonstrate that multistep tunneling cannot explain long-range ET in Ru-azurin. Instead, the data shown in Fig. 1-1 provide a benchmark for superexchange-mediated electron tunneling in proteins.

The rates of high-driving-force ET reactions have been measured for more than 30 Ru(diimine)-labeled metalloproteins.[12,13,80,81] Only modest corrections are required to scale these rates to driving-force-optimized values, permitting comparisons of ET in different proteins. The results are summarized in the electron tunneling timetable of Fig. 1-2.

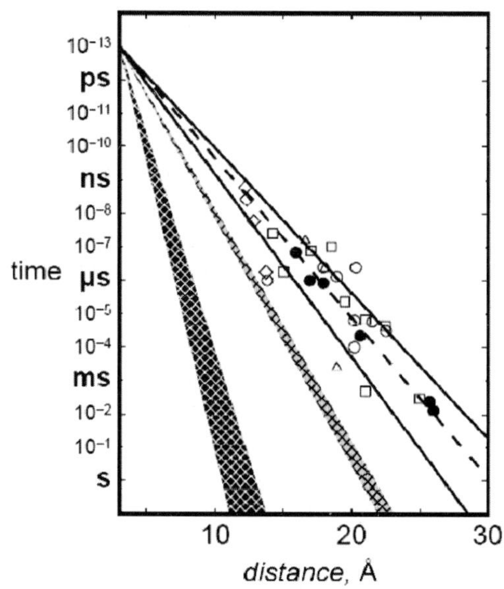

Figure 1-2. Tunneling timetable for water and Ru-modified proteins: azurin (•); cytochrome c (○); myoglobin (Δ); cytochrome b_{562} (□); and HiPIP (◊). Rates for ET reactions in crystals of tuna cytochrome *c* doped with Zn-cytochrome *c* are indicated by solid circles. The solid lines illustrate the TP predictions for coupling along β-strands ($\beta = 1.0$ Å^{-1}) and α-helices ($\beta = 1.3$ Å^{-1}); the dashed line illustrates a 1.1 Å^{-1} distance decay. Distance decay for electron tunneling through water is shown as a gray wedge. Estimated distance dependence for tunneling through vacuum is shown as the black wedge. Adapted from ref. 66.

The reported distances are all metal-to-metal measures; in the case of metal clusters, the closest metal was chosen. Tunneling times range from a few nanoseconds (12.2 Å ET in the high-potential iron sulfur protein from *C. vinosum*) to 10 milliseconds (26 Å ET in *P. aeruginosa* azurin).

The Ru-protein rates in Fig. 1-2 are scattered around the Ru-azurin $\beta = 1.1$ $Å^{-1}$ exponential distance decay. Rates at a single distance can differ by as much as a factor of 10^3 and **D/A** distances that differ by as much as 5 Å can produce virtually identical rates. Clearly, the absence of a simple exponential distance dependence in the Ru-protein rate data is a reflection of the heterogeneity of the coupling medium. The efficiency of the coupling between redox centers is determined by the three-dimensional structure of the intervening polypeptide.

4.2 Water

In 1984, Larsson suggested that long-range ET in water would be inefficient ($\beta = 2.4$ $Å^{-1}$) because of the large energy gap between the hole states of the medium and those of **D** and **A**.[82] More recent theoretical treatments, however, have produced β values in the 1.0 to 1.8 $Å^{-1}$ range.[83,84]

Investigations of $Ru(tpy)_2^{2+}$ (tpy = 2,2':6,2''-terpyridine) luminescence quenching by $Fe(OH_2)_6^{3+}$ in aqueous acidic glasses have produced a distance decay constant of 1.59 ± 0.05 $Å^{-1}$,[20,85] a value substantially larger than those for electron tunneling through saturated alkane spacers (0.9 $Å^{-1}$)[70] or proteins (1.0–1.3 $Å^{-1}$), Fig. 1-2. The region representing the distance decay for coupling through water (β = 1.6–1.7 $Å^{-1}$) demonstrates that, although better than a vacuum (β = 3–4 $Å^{-1}$), tunneling 20 Å through water is at least 100 times slower than tunneling through protein or hydrocarbon bridges. At **D-A** distances in the 10–Å range, however, ET reactions mediated by intervening water bridges could be key steps in the catalytic cycles of certain redox enzymes.[86]

5. ELECTRON TUNNELING WIRES

5.1 Ru-Wires for Cytochrome $P450_{cam}$

The original impetus for the construction of substrates linked to photoredox units was the desire to observe the fleeting intermediates in P450 catalysis, Fig. 1-3.[87–89] Under biological conditions the rate determining step (RDS) in catalysis is the addition of a second electron to the dioxygen-bound ferrous heme by the enzyme putidaredoxin (putd). Subsequent catalytic events occur far too rapidly to be observed under normal conditions.

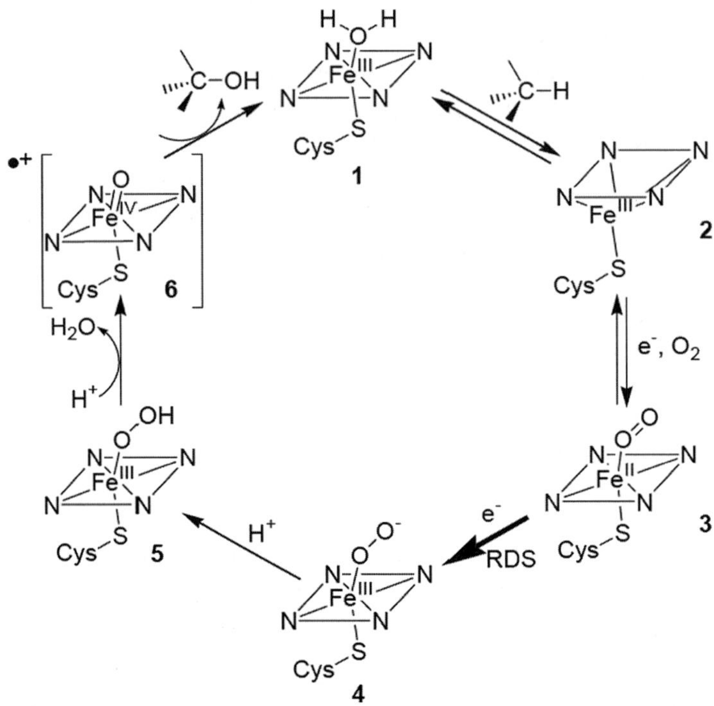

Figure 1-3. The cytochrome $P450_{cam}$ catalytic cycle. Upon binding, the substrate displaces water, converting the heme from six-coordinate, low-spin (**1**) to five-coordinate, high-spin (**2**). Subsequent reduction by putd produces a ferrous heme, which binds dioxygen (**3**). Reduction of **3** produces a ferrous, peroxide-bound heme (**4**), which rapidly protonates (**5**). In the prevalent model, the peroxide then undergoes heterolysis to produce water and a ferryl $[Fe(IV){=}O]^{\bullet+}$ species (compound I, **6**), which oxidizes the substrate. See refs. 87–89.

In order to observe intermediates **5** and **6**, we have sought to use ruthenium tris-bipyridyl complexes linked to $P450_{cam}$ substrates or heme ligands (Ru-wires) to photochemically reduce **3**, thus replacing the sluggish reduction by putd with a rapid photochemical trigger. The Ru-wires investigated consist of a Ru-diimine moiety connected to adamantane, imidazole, or ethylbenzene by either an alkyl[90] or perfluorobiphenyl[91] bridge, Fig. 1-4. The Ru-wires bind $P450_{cam}$ with micromolar K_d's, as evidenced by energy transfer from the Ru-diimine excited state to the heme, and changes to the heme absorption spectrum, Table 1-1.

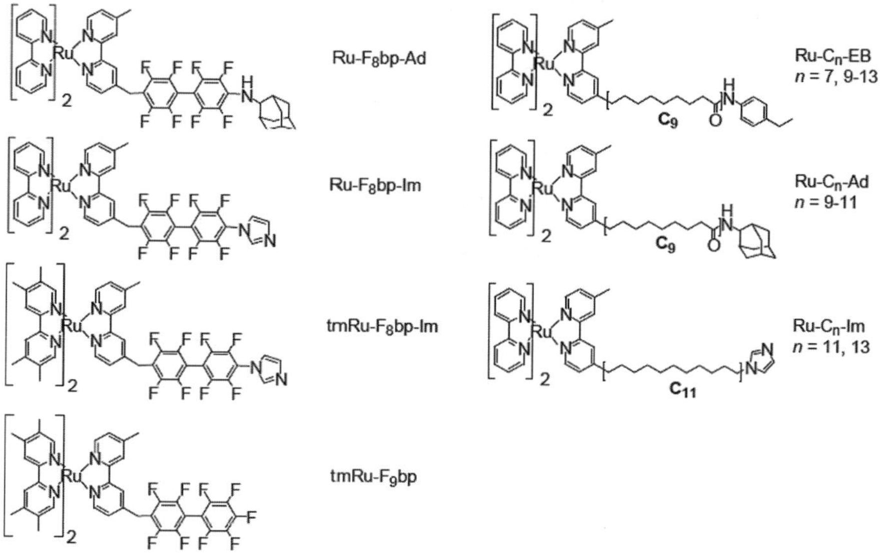

Figure 1-4. Ru-wires. Variation of the linkage connecting the Ru-diimine to the enzyme substrate or heme ligand modifies the electronic coupling and hence the rate of electron transfer. The tetramethylated ancillary bipyridine ligands in tmRu-F_8bp-Im and tmRu-F_9bp decrease the reduction potential of the excited Ru-diimine by ~130 mV. See ref. 91.

Interestingly, neither adamantyl nor imidazole is necessary for binding: Ru-wires terminating in either an alkyl chain or a perfluorobiphenyl bind to the enzyme. This observation suggests that interactions of the protein with the Ru-diimine and linker moieties provide the bulk of the binding energy.

The Ru-heme distances for a series of ethylbenzene Ru-wires remain roughly constant with varying alkyl chain lengths, Table 1-1,[92] indicating that an optimal Ru-heme distance exists in the Ru-wire:P450 conjugate. In contrast, Ru-C_{13}-Im binds $P450_{cam}$ (K_d = 4.1 μM), while Ru-C_{11}-Im does not. Evidently, the imidazole tip must ligate the heme iron in order for binding to occur, suggesting a substantial energetic penalty for its sequestration in the hydrophobic P450 active site. This result demonstrates that sensitive binding discrimination is possible with properly designed probe molecules. The Δ and Λ isomers of Ru-C_9-Ad bind $P450_{cam}$ with similar dissociation constants ($K_d(\Delta)$ = 190 nM; $K_d(\Lambda)$ = 90 nM), corresponding to a difference in binding energies of 0.46 kcal mol^{-1}.[93] Detailed analysis shows that the apparent K_d for the racemate is not the average of the stereoisomer K_d's.[94]

Photogenerated reduced or oxidized alkyl Ru-wires transfer electrons or holes to the $P450_{cam}$ ferric heme with time constants of around 50 ms, Fig. 1-5.

Table 1-1. Ru-wire dissociation constants and Ru-Fe distances derived from FET measurements.

Compound	K_d, µM	Ru-Fe, Å
Ru-C_{13}-EB	1.7 ± 0.4	20.6 ± 0.2
Ru-C_{12}-EB	1.5 ± 0.3	20.5 ± 0.2
Ru-C_{11}-EB	0.9 ± 0.4	20.1 ± 0.3
Ru-C_{10}-EB	0.9 ± 0.4	19.9 ± 0.1
Ru-C_9-EB	0.7 ± 0.1	19.4 ± 0.1
Ru-C_7-EB	6.5 ± 1.3	19.5 ± 0.1
Ru-C_9-Ad	0.8 ± 0.3	21.0 ± 0.3
Ru-C_{11}-Ad	0.7 ± 0.2	21.4 ± 0.2
Ru-C_{13}-Im	4.1 ± 1.1	21.2 ± 0.1
Ru-C_{11}-Im	>50	NA
Ru-F_8bp-Ad	0.08 ± 0.01	22.1
Ru-F_8bp-Im	3.7 ± 0.5	18.1[a]
Ru-F_8bp-Im	0.48 ± 0.18	18.1
Ru-F_9bp	2.1 ± 1.3	17.0

[a]In accord with the Ru-Fe distance calculated for tmRu-F_8bp-Im.

These rates are typical for ET through saturated bonds over comparable distances. In contrast, Ru-F_8bp-Im and tmRu-F_8bp-Im reduce the ferric heme directly from the excited state, Fig. 1-6. In particular, tmRu-F_8bp-Im photoreduces P450$_{cam}$ with a time constant of ~40 ns, five orders of magnitude faster than putd.

Our results demonstrate that it is possible to trigger reactions in the buried active sites of proteins on micro- to nanosecond timescales. In particular, the direct photoreduction observed with tmRu-F_8bp-Im occurs on timescales that cannot be accessed using stopped-flow techniques. The preassociation of the Ru-wire and the enzyme also circumvents the time limits otherwise imposed by diffusion.

Ru-wires allow the ET pathway between the Ru-diimine and the heme to be altered while retaining the surrounding protein matrix. Ru-wire:P450$_{cam}$ conjugates thus serve as a useful complement to synthetic[70,95–100] and protein[12,13] ET systems.

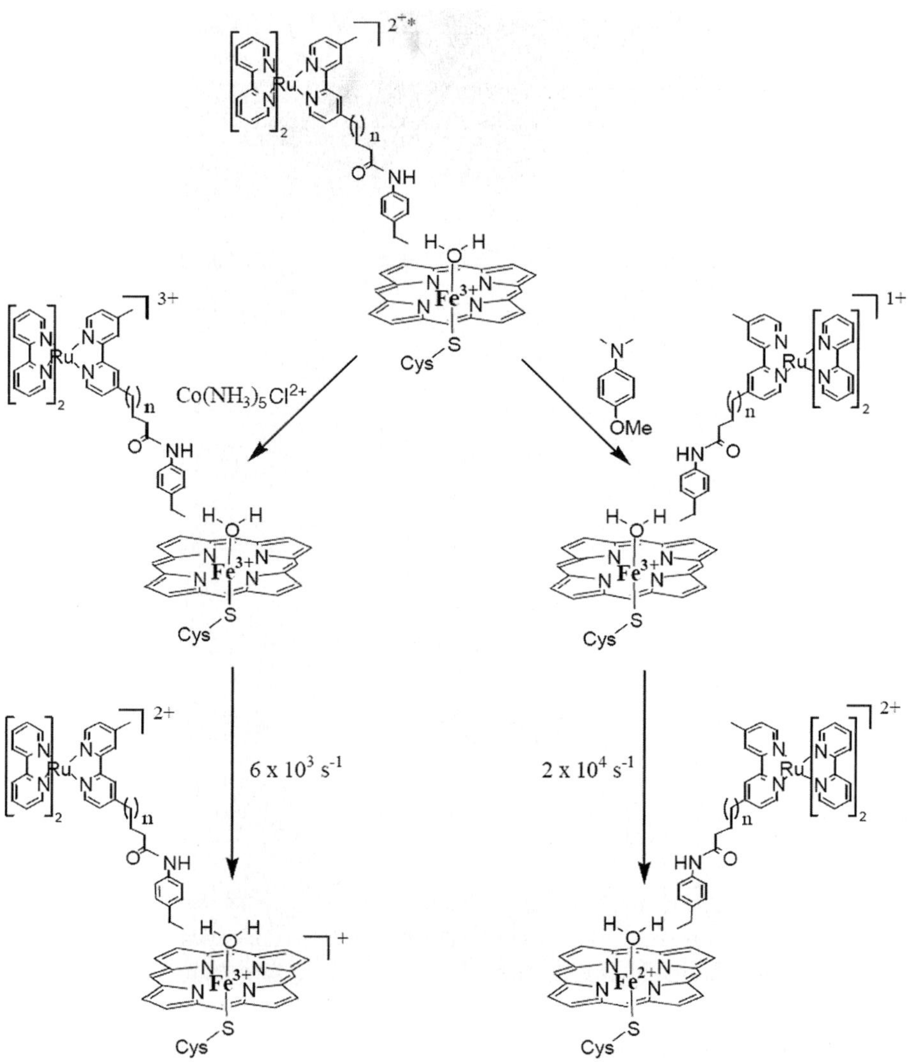

Figure 1-5. The flash-quench sequence for delivering electrons or holes to the active site of $P450_{cam}$. The Ru-wire is excited with 470-nm light (Ru*), and intercepted with either $Co(NH_3)_5Cl^{2+}$ or *p*-methoxy-*N*,*N*-dimethylaniline (PMDA) to generate the oxidized (Ru^{3+}) or reduced (Ru^{+}) Ru-wire. The photochemically generated hole or electron tunnels to the heme on the millisecond timescale, forming a heme cation radical or ferrous heme. See ref. 90.

ET through the perfluorobiphenyl bridge is 1500 times faster than through the alkyl chain, demonstrating that the nature of the intervening medium clearly mediates **D-A** coupling. Phototriggered ET is not observed in the tmRu-F_9bp and Ru-F_8bp-Ad:$P450_{cam}$ conjugates, again confirming the importance of the intervening pathway.

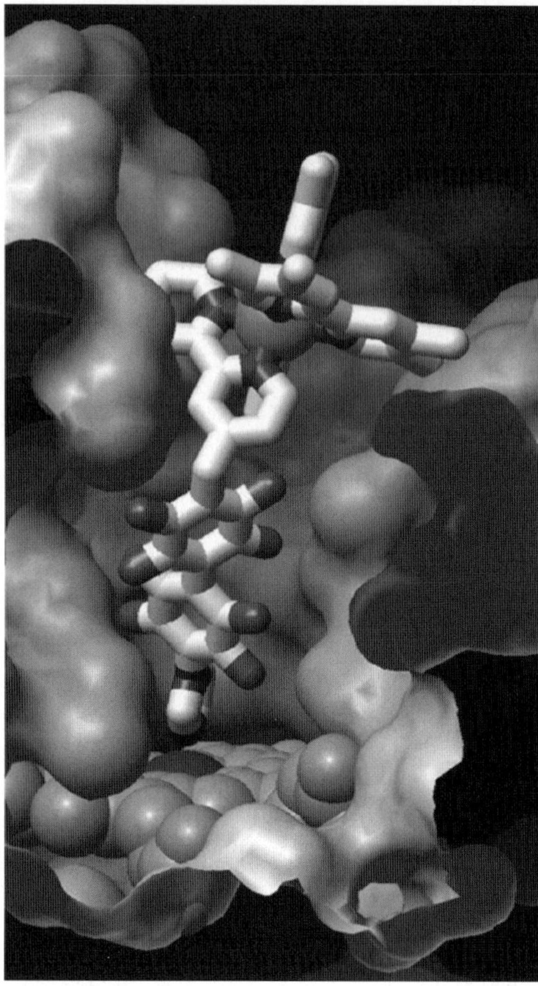

Figure 1-6. Structural model of tmRu-F_8bp-Im bound in the active site of $P450_{cam}$. The perfluorobiphenyl bridge couples $^*Ru^{2+}(L_2)L'$ to the heme, facilitating direct photoreduction ($-\Delta G^o \sim 0.45$ eV) with a time constant of 36 ns. Adapted from ref. 91.

Kurnikov has calculated molecular trajectories for Ru-wire docking in the $P450_{cam}$ active-site channel using molecular dynamics methods.[101] The couplings mediated by a perfluorobiphenyl bridge are dominated by superexchange through extra electron states, while those through hydrocarbon bridges are dominated by hole-mediated superexchange interactions. An important finding is that a small number of wire conformations with strong donor-acceptor couplings can account for the rapid ET rates in the Ru-wire:$P450_{cam}$ conjugates. Unusually rapid wire to heme ET in conjugates of inducible nitric oxide synthase with Re(perfluorobiphenyl)-wires[102] also has been observed.[103]

$P450_{cam}$ hydroxylates Ru-C_9-Ad when supplied with electrons via the natural NADH/putidaredoxin reductase/putd reduction relay.[104] Ru-C_9-Ad hydroxylation occurs at only 1.6% the rate of camphor hydroxylation, and only 10% of the electrons supplied by NADH go to product formation. Presumably the rest are diverted to the formation of reduced oxygen species such as superoxide or hydrogen peroxide.[105] The remarkable ability of $P450_{cam}$ to hydroxylate a molecule so structurally different from camphor supports the hypothesis that the structural flexibility inherent in the P450 fold allows these enzymes to hydroxylate structurally diverse substrates.[106,107]

5.2 Amine Oxidase

The conversion of amines to aldehydes and ammonia takes place at an amine oxidase (AO) active site that contains both copper and topaquinone (TPQ) cofactors.[108–111] The role of copper (if any) in catalysis remains a matter of debate.[112] The reduction potentials of the deeply buried TPQ and copper cannot be accurately measured using conventional electrochemical techniques. Instead, we measured the TPQ potential using gold electrodes functionalized with a phenyl-alkynyl bridge (DEA-OPE-SH, Fig. 1-8)[113] designed to bind in the active site of the *Arthrobacter globiformis* enzyme (AGAO),[114] thus providing an ET conduit from the electrode to the TPQ, Figs. 1-7 and 1-8. The TPQ potential was found to be –140 mV *vs.* SCE (phosphate, pH 7),[113] in accord with those of quinone model complexes.[115] Addition of the AGAO substrate phenethylamine quenches the electrochemical response, indicating that the DEA-OPE-SH bridge binds in the active site.

The TPQ reduction potential shows a linear variation with pH of –60 mV/pH, indicating a $2e^-,2H^+$ ET process, in contrast to the $2e^-,3H^+$ ET observed with model compounds.[113] Presumably the protein matrix or nearby Cu(II) stabilizes the deprotonated topaquinol. No copper electrochemistry was observed. It is not clear whether the copper potential is anomalously low or if it could not be measured due to weak electronic coupling with the phenyl-alkynyl bridge.

A distance decay constant of ~0.5 $Å^{-1}$ has been extracted from work on phenyl-ethynyl oligiomers.[98,116] In accord with this value, the CVs obtained at scan rates up to 1 V s^{-1} place a lower limit of 10^3 s^{-1} for tunneling to TPQ through the 22 Å wire; the corresponding rates through protein (1.1 $Å^{-1}$)[12,13] and water (1.6 $Å^{-1}$)[20] would be about 3 and 10^{-4} s^{-1}. It is very likely, therefore, that the DEA-wire is the coupling element between the gold electrode and TPQ, and, importantly, that tunneling timetables can play a key role in the design of functionalized electrodes for redox-active enzymes.

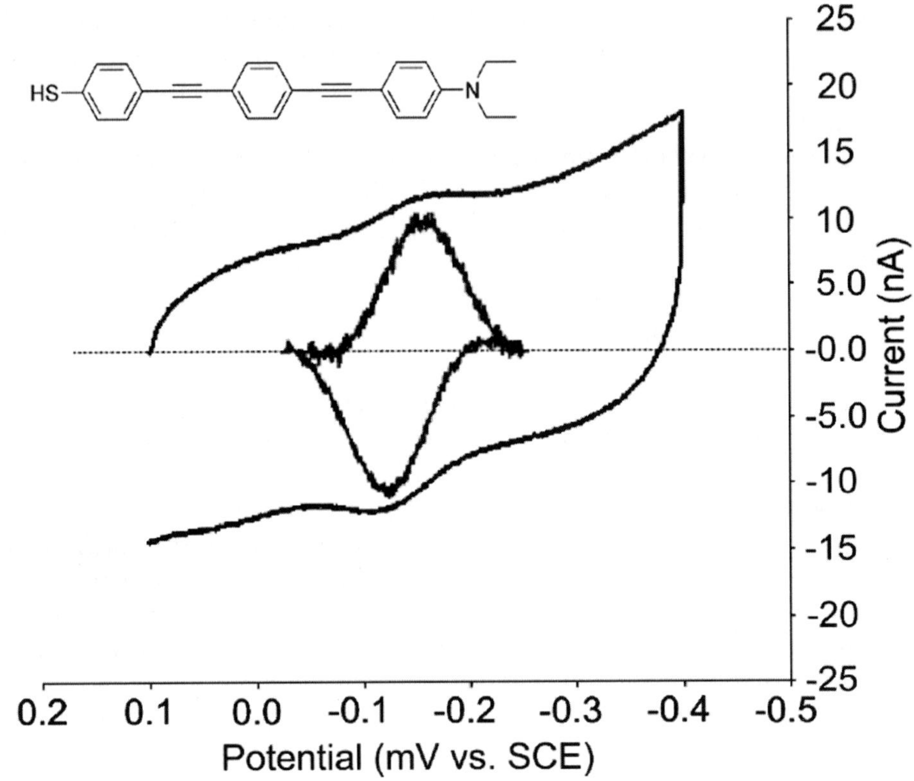

Figure 1-7. (Inset) DEA-OPE-SH. CV (black) and background-subtracted voltammogram (gray; not to scale) for AGAO on Au-bead electrodes modified with DEA-OPE-SH in 10 mM potassium phosphate, pH 7 (scan rate 100 mV s^{-1}). Gold-bead electrodes were soaked in mM solutions of DEA-OPE-SH for 24 h. Reductive stripping analyses of the resulting films indicated ~70% coverage. The modified electrodes were then incubated with AGAO for 24–48 h to allow binding to the adsorbed wires. Adapted from ref. 113.

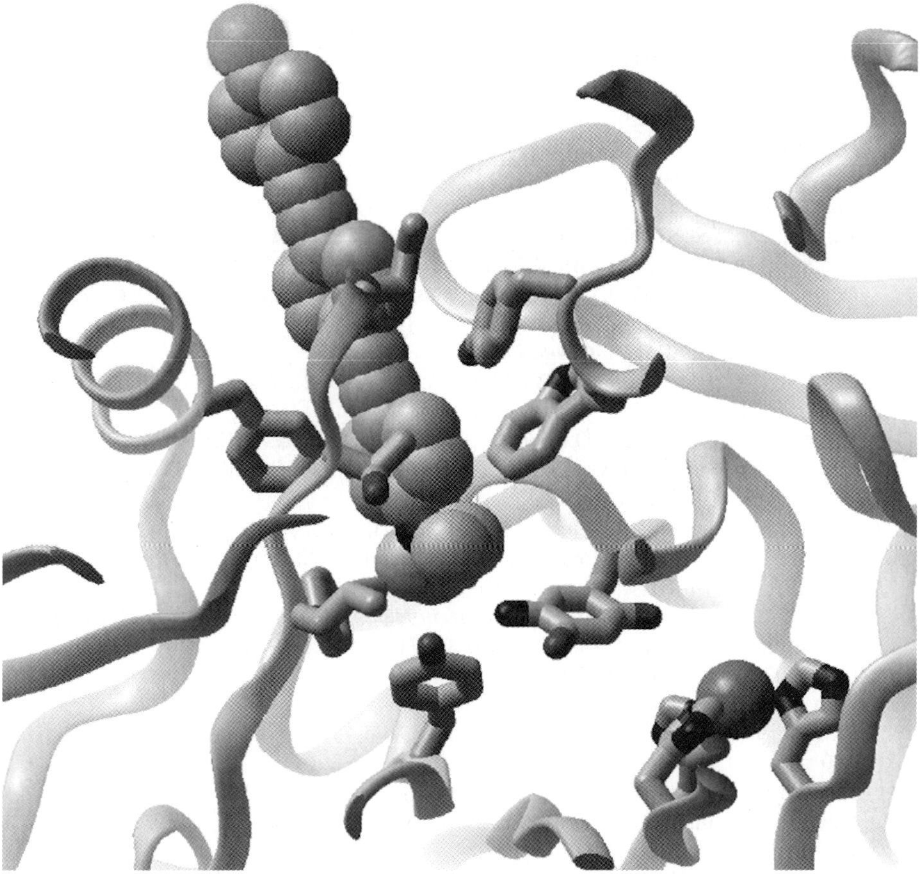

Figure 1-8. DEA-OPE-SH modeled into the substrate channel of AGAO. Key channel residues as well as TPQ and the copper center are highlighted.

ACKNOWLEDGMENT

Our research on electron tunneling through proteins is supported by the National Institutes of Health, the National Science Foundation, and the Arnold and Mabel Beckman Foundation.

REFERENCES

1. Kuznetsov, A.M.; Ulstrup, J., *Electron Transfer in Chemistry and Biology: An Introduction to the Theory*. John Wiley & Sons, Inc.: Hoboken, NJ, 1998; p. 357.
2. Nocek, J.M.; Zhou, J.S.; DeForest, S.; Priyadarshy, S.; Beratan, D.N.; Onuchic, J.N.; Hoffman, B.M., *Chem. Rev.* **1996,** *96*, 2459–2489.
3. McLendon, G.; Hake, R., *Chem. Rev.* **1992,** *92*, 481–490.
4. Marcus, R.A.; Sutin, N., *Biochim. Biophys. Acta* **1985,** *811*, 265–322.
5. Beratan, D.N.; Skourtis, S.S., *Curr. Opin. Chem. Biol.* **1998,** *2*, 235–243.
6. Regan, J.J.; Onuchic, J.N., *Adv. Chem. Phys.* **1999,** *107*, 497–553.
7. Mayo, S.L.; Ellis, W.R.; Crutchley, R.J.; Gray, H.B., *Science* **1986,** *233*, 948–952.
8. Gray, H.B.; Malmström, B.G., *Biochemistry* **1989,** *28*, 7499–7505.
9. Winkler, J.R.; Gray, H.B., *Chem. Rev.* **1992,** *92*, 369–379.
10. Onuchic, J.N.; Beratan, D.N.; Winkler, J.R.; Gray, H.B., *Annu. Rev. Biophys. Biomol. Struct.* **1992,** *21*, 349–377.
11. Bjerrum, M.J.; Casimiro, D.R.; Chang, I.-J.; Di Bilio, A.J.; Gray, H.B.; Hill, M.G.; Langen, R.; Mines, G.A.; Skov, L.K.; Winkler, J.R.; Wuttke, D.S., *J. Bioenerg. Biomemb.* **1995,** *27*, 295–302.
12. Gray, H.B.; Winkler, J.R., *Annu. Rev. Biochem.* **1996,** *65*, 537–561.
13. Winkler, J.R.; Di Bilio, A.; Farrow, N.A.; Richards, J.H.; Gray, H.B., *Pure Appl. Chem.* **1999,** *71*, 1753–1764.
14. Gray, H.B.; Winkler, J.R., In *The Porphyrin Handbook – Bioinorganic and Bioorganic Chemistry*, Kadish, K.M.; Smith, K.M.; Guilard, R., Eds. Academic Press: San Diego, CA, 2003; Vol. 11, pp. 51–73.
15. Gray, H.B., *Chem. Soc. Rev.* **1986,** *15*, 17–30.
16. Balzani, V., *Electron Transfer in Chemistry*. Wiley-VCH: Weinheim, 2001; Vol. III, p. 710.
17. Gray, H.B.; Winkler, J.R., *Quarterly Rev. Biophys.* **2003,** *36*, 341–372.
18. Gray, H.B.; Winkler, J.R., *Proc. Natl. Acad. Sci. USA* **2005,** *102*, 3534–3539.
19. Hopfield, J.J., *Natl. Acad. Sci. USA* **1974,** *71*, 3640–3644.
20. Ponce, A.; Gray, H.B.; Winkler, J.R., *J. Am. Chem. Soc.* **2000,** *122*, 8187–8191.
21. Winkler, J.R.; Wittung-Stafshede, P.; Leckner, J.; Malmström, B.G.; Gray, H.B., *Proc. Natl. Acad. Sci. USA* **1997,** *94*, 4246–4249.
22. Gray, H.B.; Malmström, B.G.; Williams, R.J.P., *J. Biol. Inorg. Chem.* **2000,** *5*, 551–559.
23. Simonson, T., *Proc. Natl. Acad. Sci. USA* **2002,** *99*, 6544–6549.
24. Closs, G.L.; Miller, J.R., *Science* **1988,** *240*, 440–447.
25. Fox, L.S.; Kozik, M.; Winkler, J.R.; Gray, H.B., *Science* **1990,** *247*, 1069–1071.
26. McLendon, G.; Miller, J.R., *J. Am. Chem. Soc.* **1985,** *107*, 7811–7816.
27. Meade, T.J.; Gray, H.B.; Winkler, J.R., *J. Am. Chem. Soc.* **1989,** *111*, 4353–4356.
28. McCleskey, T.M.; Winkler, J.R.; Gray, H.B., *J. Am. Chem. Soc.* **1992,** *114*, 6935–6937.
29. Mines, G.A.; Bjerrum, M.J.; Hill, M.G.; Casimiro, D.R.; Chang, I.-J.; Winkler, J.R.; Gray, H.B., *J. Am. Chem. Soc.* **1996,** *118*, 1961–1965.
30. Hoffman, B.M.; Ratner, M.A., *J. Am. Chem. Soc.* **1987,** *109*, 6237–6243.
31. Brunschwig, B.S.; Sutin, N., *J. Am. Chem. Soc.* **1989,** *111*, 7454–7465.
32. Skov, L.K.; Pascher, T.; Winkler, J.R.; Gray, H.B., *J. Am. Chem. Soc.* **1998,** *120*, 1102–1103.
33. Marcus, R.A.; Sutin, N., *Inorg. Chem.* **1975,** *14*, 213–216.
34. Winkler, J.R.; Nocera, D.G.; Yocom, K.M.; Bordignon, E.; Gray, H.B., *J. Am. Chem. Soc.* **1982,** *104*, 5798–5800.

35. Nocera, D.G.; Winkler, J.R.; Yocom, K.M.; Bordignon, E.; Gray, H.B., *J. Am. Chem. Soc.* **1984,** *106*, 5145–5150.
36. Moore, G.R.; Pettigrew, G.W., *Cytochromes c: Evolutionary, Structural, and Physicochemical Aspects*. Springer-Verlag: New York, 1990; p. 478.
37. Scott, R.A.; Mauk, A.G., *Cytochrome c - A Multidisciplinary Approach*. University Science Books: Sausalito, CA, 1996; p. 738.
38. Elias, H.; Chou, M.H.; Winkler, J.R., *J. Am. Chem. Soc.* **1988,** *110*, 429–434.
39. Brown, G.M.; Sutin, N., *J. Am. Chem. Soc.* **1979,** *101*, 883–892.
40. Chang, I.-J.; Gray, H.B.; Winkler, J.R., *J. Am. Chem. Soc.* **1991,** *113*, 7056–7057.
41. Chang, T.K.; Iverson, S.A.; Rodrigues, C.G.; Kiser, C.N.; Lew, A.Y.; Germanas, J.P.; Richards, J.H., *Proc. Natl. Acad. Sci. USA* **1991,** *88*, 1325–1329.
42. Adman, E.T.; Jensen, L.H., *Isr. J. Chem.* **1981,** 21, 8–12.
43. Pascher, T.; Karlsson, B.G.; Nordling, M.; Malmström, B.G.; Vanngard, T., *Eur. J. Biochem.* **1993,** *212*, 289–296.
44. Di Bilio, A.J.; Hill, M.G.; Bonander, N.; Karlsson, B.G.; Villahermosa, R.M.; Malmström, B.G.; Winkler, J.R.; Gray, H.B., *J. Am. Chem. Soc.* **1997,** *119*, 9921–9922.
45. Crane, B.R.; Di Bilio, A.J.; Winkler, J.R.; Gray, H.B., *J. Am. Chem. Soc.* **2001,** *123*, 11623–11631.
46. Newton, M.D., *J. Phys. Chem.* **1988,** *92*, 3049–3056.
47. Wasielewski, M.R., In *Photoinduced Electron Transfer. Part A: Conceptual Basis for Electron Transfer*, Fox, M. A.; Chanon, M., Eds. Elsevier: Amsterdam, 1988.
48. Moser, C.C.; Keske, J.M.; Warncke, K.; Farid, R.S.; Dutton, P.L., *Nature* **1992,** *355*, 796–802.
49. McConnell, H.M., *J. Chem. Phys.* **1961,** *35*, 508–515.
50. Beratan, D.N.; Onuchic, J.N.; Hopfield, J.J., *J. Chem. Phys.* **1987,** *86*, 4488–4498.
51. Beratan, D.N.; Onuchic, J.N., *Photosynth. Res.* **1989,** *22*, 173–186.
52. Onuchic, J.N.; Beratan, D.N., *J. Chem. Phys.* **1990,** *92*, 722–733.
53. Onuchic, J.N.; Andrade, P.C.P.; Beratan, D.N., *J. Chem. Phys.* **1991,** *95*, 1131–1138.
54. Beratan, D.N.; Betts, J.N.; Onuchic, J.N., *Science* **1991,** *252*, 1285–1288.
55. Betts, J.N.; Beratan, D.N.; Onuchic, J.N., *J. Am. Chem. Soc.* **1992,** *114*, 4043–4046.
56. Beratan, D.N.; Betts, J.N.; Onuchic, J.N., *J. Phys. Chem.* **1992,** *96*, 2852–2855.
57. Skourtis, S.S.; Beratan, D.N., *J. Phys. Chem. B* **1997,** *101*, 1215–1234.
58. Kumar, K.; Kurnikov, I.V.; Beratan, D.N.; Waldeck, D.H.; Zimmt, M.B., *J. Phys. Chem. A* **1998,** *102*, 5529–5541.
59. Roitberg, A.E.; Holden, M.J.; Mayhew, M.P.; Kurnikov, I.V.; Beratan, D.N.; Vilker, V.L., *J. Am. Chem. Soc.* **1998,** *120*, 8927–8932.
60. de Andrade, P.C.P.; Onuchic, J.N., *J. Chem. Phys.* **1998,** *108*, 4292–4298.
61. Stuchebrukhov, A.A.; Marcus, R.A., *J. Phys. Chem.* **1995,** *99*, 7581–7590.
62. Daizadeh, I.; Medvedev, E.S.; Stuchebrukhov, A.A., *Proc. Natl. Acad. Sci. USA* **1997,** *94*, 3703–3708.
63. Daizadeh, I.; Guo, J.-X.; Stuchebrukhov, A.A., *J. Chem. Phys.* **1999,** *110*, 8865–8868.
64. Dinner, A.R.; Karplus, M., *Nature Struct. Biol.* **2001,** *8*, 21–22.
65. Page, C.C.; Moser, C.C.; Chen, X.; Dutton, P.L., *Nature* **1999,** *402*, 47–52.
66. Tezcan, F.A.; Crane, B.R.; Winkler, J.R.; Gray, H.B., *Proc. Natl. Acad. Sci. USA* **2001,** *98*, 5002–5006.
67. Langen, R.; Chang, I.-J.; Germanas, J.P.; Richards, J.H.; Winkler, J.R.; Gray, H.B., *Science* **1995,** *268*, 1733–1735.
68. Regan, J.J.; Di Bilio, A.J.; Langen, R.; Skov, L.K.; Winkler, J.R.; Gray, H.B.; Onuchic, J.N., *Chem. Biol.* **1995,** *2*, 489–496.
69. Adman, E.T., *Adv. Protein Chem.* **1991,** *42*, 145–197.

70. Smalley, J.F.; Feldberg, S.W.; Chidsey, C.E.D.; Linford, M.R.; Newton, M.D.; Liu, Y.-P., *J. Phys. Chem.* **1995,** *99*, 13141–13149.
71. Smalley, J.F.; Finklea, H.O.; Chidsey, C.E.D.; Linford, M.R.; Creager, S.E.; Ferraris, J.P.; Chalfant, K.; Zawodzinsk, T.; Feldberg, S.W.; Newton, M.D., *J. Am. Chem. Soc.* **2003,** *125*, 2004–2013.
72. Jovanic, S.V.; Harriman, A.; Simic, M.G., *J. Phys. Chem.* **1986,** *90*, 1935–1939.
73. Harriman, A., *J. Phys. Chem.* **1987,** *91*, 6102–6104.
74. DeFelippis, M.R.; Murthy, C.P.; Faraggi, M.; Klapper, M.H., *Biochemistry* **1989,** *28*, 4847–4853.
75. Tommos, C.; Skalicky, J.J.; Piloud, D.L.; Wand, A.J.; Dutton, P.L., *Biochemistry* **1999,** *38*, 9495–9507.
76. Stubbe, J.; van der Donk, W.A., *Chem. Rev.* **1998,** *98*, 705–762.
77. Bollinger, J.M., Jr.; Edmondson, D.E.; Huynh, B.H.; Filley, J.; Norton, J.R.; Stubbe, J., *Science* **1991,** *253*, 292–298.
78. Hofbauer, W.; Zouni, A.; Bittl, R.; Kern, J.; Orth, P.; Lendzian, F.; Fromme, P.; Witt, H.T.; Lubitz, W., *Proc. Natl. Acad. Sci. USA* **2001,** *98*, 6623–6628.
79. Sucheta, A.; Szundi, I.; Einarsdottir, O., *Biochemistry* **1998,** *37*, 17905–17914.
80. Wuttke, D.S.; Bjerrum, M.J.; Winkler, J.R.; Gray, H.B., *Science* **1992,** *256*, 1007–1009.
81. Babini, E.; Bertini, I.; Borsari, M.; Capozzi, F.; Luchinat, C.; Zhang, X.; Moura, G.L.C.; Kurnikov, I.V.; Beratan, D.N.; Ponce, A.; Di Bilio, A.J.; Winkler, J.R.; Gray, H.B., *J. Am. Chem. Soc.* **2000,** *122*, 4532–4533.
82. Larsson, S., *J. Phys. Chem.* **1984,** *88*, 1321–1323.
83. Miller, N.E.; Wander, M.C.; Cave, R.J., *J. Phys. Chem. A* **1999,** *103*, 1084–1093.
84. Benjamin, I.; Evans, D.; Nitzan, A., *J. Chem. Phys.* **1997,** *106*, 6647–6654.
85. Wenger, O.S.; Leigh, B.S.; Villahermosa, R.M.; Gray, H.B.; Winkler, J.R., *Science* **2005,** *307*, 99–102.
86. Lin, J.P.; Balabin, I.A.; Beratan, D.N., *Science* **2005,** *310*, 1311–1313.
87. Schlichting, I.; Berendzen, J.; Chu, K.; Stock, A.M.; Maves, S.A.; Benson, D.E.; Sweet, B.M.; Ringe, D.; Petsko, G.A.; Sligar, S.G., *Science* **2000,** *287*, 1615–1622.
88. Davydov, R.; Ledbetter-Rogers, A.; Martásek, P.; Larukhin, M.; Sono, M.; Dawson, J.H.; Masters, B.S.S.; Hoffman, B.M., *Biochemistry* **2002,** *41*, 10375–10381.
89. Groves, J.T.; McClusky, G.A.; White, R.E.; Coon, M.J., *Biochem. Biophys. Res. Commun.* **1978,** *81*, 154–160.
90. Wilker, J.J.; Dmochowski, I.J.; Dawson, J.H.; Winkler, J.R.; Gray, H.B., *Angew. Chem., Int. Ed. Eng.* **1999,** *38*, 89–92.
91. Dunn, A.R.; Dmochowski, I.J.; Winkler, J.R.; Gray, H.B., *J. Am. Chem. Soc.* **2003,** *125*, 12450–12456.
92. Dmochowski, I.J.; Crane, B.R.; Wilker, J.J.; Winkler, J.R.; Gray, H.B., *Proc. Natl. Acad. Sci. USA* **1999,** *96*, 12987–12990.
93. Dmochowski, I.J.; Winkler, J.R.; Gray, H.B., *J. Inorg. Biochem.* **2000,** *81*, 221–228.
94. Dmochowski, I.J. Ph.D. Thesis, California Institute of Technology, Pasadena, 2000.
95. Yonemoto, E.H.; Saupe, G.B.; Schmehl, R.H.; Hubig, S.M.; Riley, R.L.; Iverson, B.L.; Mallouk, T.E., *J. Am. Chem. Soc.* **1994,** *116*, 4786–4795.
96. Paddon-Row, M.N.; Oliver, A.M.; Warman, J.M.; Smit, K.J.; de Haas, M.P.; Oevering, H.; Verhoeven, J.W., *J. Phys. Chem.* **1988,** *92*, 6958–6962.
97. Helms, A.; Heiler, D.; McLendon, G., *J. Am. Chem. Soc.* **1992,** *114*, 6227–6238.
98. Creager, S.; Yu, C.J.; Bamdad, C.; O'Connor, S.; MacLean, T.; Lam, E.; Chong, Y.; Olsen, G.T.; Luo, J.; Gozin, M.; Kayyem, J.F., *J. Am. Chem. Soc.* **1999,** *121*, 1059-1064.
99. Sikes, H.D.; Smalley, J.F.; Dudek, S.P.; Cook, A.R.; Newton, M.D.; Chidsey, C.E.D.; Feldberg, S.W., *Science* **2001,** *291*, 1519–1523.

100. Davis, W.B.; Svec, W.A.; Ratner, M.A.; Wasielewski, M.R., *Nature* **1998,** *396*, 60–63.
101. Kurnikov, I.V.; Dunn, A.R.; Gray, H.B.; Ratner, M.A., unpublished results.
102. Dunn, A.R.; Belliston-Bittner, W.; Winkler, J.R.; Getzoff, E.D.; Stuehr, D.J.; Gray, H.B., *J. Am. Chem. Soc.* **2005,** *127*, 5169–5173.
103. Belliston-Bittner, W.; Dunn, A.R.; Nguyen, Y.H.L.; Stuehr, D.J.; Winkler, J.R.; Gray, H.B., *J. Am. Chem. Soc.* **2005,** *127*, 15907–15915.
104. Dmochowski, I.J.; Dunn, A.R.; Wilker, J.J.; Crane, B.R.; Green, M.T.; Dawson, J.H.; Sligar, S.G.; Winkler, J.R.; Gray, H.B., *Meth. Enzymol.* **2002,** *357*, 120–133.
105. Sligar, S.G.; Lipscomb, J.D.; Debrunne, P.G.; Gunsalus, I.C., *Biochem. Biophys. Res. Commun.* **1974,** *61*, 290–296.
106. Dunn, A.R.; Dmochowski, I.J.; Bilwes, A.M.; Gray, H.B.; Crane, B.R., *Proc. Natl. Acad. Sci. USA* **2001,** *98*, 12420–12425.
107. Dunn, A.R.; Hays, A.-M.A.; Goodin, D.B.; Stout, C.D.; Chiu, R.; Winkler, J.R.; Gray, H.B., *J. Am. Chem. Soc.* **2002,** *124*, 10254–10255.
108. Halcrow, M.; Phillips, S.E.V.; Knowles, P.F., In *Enzyme-Catalyzed Electron and Radical Transfer*, Scrutton, N.S.; Holzenburg, A.K.H., Eds. Kluwer Academic/Plenum Publishers: London, 2000; Vol. 35, pp. 183–231.
109. McGuirl, M.A.; Dooley, D.M., *Curr. Opin. Chem. Biol.* **1999,** *3*, 138–144.
110. Klinman, J.P., *Chem. Rev.* **1996,** *96*, 2541–2561.
111. Dooley, D.M., *J. Biol. Inorg. Chem.* **1999,** *4*, 1–11.
112. Mure, M.; Mills, S.A.; Klinman, J.P., *Biochemistry* **2002,** *41*, 9269–9278.
113. Hess, C.R.; Juda, G.A.; Dooley, D.M.; Amii, R.N.; Hill, M.G.; Winkler, J.R.; Gray, H.B., *J. Am. Chem. Soc.* **2003,** *125*, 7156–7157.
114. Wilce, M.C.J.; Dooley, D.M.; Freeman, H.C.; Guss, J.M.; Matsunami, H.; McIntire, W.S.; Ruggiero, C.E.; Tanizawa, K.; Yamaguchi, H., *Biochemistry* **1997,** *36*, 16116–16133.
115. Mure, M.; Klinman, J.P., *J. Am. Chem. Soc.* **1993,** *115*, 7117–7127.
116. Sachs, S.B.; Dudek, S.P.; Hsung, R.P.; Sita, L.R.; Smalley, J.F.; Newton, M.D.; Feldberg, S.W.; Chidsey, C.E.D., *J. Am. Chem. Soc.* **1997,** *119*, 10563–10564.

Chapter 2

THE RESPIRATORY ENZYME AS AN ELECTROCHEMICAL ENERGY TRANSDUCER

MÅRTEN WIKSTRÖM
Helsinki Bioenergetics Group, Institute of Biotechnology, University of Helsinki, PB 65, 00014 University of Helsinki, Finland.

1. INTRODUCTION

Cell respiration is essential for all higher forms of life, which have evolved to make use of the high oxidation potential of the O_2/H_2O redox couple to oxidise foodstuff hydrocarbon derivatives, and thereby make full use of their available energy. Cell respiration closes the biological dioxygen/water cycle that is initiated by oxygenic photosynthesis in green plants and algae (cyanobacteria) and that is driven by light energy from the sun. The primary biological energy converters comprise the photosynthetic molecular machineries of plants and bacteria, bacteriorhodopsin, the special light-driven proton pump in *Halobacteria*, and the respiratory chain complexes in both bacteria and mitochondria of eukaryotic cells. All these membrane proteins convert either light energy or electron transfer across a redox potential difference to a electrochemical gradient of protons across the membrane in question. Thus Peter Mitchell's *protonmotive force* has a central position in primary bioenergetics. In this account, I will briefly review the present status of our knowledge of one of the key players in this scenario, *viz.* the respiratory enzyme (Otto Warburg's Atmungsferment, or David Keilins cytochrome *c* oxidase), which is not only responsible for catalysing the intriguing chemistry of respiratory O_2 reduction, but which simultaneously couples this reaction to translocation of protons across the bacterial or mitochondrial membrane, thereby contributing to formation of protonmotive force. The mechanism by which the redox reaction of

O. Hammerich and J. Ulstrup (eds.), Bioinorganic Electrochemistry, 25–35.

cytocrome *c* oxidase is coupled to proton translocation has long been enigmatic, and several seemingly paradoxical observations have been difficult to comprehend. It now seems that many of these paradoxes can be explained.

2. THE REDOX CENTRES AND THE OVERALL REACTION

In cytochrome *c* oxidase, all redox-active metal centres are found in subunits I and II. Subunit I comprises 12 transmembrane helices that are arranged in a propeller-like fashion with three redox-active metal centres.[1,2] The binuclear haem a_3-Cu_B site lies one third into the membrane from the positively charged *P*-side; this is the site of O_2 binding and reduction, Fig. 2-1.

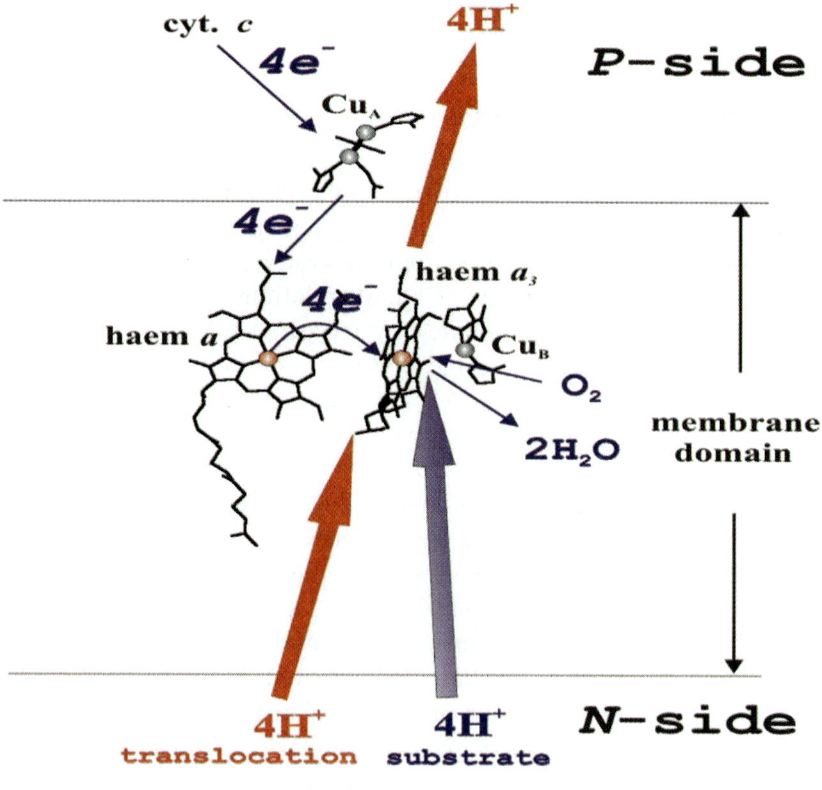

Figure 2-1. Schematic outline of the function of cytochrome *c* oxidase. Blue arrows denote the reduction of dioxygen catalysed by the enzyme. Red arrows denote the proton-pumping function that is linked to the redox reaction.

The low spin haem *a* lies next to haem a_3, at the same depth and almost at a right angle between the porphyrin planes, both of which lie orthogonally to the membrane plane. This arrangement of the two haem groups and Cu_B in subunit I is a completely conserved feature within the large family of otherwise variable haem-copper oxidases. Haem a_3 has a proximal histidine ligand; the distal coordination site binds O_2. Cu_B lies <5Å from the haem a_3 iron on the distal side and is coordinated by three histidines. The fourth coordination site points towards haem a_3 and is either unoccupied, or binds water or hydroxide[3]. Haem *a* with two axial histidine ligands functions as the electron donor to the binuclear haem a_3-Cu_B site, which therefore receives one electron at the time during function. Haem *a*, in turn, receives one electron at the time from the bimetallic Cu_A centre which is bound to a β-barrel domain of subunit II on the *P*-side of the membrane, and which covers the haem domain from that side as a cap.

Figure 2-1 summarises the overall protonmotive function of cytochrome *c* oxidase. Note that the redox reaction (*i.e.* reduction of O_2 to water by ferrocytochrome *c*) is itself organised vectorially with respect to the membrane. Electrons are derived from cytochrome *c* on the *P*-side of the membrane and the substrate protons from the *N*-side, so that when their electric charges are annihilated to reduce O_2 to two water molecules at the haem a_3-Cu_B centre, one charge equivalent is translocated across the membrane contributing to the formation of protonmotive force. An additional electrical charge equivalent is translocated across the membrane per transferred electron due to the enzyme functioning as a proton pump[4], which doubles the energy-conserving efficiency of this device.

3. PROTON TRANSFER PATHWAYS

Six out of the eight protons taken up per catalytic cycle from the negatively charged *N*-side of the membrane are transferred via the same so-called D-pathway[5,6]. Two of the four substrate protons is probably transferred to the binuclear site by another pathway, the so-called K-channel that involves a conserved lysine residue and ends up at a tyrosine group within the site. The D-pathway starts with an aspartic acid near the *N*-side and continues into the membrane domain with several hydrophilic residues and bound water molecules[7–9], and reaches a conserved glutamic acid which lies ca. 10 Å from the haem groups, Fig. 2-2. The domain between the glutamic acid and the haem groups comprises a narrow cavity lined by mostly hydrophobic residues, but with no obvious means for transferring the protons any further, as judged from the X-ray structures. Since all pumped protons, and most substrate protons that are consumed at the binuclear site, must pass through this pathway, some kind of "switching" between proton transfer paths is necessary beyond the glutamic acid[10] (see below).

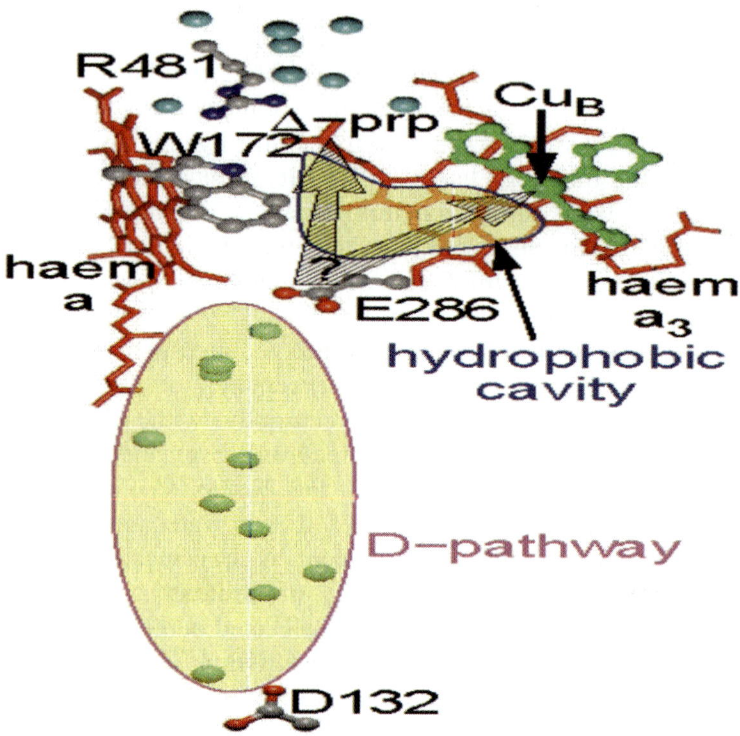

Figure 2-2. Proton transfer via the D-pathway in cytochrome *c* oxidase. The view is similar to that in Fig. 2-1, parallel to the membrane. The D-pathway starts with an aspartic acid (D132) near the *N*-side of the membrane and continues towards a conserved glutamic acid (E286). D-pathway water molecules seen in the crystal structure[9] of the enzyme from *Rh. sphaeroides* are shown in light green, and water molecules above the haem centres in light blue. Δ-prp denotes the Δ-propionate of haem a_3. The haem groups are shown in red and Cu_B with its three imidazole ligands in green. A hydrophobic cavity is depicted above E286, with arrows indicating proton transfer from E286 further to the propionate and the binuclear site (see the text).

Statistical-mechanical calculations have suggested that the hydrophobic cavity may contain 3–4 water molecules[7], at least transiently, and these may be crucial not only for proton transfer as such, but also in the mechanism of coupling between electron and proton transfer, as outlined below. Finally, it should be emphasised that the hydrophobic cavity is next to the binuclear oxygen reduction site, and that it is likely that the product of the redox reaction, water, is initially ejected into the hydrophobic cavity during the reaction. We propose here that the product water is not wasted, but may have an important function in the enzyme mechanism before being disposed of.

4. CHEMISTRY OF O_2 REDUCTION AND PROTON TRANSLOCATION

Figure 2-3 summarises the states of the binuclear centre during the catalytic cycle. It is especially noteworthy that after the reduced ferrous/cuprous site (**R**) has bound dioxygen to form state **A**, there is next a full four-electron reduction of the bound dioxygen is one step[11,12], leading to state $\mathbf{P_M}$ with ferryl haem iron and cupric Cu_B.

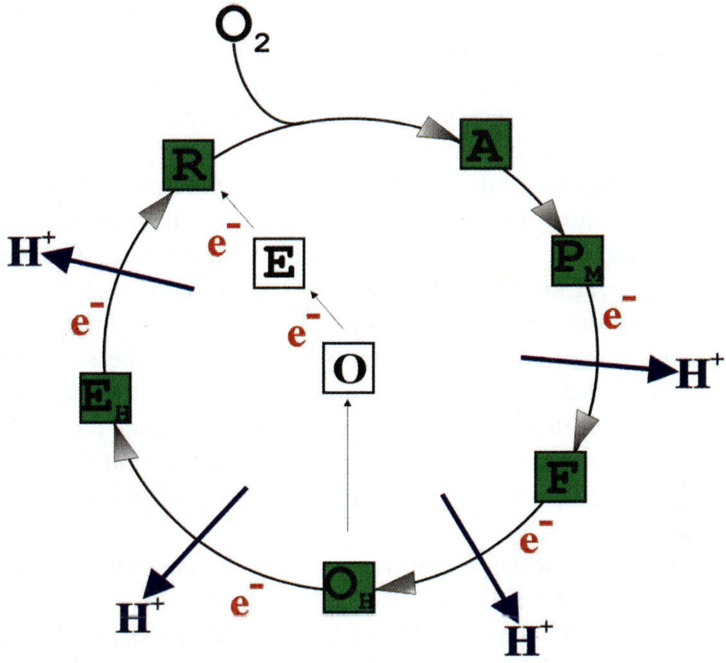

Figure 2-3. Catalytic cycle of cytochrome *c* oxidase. The figure depicts different states of the binuclear haem a_3-Cu_B centre (squares, see the text), and shows the reaction steps where an electron is transferred to the centre from cytochrome *c* via haem a. If the enzyme is deprived of an electron donor, the oxidised state $\mathbf{O_H}$ decays into a relaxed form **O**. The latter may be reduced back to state **R** with uptake of two substrate protons (not shown), but this is not coupled to proton translocation (blue arrows). Note that otherwise each electron transfer into the binuclear site is coupled to proton translocation, and to uptake of a substrate proton (not shown). Thus each blue arrow represents uptake of two protons from the *N*-side and release of one proton to the *P*-side of the membrane (see also Fig. 2-1).

The product "water" remains partially unprotonated in the form of the oxide of the ferryl iron and a hydroxide ligand of cupric Cu_B. The fourth reducing equivalent probably derives from a conserved tyrosine within the site that is covalently bonded to one of the histidine ligands of Cu_B, possibly forming a neutral tyrosine radical in $\mathbf{P_M}$. However, the precise location of the unpaired spin is not known with certainty, and it is possible that it is distributed between the tyrosine, Cu_B, and its histidine ligands. At any rate, the crux of this mechanism is that the entire oxidation potential of dioxygen is transferred to the enzyme's binuclear site in a single reaction that is not coupled to proton or charge translocation across the membrane. Hence, this is the prelude to exergonic electron transfer from cytochrome *c* to the now highly oxidising acceptor site, which is coupled to proton-pumping.

Input of the next electron into the site reduces the radical and leads to the intermediate **F** which has taken up an additional proton, presumably to protonate the Cu_B[II] hydroxide ligand to water. This step is the first one that is coupled to translocation of a proton across the membrane. The next electron (and proton) enters the site, again coupled to translocation of a proton, which yields a ferric-cupric intermediate $\mathbf{O_H}$ (previously called O~ and H, Fig. 2-3). $\mathbf{O_H}$ is a metastable form of the binuclear site, which relaxes to the more stable form **O** in the absence of further electron donation[13]. **O** is the state of the enzyme "as isolated". The chemical difference between states $\mathbf{O_H}$ and **O** is not yet known (their redox states are the same), but it is likely that the redox potential of Cu_B is considerably higher in $\mathbf{O_H}$ than it is in **O**.

The oxidative half of the catalytic cycle ends with the states $\mathbf{O_H}$, or **O**. Both these states of the binuclear site may be reduced in the reductive half of the cycle by two electrons to form state **R**, which initiates the next cycle by binding O_2. However, only the reductive route via $\mathbf{O_H}$ and the one-electron reduced $\mathbf{E_H}$ state is coupled to proton translocation, whereas the route via **O** and **E** is not,[14] Fig. 2-3.

Michel and co-workers[15,16] have recently dismissed the notion of a metastable state of the fully oxidised enzyme, $\mathbf{O_H}$. In their experiments[16] they produced the **F** state by treating the oxidised enzyme with H_2O_2. Subsequently, they exposed the enzyme in the **F** state to carbon monoxide, which is known to react with **F** by a strict two-electron reduction. Consequently, this procedure yields an **E** state without forming the metastable intermediate $\mathbf{O_H}$ (see Fig. 2-3). Hence, their report that photo-induced reduction of the **E** state produced this way (we would ascribe this state to $\mathbf{E_H}$) is linked to proton translocation does not contradict our notion, but rather strengthens it.

5. ROLE OF WATER MOLECULES WITHIN THE ENZYME

We have recently proposed a redox-linked proton translocation mechanism[17] that relies on the unique behaviour of water molecules in a restricted hydrophobic space, such as the cavity in subunit I between the conserved glutamic acid and the binuclear site, Fig. 2-2. Single file arrays of water molecules are excellent unidirectional proton transfer devices, or "wires"[18]. Proton transfer through such wires is often termed the Grotthuss mechanism[19] after a proposal of C.J.T. de Grotthuss[20] in 1806, and is 40 times faster than proton transfer in bulk water[21]. Water dipoles confined to narrow hydrophobic cavities exhibit strong water-water hydrogen bonds in the absence of interactions with the surroundings, and tend to form single files that behave cooperatively to sustain these bonds. The occupancy of narrow cavities with water molecules is a sensitive function that is strongly dependent on the strength of the interactions with the cavity wall. Thus even subtle changes in these interactions can cause transitions between an empty and a full cavity[22]. Such behaviour may well explain why the cavity in cytochrome *c* oxidase appears empty by X-ray crystallographic criteria, but appears to be filled at least transiently with 3–4 water molecules during dynamic turnover.

Kornblatt[23] has shown in elegant experiments that the electron transfer from haem *a* to the binuclear site in cytochrome *c* oxidase depends on the presence of water molecules within the protein. It has long been known that oxidised cytochrome oxidase, "as isolated", is in a notoriously slow-reacting "resting" state, which is typically reflected as a limitation of electron transfer between haem *a* and the binuclear site. This limitation is removed by enzyme turnover, or by pre-reducing the enzyme and reoxidising it with O_2 ("pulsed" oxidase)[24–26]. It is possible, therefore, that "resting" oxidase represents a state in which key domains of the structure are "dry", and that the water molecules produced during turnover convert the enzyme to its "pulsed" form.

Electron transfer *per se* from haem *a* to the binuclear site, over a distance of ca. 7–9 Å between the edges of the two haem groups, is expected to be very fast (τ = a few nanoseconds)[27,28]. Although haem-haem electron transfer with a time constant of ca. 3 μs has been measured in several laboratories, there are indications of a much faster rate of electron tunneling[29] (but see also ref. 30). Yet, the electron transfer rate measured is already about three orders of magnitude faster than V_{max}, and can hence hardly be the cause of the limitation in the "resting" enzyme. However, the observed 3 μs electron transfer is an electron equilibration between haems a_3 and *a* after reduction of haem a_3 in the presence of CO, followed by flash photolysis of CO.

Normally, the reduction of haem a_3 is limited by the rate of accompanying proton uptake; as formulated by Peter Rich[31], reduction of the binuclear site is governed by the electroneutrality principle. In essence, this means that whilst electron tunneling is very fast, the redox potential of the binuclear site is very low in the absence of a charge-compensating proton. The overall rate is thus determined by the product of the rate constant for proton transfer and the (low) occupancy of the reduced unprotonated binuclear centre. With this in mind, it seems plausible that the slow rate of electron transfer from haem *a* to the binuclear site in the "resting" enzyme is due to slow proton transfer, which in turn may be caused by the absence of suitably arranged water molecules transmitting it.

6. A POSSIBLE ELECTROCHEMICAL MECHANISM OF PROTON TRANSLOCATION

Since protons to be pumped across the membrane, or to be consumed to form water at the binuclear site, are initially transferred via the common D-pathway to the conserved glutamic acid, Fig. 2-2, some kind of molecular "switch" would be required beyond the glutamic acid to guide protons in a controlled fashion to the correct destination. It occurred to us that water molecules in the cavity between the glutamic acid and the binuclear centre might not only provide a pathway for proton transfer, but could also function as a switch, and simultaneously provide the linkage between proton transfer and the redox reaction that is essential in a redox-coupled proton pump[17]. This idea was further nourished by the unique and fully conserved structural arrangement of haem *a* and the binuclear site, relative to one another and to the membrane. The proximity of haem *a* to the binuclear site and to the hydrophobic cavity, and the arrangement of the two haem groups at the same depth within the membrane are particularly intriguing features.

As summarised above, the key reaction step in the catalytic cycle that is coupled to proton translocation is the electron transfer from haem *a* to the binuclear site, whether the latter is in the $\mathbf{P_M}$, **F**, $\mathbf{O_H}$ or $\mathbf{E_H}$ state, Fig. 2-3. Prior to this electron transfer, haem *a* is more negatively charged than the binuclear site. Thus there is a corresponding electric field parallel to the membrane between the electron donor and the acceptor sites. We conjectured that 3–4 water dipoles in the hydrophobic cavity forming a single file, will orientate in this field to provide a Grotthuss path from the glutamic acid to the Δ-propionate of haem a_3, which has been inferred by site-directed mutagenesis experiments to be at the beginning of the exit path for pumped protons[32]. Notably, these water molecules do not provide a conduction path to the binuclear site in this situation, which prevents completion of the oxygen reduction chemistry. Instead, proton transfer occurs from glutamic acid via the propionate group to the relatively hydrophilic domain above haem a_3. This initial proton transfer raises the redox potential of the binuclear site sufficiently to allow electron transfer. Meanwhile, the glutamate is reprotonated via the D-pathway. The transfer of the electron from haem *a* to the binuclear centre switches the electric field into the opposite orientation. As a result, the array of water molecules reorientates to provide a proton-conducting path from the glutamic acid to the binuclear site. Finally, a second proton is transferred from the glutamic acid to this site with complete neutralisation of its negative charge, whereby the "prepumped" proton above the haem group is ejected towards the *P*-side of the membrane by electrostatic repulsion. Figure 2-4 provides a schematic summary of these events.

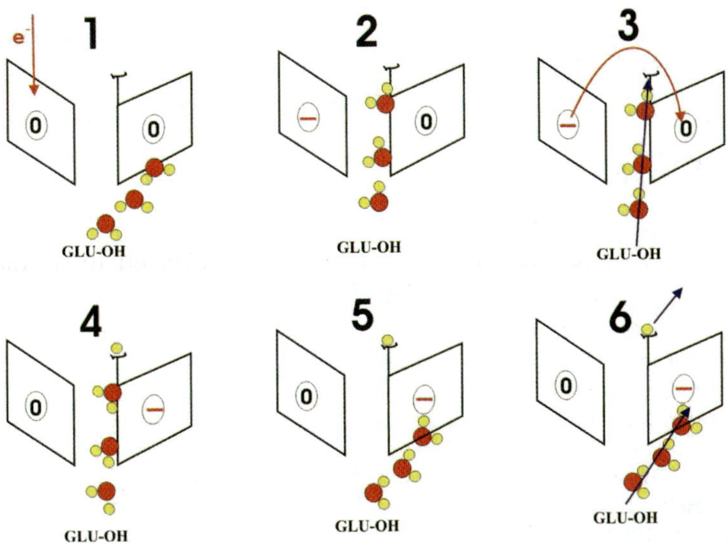

Figure 2-4. A simplified scheme of the proposed water-gated mechanism of proton translocation. Each numbered state shows haem *a* and the binuclear site (left and right rectangles, respectively); the Δ-propionate of haem a_3 is shown schematically. Three water molecules (oxygen in red; hydrogen in yellow) are shown to mediate Grotthuss proton transfer from the glutamic acid (GLU-OH) to the propionate or the binuclear site, respectively. In state 1, an electron is transferred to haem *a*. The formed electric field between the redox sites orientates the water molecules towards the propionate (state 2). In state 3, electron transfer to the binuclear site is accompanied by proton transfer via the propionate; a proton is deposited above haem a_3 and the glutamate is reprotonated via the D-pathway (state 4). The switch of electric field orientation reorientates the water array towards the binuclear site (state 5). Finally, a proton is transferred to this site, and the first proton is ejected (state 6). Reprotonation of the glutamate transfers the system back to state 1. For details, see the text and ref. 17.

This water-gated mechanism is consistent with the strength of the charge-dipole interactions expected between the electron donor (acceptor), and the water dipoles in the cavity, as well as with the large amount of biochemical data available to date. It is also strongly supported by molecular dynamics simulations[17], which showed a remarkably consistent behaviour of water molecules in the cavity in response to the redox states of haem *a* and the binuclear site. Finally, this mechanism provides a satisfactory explanation of the unique structural arrangement of the redox-active centres in subunit I of all haem-copper oxidases.

ACKNOWLEDGEMENTS

I am grateful to Gerhard Hummer and Michael I. Verkhovsky for numerous discussions and help with the figures, and to the Sigrid Juselius Foundation and the Academy of Finland for financial support.

REFERENCES

1. Iwata, S.; Ostermeier, C.; Ludwig, B.; Michel, H., *Nature* **1995**, *376*, 660–669.
2. Tsukihara, T.; Aoyama, H.; Yamashita, E.; Tomizaki, T.; Yamaguchi, H.; Shinzawa-Itoh, K.; Nakashima, R.; Yaono, R.; Yoshikawa, S., *Science* **1995**, *269*, 1069–1094.
3. Fann, Y.C.; Ahmed, I.; Blackburn, N.J.; Boswell, J.S.; Verkhovskaya, M.L.; Hoffman, B.M.; Wikström, M., *Biochemistry* **1995**, *34*, 10245–10255.
4. Wikström, M., *Nature* **1977**, *266*, 271–273.
5. Wikström, M.; Jasaitis, A.; Backgren, C.; Puustinen, A.; Verkhovsky, M.I., *Biochim. Biophys. Acta* **2000**, *1459*, 514–520.
6. Gennis, R. B., *Biochim. Biophys. Acta* **1998**, *1365*, 241–248.
7. Riistama, S.; Hummer, G.; Puustinen, A.; Dyer, R.B.; Woodruff, W.H.; Wikström, M., *FEBS Lett.* **1997**, *414*, 275–280.
8. Hofacker, I.; Schulten, K., *Proteins: Structure, Function and Genetics* **1998**, *30*, 100–107.
9. Svensson-Ek, M.; Abramson, J.; Larsson, G.; Tornroth, S.; Brzezinski, P.; Iwata S., *J. Mol. Biol.* **2002**, *321*, 329–339.
10. Pawate, A.S.; Morgan, J.; Namslauer, A.; Mills, D.; Brzezinski, P.A.; Ferguson-Miller, S.; Gennis, R.B., *Biochemistry* **2002**, *41*, 13417–13423.
11. Babcock, G.T., *Proc. Natl. Acad. Sci. USA* **1999**, *96*, 12971–12973.
12. Proshlyakov, D.A.; Pressler, M.A.; DeMaso, C.; Leykam, J.F.; DeWitt, D.L.; Babcock, G.T., *Science* **2000**, *290*, 1588–1591.
13. Verkhovsky, M.I.; Jasaitis, A.; Verkhovskaya, M.L.; Morgan, J.E.; Wikström, M., *Nature* **1999**, *400*, 480–483.
14. Wikström, M.; Verkhovsky, M.I., *Biochim. Biophys. Acta* **2002**, *1555*, 128–132.
15. Michel, H., *Biochemistry* **1999**, *38*, 15129–15140.
16. Ruitenberg, M.; Kannt, A.; Bamberg, E.; Fendler, K.; Michel, H., *Nature* **2002**, *417*, 99–102.

17. Wikström, M.; Verkhovsky, M.I.; Hummer, G., *Biochim. Biophys. Acta* **2003**, *1604*, 61–65.
18. Nagle, J.F., Tristram-Nagle, S., *J. Membrane Biol.* **1983**, *74*, 1–14.
19. Agmon, N., *Chem. Phys. Lett.* **1995**, *244*, 456–462.
20. de Grotthuss, C.J.T., *Ann. Chim.* **1806**, *58*, 54–74.
21. Dellago, C.; Naor, M.M.; Hummer, G., *Phys. Rev. Lett.* **2003**, *90*, 105902.
22. Hummer, G.; Rasaiah, J.C.; Noworyta, J.P., *Nature* **2001**, *414*, 188–190.
23. Kornblatt, J.A., *Biophys. J.* **1998**, *75*, 3127–3134.
24. Antonini, E.; Brunori, M.; Colosimo, A.; Greenwood, C.; Wilson, M.T., *Proc. Natl. Acad. Sci. USA* **1977**, *74*, 3128–3132.
25. Brunori, M.; Sarti, P.; Malatesta, F.; Antonini, G.; Wilson, M.T., In *Cytochrome systems: Molecular biology and bioenergetics*, Papa, S.; Chance, B.; Ernster, L., Eds. Plenum: N.Y., 1987, pp. 689–695.
26. Brunori, M.; Antonini, G.; Malatesta, F.; Sarti, P.; Wilson, M.T., *Eur. J. Biochem.* **1987**, *169*, 1–8.
27. Moser, C.C.; Keske, J.M.; Warncke, K.; Farid, R.S.; Dutton, P.L., *Nature* **1992**, *355*, 796–802.
28. Moser, C.C.; Page, C.C.; Farid, R; Dutton, P.L., *J. Bioenerg. Biomemb.* **1995**, *27*, 263–274.
29. Verkhovsky, M.I.; Jasaitis, A.; Wikström, M., *Biochim. Biophys. Acta* **2001**, *1506*, 143–146.
30. Namslauer, A.; Branden, M., Brzezinski, P., *Biochemistry* **2002**, *41*, 10369–10374.
31. Rich, P.R., *Aust. J. Plant Physiol.* **1995**, *22*, 479–486.
32. Puustinen, A., Wikström, M., *Proc. Natl. Acad. Sci. USA* **1999**, *96*, 35–37.

Chapter 3

RECONSTITUTED REDOX PROTEINS ON SURFACES FOR BIOELECTRONIC APPLICATIONS

BILHA WILLNER AND ITAMAR WILLNER
Institute of Chemistry, The Hebrew University of Jerusalem, Jerusalem 91904, Israel

1. INTRODUCTION

The electrical contacting of redox proteins with electrodes is a central issue in bioelectronics. Redox proteins usually lack direct electrical communication with electrodes, and hence the direct electrochemical activation of the proteins is prohibited. The lack of electrical contact between the redox centers of proteins and electrodes can be explained by the electron transfer (ET) Marcus theory.[1] The ET rate constant between a donor and acceptor pair is given by eqn. 3-1 where d and d_o are the distance separating the electron and donor, and the van der Waals distance, respectively, β is the electron-coupling constant and $\Delta G°$ and λ are the free energy change and the reorganization energy accompanying the electron-transfer process, respectively.

$$k_{ET} \propto \exp[-\beta(d-d_o)]\times\exp[-(\Delta G° + \lambda)^2/(4RT\lambda)] \qquad (3\text{-}1)$$

Since the electrode and the protein redox center may be considered as a donor-acceptor pair, the distance, or spatial, separation of the enzyme redox center from the electrode by means of the protein shell prohibits the direct electrical communication between the redox site and the electrode. Substantial research efforts were directed during the last two decades to overcome the insulating protein barrier surrounding the redox centers and

O. Hammerich and J. Ulstrup (eds.), Bioinorganic Electrochemistry, 37–90.

developing means to electrically contact redox proteins and electrodes.[2] Diffusional molecular electron transfer mediators such as ferricyanides,[3] quinones,[4] bipyridinium salts[5] or transition metal complexes,[6] were widely applied to contact redox proteins with the electrodes by a diffusional path. A different approach employed molecular modifiers on electrodes (promoters) that interact with the proteins and align them on the surface in a configuration that facilitates electrical contact.[7] Using this method, hemoproteins such as cytochrome *c*[7] or myoglobin[8] were activated towards ET communication with conductive supports. A different method to establish ET between redox enzymes and electrodes involved the chemical modification of the protein with electron relay units that shorten the ET distances.[9] The tethered relay units activate the electrical contacting of the redox centers and the electrodes by an intra-protein ET hopping process. This method was successfully applied to develop integrated, electrically contacted, layered enzyme electrodes that included tethered relay units.[10] The loading of the protein by the relay groups is, however, important to control the ET effectiveness. Increased loading of the relay units enhances the ET hopping probability, the introduction of the foreign relay units adversely affect the protein structure and its catalytic activity, and hence an appropriate balance of the extent of loading has to be optimized for the systems. A further method to electrically contact redox enzymes with electrode surfaces involves the immobilization of the biocatalysts in conductive polymer matrices[11,12] or redox-active polymers and, specifically, hydrogels.[13,14] The entanglement of the polymer chains within the protein structure facilitates the charge transport between the electrode and the biocatalytic redox centers by means of the electroactive polymer wires.

Although the different methods discussed so far are successful in establishing ET communication between the redox centers and electrodes, they suffer from intrinsic limitations. The properties of the macroscopic integrated enzyme-electrodes represent the collective functions of numerous *different* enzymes that include various degrees and positions of the anchored relay units, and different alignment of the proteins in respect to the polymer backbone and conductive support. These difficulties limit the ET communication of the biocatalysts and the conductive supports, as compared to the ET efficiency between the enzyme redox sites and their natural ET substrates or cofactors. For example, the most efficient electrically-contacted ferrocene-tethered glucose oxidase revealed an ET exchange rate that corresponded to 2 s^{-1} between the redox center and the electrode,[15] while the ET turnover rate between the glucose oxidase redox center and its native oxygen electron acceptor[16] is ca. 650 s^{-1}. The effectiveness of ET communication between the enzyme redox centers and the electrodes has important implications on enzyme electrodes performances and their practical utility in bioelectronic devices. The efficiency of ET communication determines the current output of the electrodes, and thus controls their sensitivities, their specificities and their sensing processes. Also, the electrical communication efficiency between redox enzymes and electrode surfaces controls the current and extractable voltage of enzyme electrodes. These are two invaluable parameters that regulate the output power of biofuel cells.[17]

The need to improve the electrical communication between redox proteins and electrodes, and the understanding that the structural orientation at the molecular level of redox proteins and electroactive relay units on the conductive surfaces is a key element to facilitate ET, introduced tremendous research efforts to nano-engineer enzyme electrodes with improved ET functionalities. The present chapter addresses recent advances in the assembly of structurally aligned enzyme layers on electrodes by means of surface reconstitution and surface crosslinking of structurally oriented enzyme/cofactor complexes on electrodes. The ET properties of the nano-structured interfaces is discussed, as well as the possible application of the systems in bioelectronic devices such as biosensors, biofuel cell elements or optical and electrical switches.

The reconstitution process is a well practiced methodology to examine the structure-function relationship of proteins. The reconstitution method, Fig. 3-1, (exemplified for myoglobin) involves the exclusion of the native active center from the protein, e.g. an ion or a cofactor, to yield the respective apo-protein (or hollow-protein). The implanting (or reconstitution) of a structurally-related cofactor or ion generates the semi-synthetic reconstituted protein that might exhibit new tailored functions that are not present in the native protein. The reconstitution process has been extensively practiced upon studying mechanistic aspects of photoinduced ET in proteins.[18–20] Hemoproteins such as cytochrome *c*[21] or myoglobin[22] were reconstituted with photoactive Zn(II)-protoporphyrin IX or different metal analog systems that enabled the systematic elucidation of the effects of free energy and distance dependencies on intra-protein ET. The reconstitution of hemoproteins with synthetically modified heme units[23] or with metal protoporphyrin analogs[24] enabled the synthesis of proteins of new designed photochemical, binding and catalytic functions. For example, Fig. 3-1 depicts several structures of the functional reconstituted proteins. The reconstitution of apo-myoglobin with the Ru(II)-*tris*-bipyridine-heme dyad, (**1**), yielded[25] a reconstituted photoactive protein that photocatalyzed the reduction of O_2.

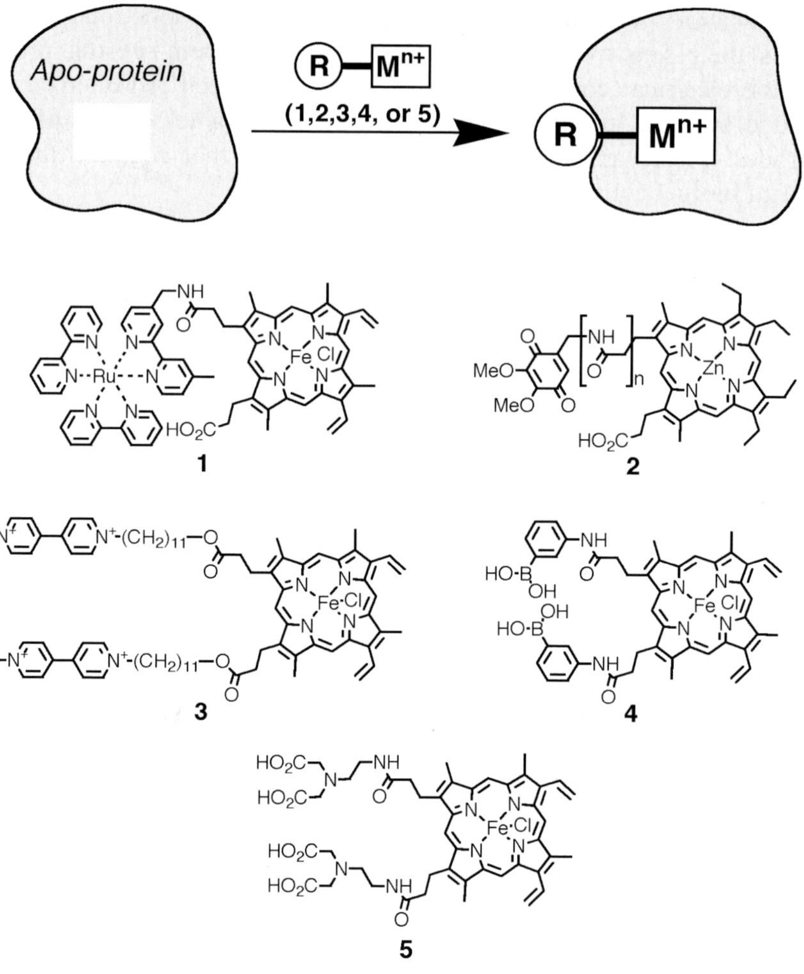

Figure 3-1. The reconstitution of apo-myoglobin with functionalized heme units.

The reconstitution of apo-myoglobin with the quinone-Zn(II)-protoporphyrin IX, (**2**),[26] or the bipyridinium-Zn(II)-protoporphyrin IX, (**3**),[27] generated photoactive reconstituted proteins that revealed protein-regulated ET properties. The reconstitution of apo-myoglobin with the heme unit tethered to the phenyl boronic acid ligand, (**4**), or the amino-diacetic acid ligand, (**5**), generated proteins with sugar binding-sites[28] and metal-ion[29] binding functions, respectively. Similarly, the reconstitution of apo-myoglobin with Co(III)-protoporphyrin IX and the covalent attachment of eosin chromophores to the protein, generated a photoactive hydrogenation biocatalyst that stimulated the light-induced hydrogenation of acetylenes.[23]

2. ELECTRODES FUNCTIONALIZED WITH RECONSTITUTED REDOX PROTEINS

Two general strategies to assemble the reconstituted enzymes on electrodes were developed, Fig. 3-2.

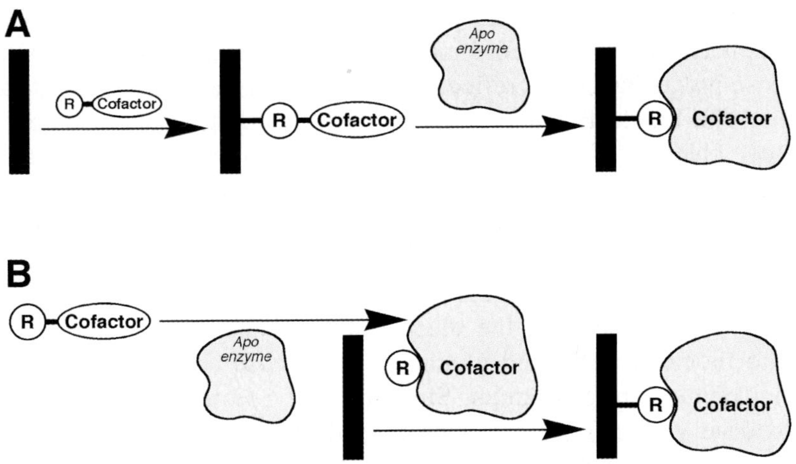

Figure 3-2. Methods for the assembly of reconstituted relay-cofactor enzyme assemblies (A) by the surface reconstitution of the apo-enzyme; (B) by the attachment of the pre-reconstituted enzyme on the electrode.

By one method a relay-cofactor dyad is assembled on the electrode, and the respective apo-protein is reconstituted on the surface to yield an aligned protein that is linked to the conductive surface by the relay component. The second method involves the synthesis of the relay-cofactor unit and the reconstitution of the apo-protein in solution. The specific immobilization of the enzyme on the electrode by the relay unit provides the structurally organized enzyme electrodes. While the first method is technically easier, the second methodology that involves tedious synthetic and separation steps, permits the fundamental structural characterization of the reconstituted protein. In the two configurations, the redox enzymes are anticipated to be electrically contacted with the electrode by means of the relay, a conductive molecular or polymer wire, a conductive nanoparticle, or even conductive carbon or doped silicon nanotubes.

2.1 Reconstituted Flavoenzyme Electrodes

The reconstitution of apo-glucose oxidase, apo-GOx, with the ferrocene-amino-FAD (FAD = flavin adenine dinucleotide) dyad, (**6**), led to the formation of an electrically contacted enzyme.[30] At a cystamine monolayer-functionalized Au-electrode, Fig. 3-3A, the reconstituted enzyme was electrochemically activated towards the oxidation of glucose. Figure 3-3B shows the cyclic voltammograms observed in the presence of the reconstituted glucose oxidase, GOx, at different concentrations of glucose. As the concentration of glucose increased, the bioelectrocatalyzed oxidation of glucose was enhanced as reflected by the increased electrocatalytic anodic currents. The bioelectrocatalytic oxidation of glucose was attributed to the mediated electron transfer from the FAD cofactor to the electrode. Electrochemical oxidation of the ferrocene units to the ferrocenylium cation is followed by the oxidation of the FAD cofactor that activates the biocatalyzed oxidation of glucose, Fig. 3-3A. The intermediary functions of the ferrocene units in the bioelectrocatalyzed oxidation of glucose is supported by the fact that the electrocatalytic anodic currents originating upon the bioelectrocatalyzed oxidation of glucose appear at the oxidation potential of the ferrocene units. Similarly, the reconstitution of apo-amino acid oxidase with (**6**) yielded a bioelectrocatalytically active enzyme that stimulated the electrochemical oxidation of L-alanine.

Integrated electrically-contacted enzyme electrodes were prepared by the surface reconstitution of different apo-enzymes on electrode surfaces.[31] The pyrroloquinoline quinone, PQQ, (**7**), was covalently linked to a cystamine monolayer associated with an Au-electrode, and N^{-6}-(2-aminoethyl-FAD), (**8**), was covalently attached to the PQQ units, Fig. 3-4A. The integrated enzyme-electrode was then prepared by the reconstitution of apo-GOx on the FAD units. The surface coverage of the PQQ-FAD units was estimated to be 5.5×10^{-10} mole cm^{-2}, whereas the surface coverage of the reconstituted enzyme was determined to be 1.5×10^{-12} mole cm^{-2}. Figure 3-4B shows the cyclic voltammogram corresponding to the integrated enzyme electrode in the presence of glucose, 80 mM. The electrocatalytic anodic current observed at $E = -0.125$ V vs. SCE (the redox potential of the PQQ units) indicates that the reconstituted enzyme is electrically-contacted with the electrode.

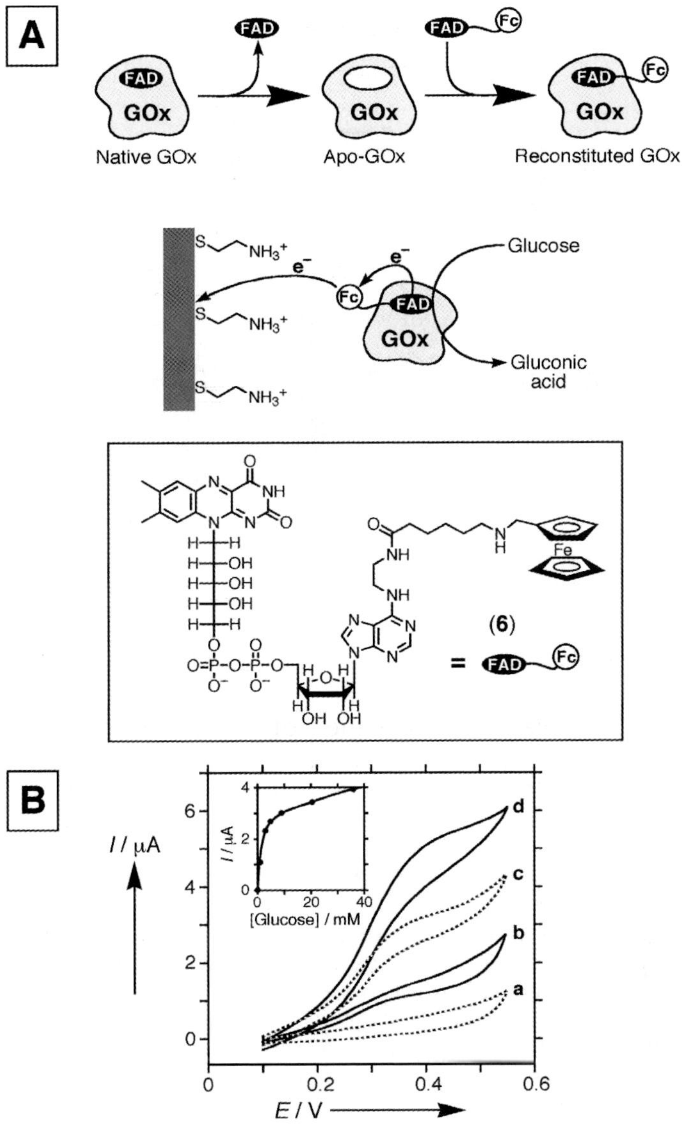

Figure 3-3. (A) The reconstitution of apo-glucose oxidase with ferrocene-functionalized FAD, (**6**), and the bioelectrocatalytic functions of the reconstituted enzyme. (B) Cyclic voltammograms corresponding to the bioelectrocatalayzed oxidation of glucose by the (**6**)-reconstituted glucose oxidase in the presence of different concentrations of glucose: (a) 0 mM, (b) 1 mM, (c) 3 mM, (d) 20.5 mM. Inset: Calibration curve corresponding to the electrocatalytic currents at different glucose concentrations of the system. Data recorded in 0.1 M phosphate buffer, pH 7.3, 35°C, scan rate 2 mV s^{-1}. Part B is adapted from ref. 30 with permission. Copyright Nature Publishing Group, 1995.

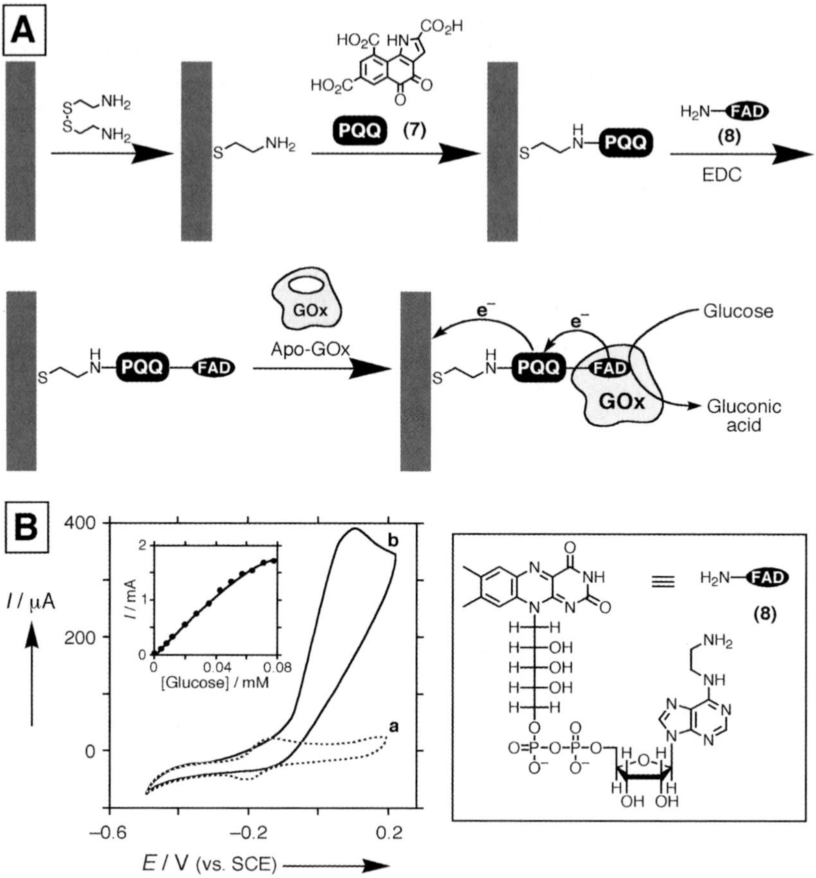

Figure 3-4. (A) Surface reconstitution of apo-glucose oxidase on a PQQ-FAD monolayer associated with an Au-electrode. (B) Cyclic voltammograms corresponding to (a) the reconstituted glucose oxidase monolayer in the absence of glucose, and (b) in the presence of glucose, 80 mM. Data recorded in 10 mM phosphate buffer, pH 7.0, at 35°C, scan rate 5 mV s^{-1}. Inset: Calibration curve corresponding to the chronoamperometric currents at $E = 0.2$ V at different glucose concentrations. Reproduced with permission from ref. 31. Copyright 1996 American Chemical Society.

Since the reconstitution of apo-GOx onto an FAD-monolayer linked to the Au-electrode did not yield an electrically contacted enzyme electrode, and since the electrocatalytic anodic current observed for the integrated electrode was observed at the redox-potential of the PQQ units, the electrical communication between the enzyme redox center and the electrode and its bioelectrocatalytic activation was attributed to the PQQ-mediated electron transfer in the system, Fig. 3-4A. The FAD centers of reconstituted GOx are reduced by glucose, and the reduced cofactors are then oxidized by the PQQ units to form $PQQH_2$. Electrochemical oxidation of $PQQH_2$ that is not shielded by the protein shell regenerates the ET mediator, and this enables the cyclic biocatalyzed oxidation of glucose as long as the $PQQH_2$ units are electrochemically oxidized. Figure 3-4B, inset, shows the calibration curve that corresponds to the anodic currents observed at variable concentrations of glucose. The current increases as the glucose concentration is elevated, and it levels off at a concentration of glucose that corresponds to 80 mM. The saturated anodic current corresponds to the maximum efficiency of the bioelectrocatalytic process at the electrode interface. Knowing the value of the saturation current, and the surface coverage of the enzyme, the ET turnover rate between the FAD center and the electrode was calculated to be $900 \pm 150\ s^{-1}$ at 35°C. This is an unprecendetedly high turnover rate that equals, and eventually exceeds, the ET turnover rate between GOx and its native electron acceptor (O_2).[16] This effective electrical contacting of the redox enzyme and the electrode has important consequences on the bioelectrocatalytic sensoric features of the electrode. The efficient ET turnover rate leads to an electrode that is unaffected by oxygen or common glucose sensing interferants such as ascorbic acid or uric acid.

The organization of the reconstituted integrated glucose sensing electrode has required the use of the scarce semi-synthetic FAD cofactor (**8**). A modified procedure depicted in Fig. 3-5A that uses the native FAD cofactor for reconstituting the enzyme, was developed to resolve this limitation.[32] The PQQ electron mediator (**7**) was coupled to the cystamine monolayer-functionalized electrode, and 3-aminophenyl boronic acid, (**9**), was covalently coupled to the PQQ units. The boronate complex generated between the boronic acid ligand and the ribose units of FAD followed by the reconstitution of apo-GOx on the FAD sites, yielded the integrated enzyme electrode. The kinetics of the reconstitution process on the cofactor-functionalized monolayer electrode was characterized by electrochemical means,[32] Fig. 3-5. Since the reconstitution of the apo-protein on the electrodes yields a hydrophobic insulating protein layer, the interfacial electron transfer resistances is expected to increase upon surface reconstitution. Figure 3-5B shows the Faradaic impedance spectra of the enzyme-electrodes at time intervals of reconstitution. The interfacial electron transfer resistances increase as reconstitution proceeds, and it reaches a saturation value after ca. 4 hours of reconstitution. (See inset, Fig. 3-5B). A similar conclusion was derived by following the bioelectrocatalytic functions of the enzyme electrode at time intervals of reconstitution in the presence of a fixed concentration of glucose, 8×10^{-2} M, Fig. 3-5C.

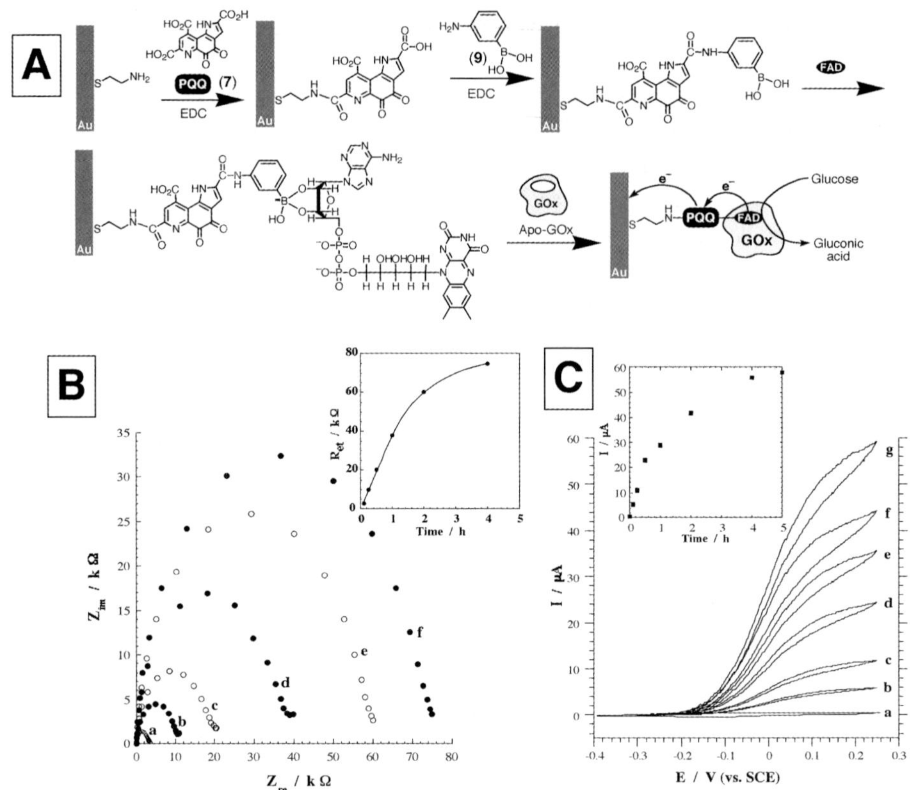

Figure 3-5. (A) Assembly of reconstituted glucose oxidase on a PQQ-FAD monolayer linked to an Au-electrode. (B) Faradaic impedance spectra of the modified electrode at time intervals of reconstitution. (a) 0.1 h, (b) 0.25 h, (c) 0.5 h, (d) 1 h, (e) 2 h, (f) 4 h. Inset: Interfacial electron transfer resistance of the modified electrode at time-intervals of reconstitution. (C) Cyclic voltammograms corresponding to the bioelectrocatalyzed oxidation of glucose, 80 mM, by the enzyme-functionalized electrode at time-intervals of reconstitution: (a) 0 h, (b) 0.1 h, (c) 0.25 h, (d) 0.5 h, (e) 1 h, (f) 2 h, (g) 4 h. Inset: Electrocatalytic currents transduced by the enzyme-modified electrode at time-intervals of reconstitution. Reproduced with permission from ref. 32. Copyright 2002 American Chemical Society.

As reconstitution proceeds, the surface coverage of the bioelectrocatalyst increased, and the resulting anodic current is enhanced. The anodic current reaches a saturation value after ca. 4 hours of reconstitution that corresponds to the time interval required to generate the optimal surface coverage of the protein at the stated conditions. The saturated surface-coverage of the enzyme was elucidated by electrochemical means and microgravimetric quartz-crystal-microbalance measurements[32] to be ca. 2×10^{-12} mole cm^{-2}. Realizing that the footprint[33] of GOx is 58 nm^2, the experimental surface coverage is translated into the value of 70% of an ordered densely packed monolayer. The electrocatalytic anodic currents of the electrically-contacted enzyme-electrode were analyzed at variable concentrations of glucose, Fig. 3-6.

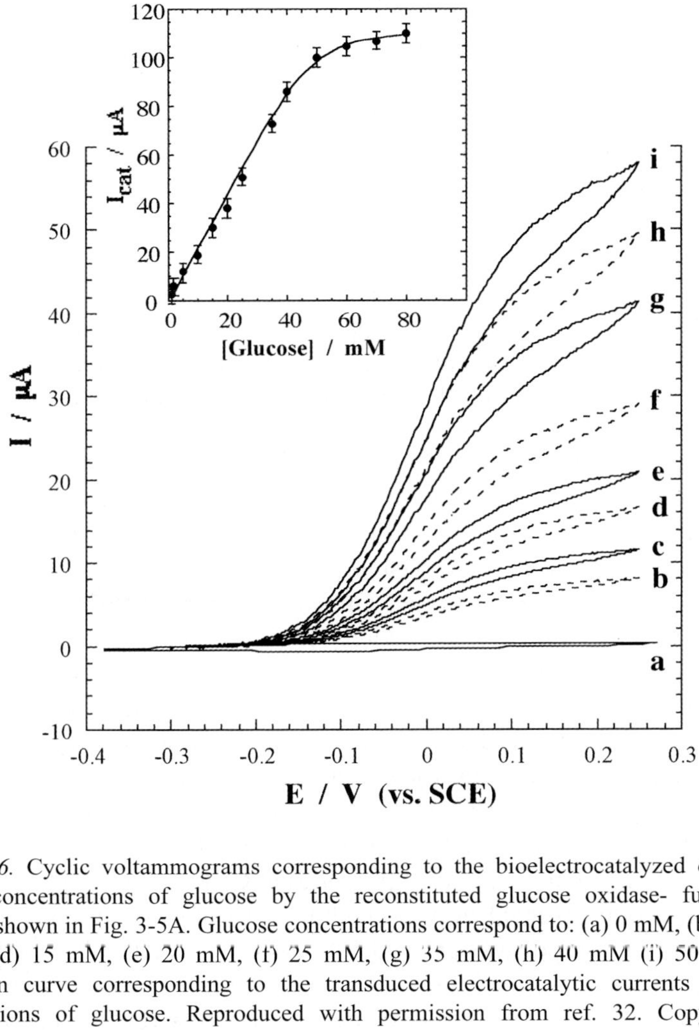

Figure 3-6. Cyclic voltammograms corresponding to the bioelectrocatalyzed oxidation of variable concentrations of glucose by the reconstituted glucose oxidase- functionalized electrode shown in Fig. 3-5A. Glucose concentrations correspond to: (a) 0 mM, (b) 5 mM, (c) 10 mM, (d) 15 mM, (e) 20 mM, (f) 25 mM, (g) 35 mM, (h) 40 mM (i) 50 mM. Inset: Calibration curve corresponding to the transduced electrocatalytic currents at different concentrations of glucose. Reproduced with permission from ref. 32. Copyright 2002 American Chemical Society.

The catalytic current levels off to saturation value at glucose concentrations higher than 6×10^{-2} M, conditions where the maximal bioelectrocatalytic performance of the reconstituted enzyme electrode is observed. Knowing the saturation value of the catalytic anodic current, I_{cat}^{sat}, the surface coverage of GOx, Γ_{GOx}, and the electrode area, A, the ET turnover rate from the active center to the electrode TR_{max}, was estimated to be ca. 700 s^{-1}, using eqn. 3-2. This ET turnover rate is similar to the ET rate between GOx and O_2.

$$TR_{max} = I_{cat}^{sat}/(F \cdot n \cdot \Gamma_{GOx} \cdot A) \tag{3-2}$$

A different approach to electrically contact reconstituted flavoenzymes on electrodes involved the use of the conductive redox-polymer polyaniline as a redox mediator for the electrical contacting of the enzyme with the electrode.[34] Aniline was electropolymerized on Au-electrodes in the presence of polyacrylic acid. Polymerization yielded a polyaniline film with entangled polyacrylic acid chains. In contrast to a bare polyaniline film that reveals redox activities only in acidic aqueous solutions, the polyaniline/polyacrylic acid composite film was reported[35] to be electrochemically active in neutral aqueous solutions, and thus the polymer film could be coupled to bioelectrocatalyzed transformations. The enzyme-electrode was constructed as depicted in Fig. 3-7. The amino-FAD semi-synthetic cofactor (**8**) was covalently linked to the polyacrylic acid chains entangled in the polymer film, and apo-glucose oxidase was reconstituted onto the FAD units to generate the integrated enzyme electrode. In the specific construction assembled on the electrode, the polyaniline/polyacrylic acid film exhibited a thickness of ca. 90 nm and the surface coverage of the FAD units and of the reconstituted GOx components was estimated to be 2×10^{-11} mole cm^{-2} and 3×10^{-12} mole cm^{-2}, respectively. The resulting enzyme electrode revealed bioelectrocatalytic activities towards the oxidation of glucose, Fig. 3-8. Knowing the enzyme content in the film and using the saturated current density observed in the system, $i = 0.3$ mA cm^{-2}, the maximum ET turnover rate for the biocatalyst was estimated to be $TR_{max} \approx 1000$ s^{-1}. In a control experiment the lysine residues of native GOx, with its naturally implanted FAD cofactor, were directly coupled to the polyacrylic acid chains of the polyaniline/polyacrylic acid film. The resulting electrode revealed very poor bioelectrocatalytic activities, and the maximum current output was ca. 100-fold lower. These results highlight the importance of alignment of the biocatalyst in respect to the redox-active polymer film. The electrostatic attraction of the polyacrylic acid chains to the oxidized polyaniline attracts simultaneously the aligned tethered enzyme components, thus enabling the mediated ET from the active site through the oxidized polyaniline to the electrode, Fig. 3-7. This mediated ET electrically contacts the biocatalyst towards the biocatalyzed oxidation of glucose.

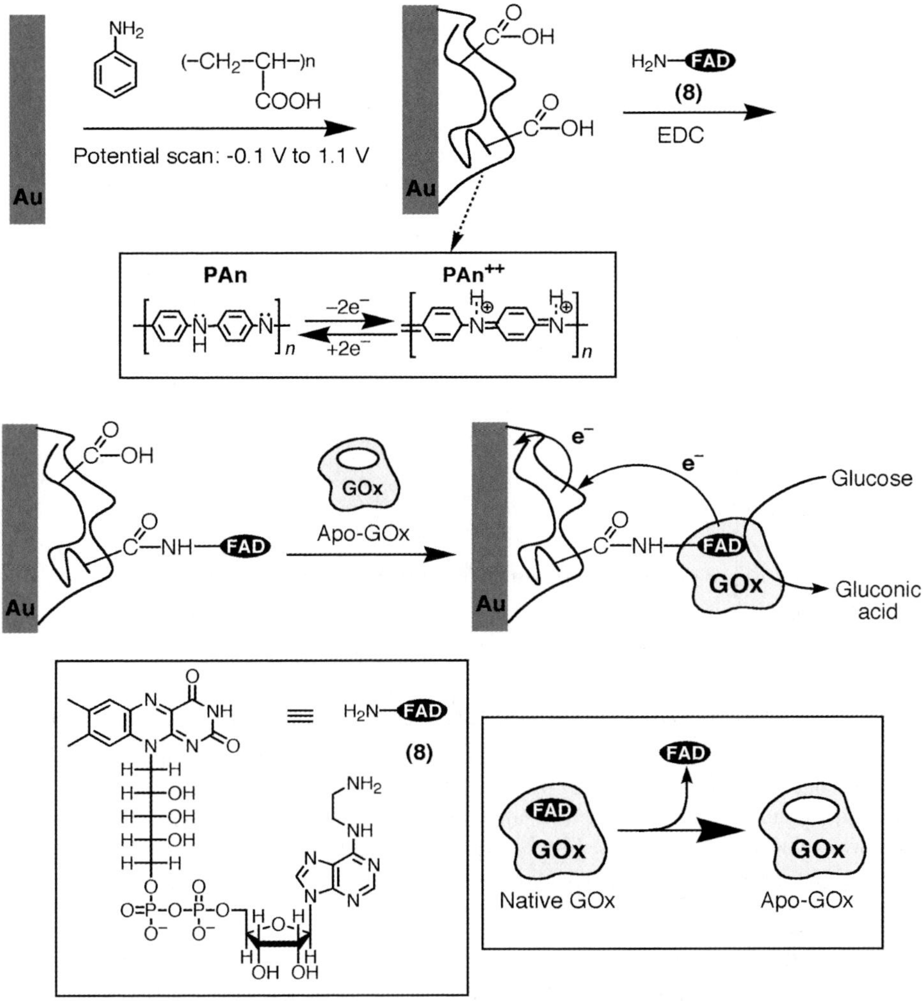

Figure 3-7. Assembly of an electrically-contacted polyaniline-reconstituted glucose oxidase electrode. Reproduced with permission from ref. 34. Copyright 2002 American Chemical Society.

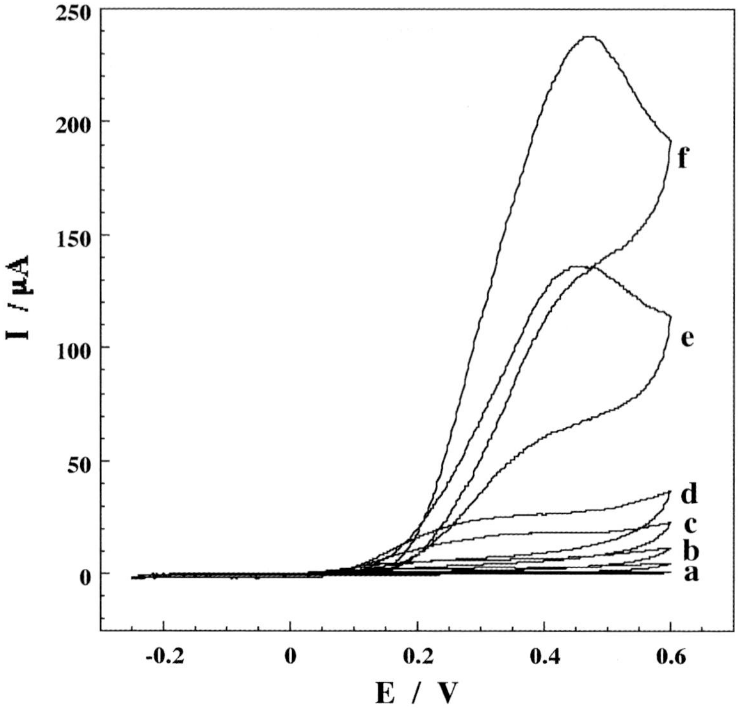

Figure 3-8. Cyclic voltammograms corresponding to the bioelectrocatalyzed oxidation of variable concentrations of glucose by the integrated, electrically-contacted polyaniline-reconstituted glucose oxidase electrode. Glucose concentrations correspond to: (a) 0 mM, (b) 5 mM, (c) 10 mM, (d) 20 mM, (e) 35 mM, (f) 50 mM. Reproduced with permission from ref. 34. Copyright 2002 American Chemical Society.

The system was further characterized by *in situ* electrochemical/surface plasmon resonance (SPR) measurements,[34] and surface plasmon resonance spectroscopy was introduced as a means to follow bioelectrocatalytic transformations. Surface plasmon resonance spectroscopy (SPR) is a useful method for the characterization of the refractive index and thickness of interfaces associated with Au surfaces (or Ag surfaces).[36,37] In fact, SPR spectroscopy is a common practice in biosensing, and numerous examples that follow the formation of biorecognition complexes on SPR active surfaces have been reported.[38,39] The redox transformations of polyaniline films on an Au-surface were examined in detail.[40] The minimum reflectivity angle of the oxidized polyaniline was found to be shifted to higher values. Thus, under conditions where the applied potential oxidizes the polyaniline, PAn, to its oxidized state PAn^{2+}, and provided that the mediated biocatalyzed oxidation of glucose is fast, a steady-state equilibrium that is controlled by the glucose concentration, between the oxidized polyaniline PAn^{2+} and reduced polyaniline, PAn, will be generated. As the concentration of glucose is elevated, the concentration of PAn will increase, and the ratio PAn^{2+}/PAn will decrease. As a result, the SPR spectra of the polymer film would shift to lower reflectivity angles that resemble the spectrum of PAn, as the concentration of glucose is elevated, even though the potential applied on the electrode is thermodynamically adequate to oxidize the polymer film. Figure 3-9A shows the SPR of the reduced polymer film, curve (a), and oxidized (PAn^{2+}) film in the absence of glucose. While the effect of addition of glucose on the SPR spectra of the reduced PAn film is minute, curves (b)-(f), addition of glucose to the film subjected to the oxidative potential, results in a shift to lower values in the minimum reflectivity angles as the concentration of glucose increases. This is consistent with the enrichment of the polymer film with the reduced state of PAn. Figure 3-9B shows the derived calibration curves that correspond to the amperometric detection of glucose and to the SPR transduction of glucose sensing by the integrated PAn/reconstituted GOx electrode.

Significant progress in tailoring electrically-contacted enzyme electrodes was accomplished by the application of Au-nanoparticles (diameter 1.2 nm) as electron mediator units for the enzyme glucose oxidase.[41] The Au-nanoparticle/reconstituted GOx electrodes were prepared by two alternative methods, Fig. 3-10, and yield integrated enzyme electrodes of similar composition and functions. By one method, route a, the amino-FAD semi-synthetic cofactor was attached to the maleimide-functionalized Au_{55}-nanoparticle (diameter 1.2 nm). Apo-GOx was then reconstituted on the particle and the resulting nanoparticle-enzyme hybride was linked to a dithiol monolayer associated with the Au-electrode. The scanning transmission electron micrograph (STEM) of the nanoparticle-enzyme conjugate is shown in Fig. 3-11A that visualizes nicely the composition of the nanoparticle-enzyme hybrid. The second method to organize the enzyme electrode, route b, has involved the attachment of the FAD-functionalized Au-nanoparticles on the dithiol monolayer, and the surface reconstitution of apo-GOx onto the FAD units. The surface coverage of the reconstituted enzyme was estimated to be 1×10^{-12} mole cm^{-1}. In contrast to reconstituted GOx units on flat Au-electrodes that lack direct electrical contact with electrodes, the Au-nanoparticles serve as electron mediating units that transport the charge between the active center and the bulk electrode surface.

Figure 3-11B shows the cyclic voltammograms that correspond to the bioelectrocatalyzed oxidation of glucose by the Au-nanoparticle-reconstituted GOx linked to the bulk Au-electrode by the benzene dithiol bridge (**10**).

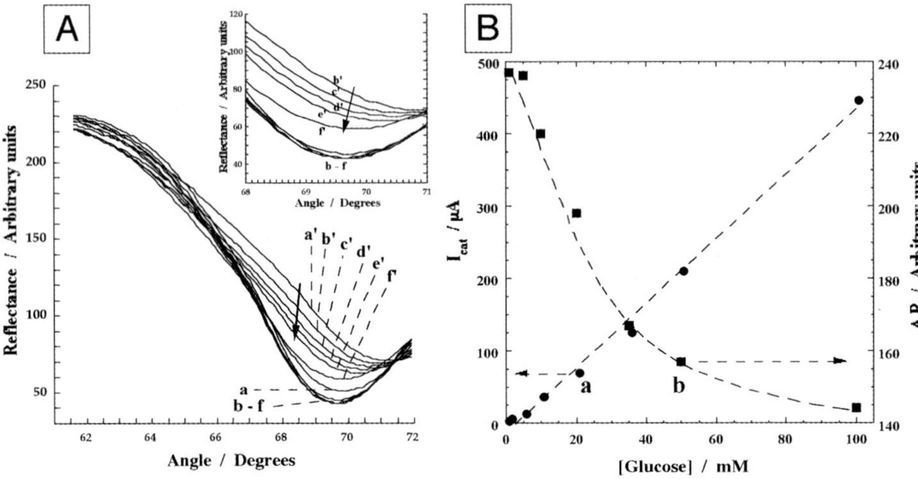

Figure 3-9. (A) Surface plasmon resonance (SPR) spectra corresponding to (a) the reduced polyaniline film with covalently linked reconstituted glucose oxidase in the absence of glucose (applied potential $E = -0.3$ V vs. SCE), and (b) to (f) the reduced polyaniline-reconstituted GOx film upon addition of 5 mM, 10 mM, 20 mM, 35 mM, 50 mM of glucose, respectively. The (a') spectrum corresponds to the oxidized polyaniline-reconstituted GOx film ($E = 0.6$ V vs. SCE) in the absence of glucose whereas spectra (b') to (f') correspond to the oxidized polyaniline-reconstituted GOx film in the presence of 5 mM, 10 mM, 20 mM, 35 mM and 50 mM, respectively. (B) Calibration curves corresponding to: (a) Transduced anodic currents in the presence of variable concentrations of glucose. (b) The changes in the reflectance intensities at the minimum reflectivity angles, ΔR, of the oxidized polyaniline-reconstituted GOx-modified Au-surface in the presence of different concentrations of glucose. Reproduced with permission from ref. 34. Copyright 2002 American Chemical Society.

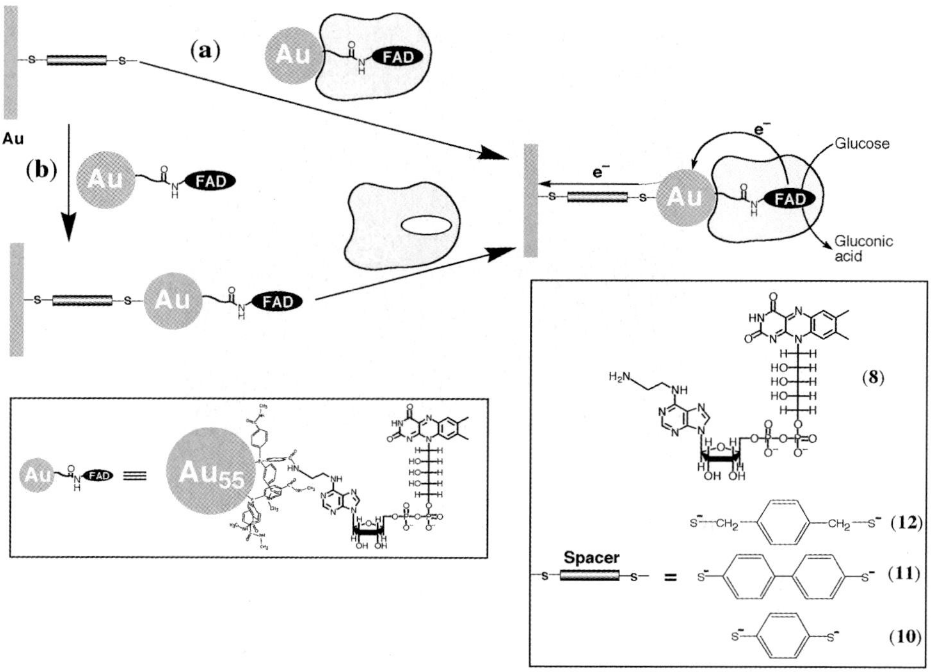

Figure 3-10. The assembly of an Au-nanoparticle (1.4 nm) electrically contacted glucose oxidase electrode by: (a) the primary reconstitution of apo-GOx on the FAD-functionalized Au-nanoparticle, and the immobilization of the enzyme-nanoparticle hybrid on an electrode surface; (b) the primary immobilization of the FAD-modified Au-nanoparticle on the electrode surface and the surface reconstitution of apo-GOx on the functionalized electrode. Reproduced with permission from ref. 41. Copyright 2003, AAAS.

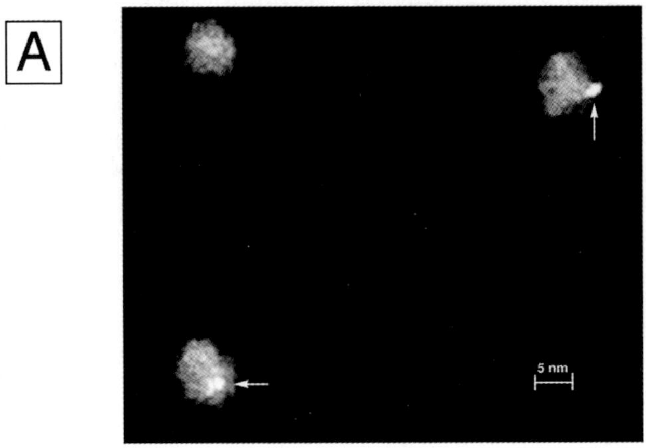

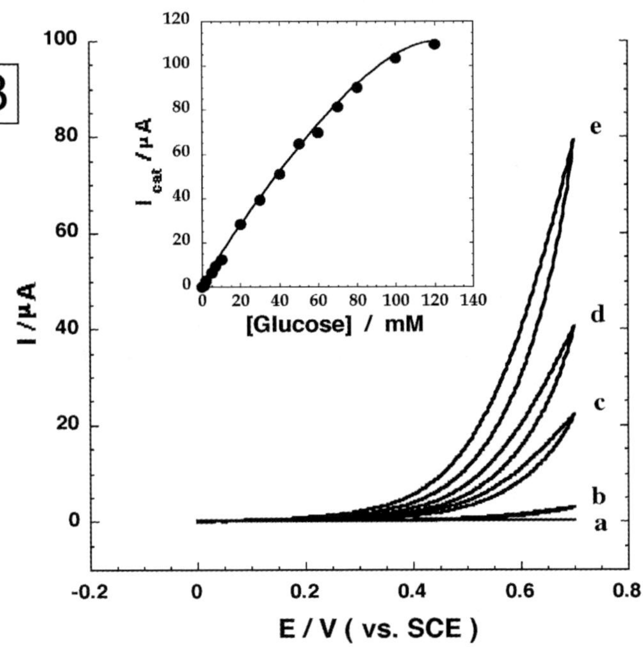

Figure 3-11. (A) Scanning transmission electron microscopy image of the Au-nanoparticle (1.4 nm)-reconstituted GOx hybrid. (The arrow indicates the Au-nanoparticle). (B) Cyclic voltammograms corresponding to the bioelectrocatalyzed oxidation of variable concentrations of glucose by the electrically-contacted Au-nanoparticle-reconstituted GOx-modified electrode. Glucose concentrations correspond to: (a) 0 mM, (b) 1 mM, (c) 2 mM, (d) 5 mM, (e) 10 mM. Inset: Calibration curve corresponding to the electrocatalytic currents at different glucose concentrations. Reproduced with permission from ref. 41. Copyright 2003, AAAS.

The inset in Fig. 3-11B corresponds to the calibration curve that corresponds to the resulting electrocatalytic anodic current at variable concentrations of glucose. From the observed saturation current, and knowing the surface coverage of the biocatalyst, the maximum ET turnover rate is calculated to be $TR_{max} = 5000\ s^{-1}$. This value is ca. six-fold higher than the electron transfer turnover rate of the enzyme with its native O_2 electron acceptor. This unprecedented effective electrical contacting of the enzyme's redox center with the electrode has important consequences on the functional properties of the electrode. The electrode was found to be insensitive to common glucose sensing interferants such as ascorbic acid or uric acid, and it was unaffected by the presence of oxygen. The mechanistic analysis of the electron transfer process in the Au-nanoparticle/reconstituted enzyme system revealed that the ET transport through the molecular bridge linking the Au-nanoparticle to the electrode is the rate limiting process. It was found that benzene dithiol, (**10**), was the most efficient electron transporting bridge, whereas the biphenyl dithiol, (**11**), and the *p*-xylylene dithiol, (**12**), were less efficient bridging elements. This was attributed to the different electron tunneling properties of the different bridges, while the conjugated planar π-system in (**11**) and the SP^3 methylene units in (**12**) increase the tunneling barriers, resulting in slower ET turnover rates. The results suggest that the use of conductive molecular wires such as oligophenyl acetylenes to bridge the functional nanoparticles to the surface could further enhance the electrical contacting efficiency between the reconstituted enzyme and the electrode. Also, the use of other electrical contacting units such as carbon nanotubes for the reconstitution of the enzyme on electrodes could be an attractive method to generate integrated electrically contacted enzyme electrodes.

2.2 Integrated Electrically-Contacted Electrodes Composed of Reconstituted Quinoproteins

Redox proteins that include quinone cofactor units play important roles in biological ET processes.[42,43] Some of the quinoproteins include the quinone cofactor in a non-covalently linked configuration, such as the pyrroloquinoline quinone, PQQ, dependent enzymes, whereas other quinoproteins include the quinone cofactor covalently-linked to the protein, for example topaquinone (2,4,6-trihydroxyphenylalanine quinone, TPQ) dependent enzymes. A number of quinoproteins include in addition to the quinone cofactor an ET cofactor unit in another protein subunit. These cofactors may be metal ions or a cytochrome-type heme cofactor such as D-fructose dehydrogenase that is a heme containing PQQ-dependent enzyme.[44]

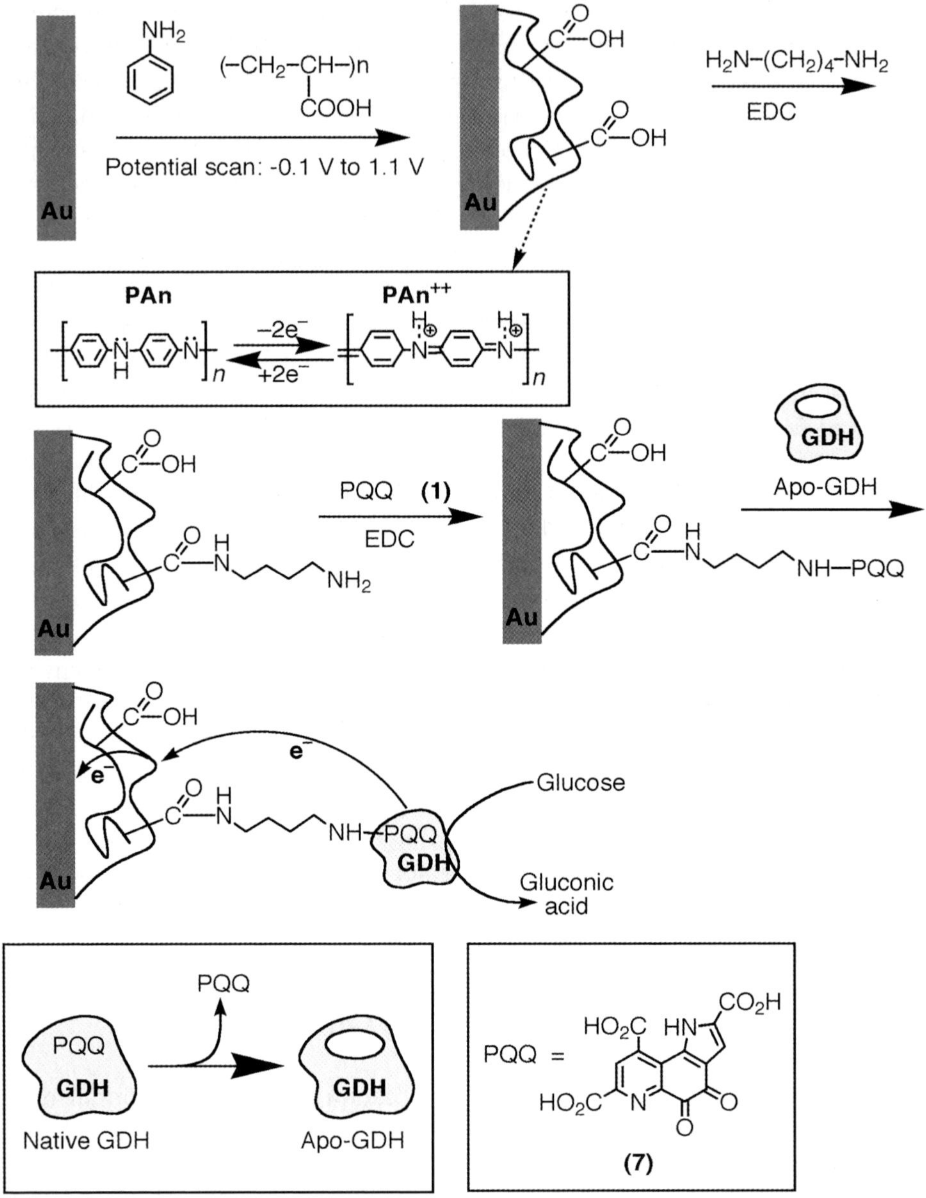

Figure 3-12. Assembly of the electrically-contacted polyaniline/PQQ-reconstituted glucose dehydrogenase electrode. Reproduced from ref. 46 by permission of the Royal Society of Chemistry (RSC).

Reconstitution of apo-glucose dehydrogenase on a PQQ-monolayer-functionalized Au-electrode led to an active configuration of the enzyme on the electrode support, but it lacked direct electrical contact with the electrode surface.[45] An integrated electrically contacted glucose dehydrogenase electrode was organized[46] by the reconstitution of the apo-enzyme on PQQ units linked to a polyaniline/polyacrylic acid redox-active polymer composite associated with an electrode, Fig. 3-12. Electropolymerization of aniline in the presence of polyacrylic acid led to the formation of a polyaniline/polyacrylic acid composite film on the electrode that exhibited redox activity at neutral pH values. The covalent coupling of 1,4-diaminobutane to the carboxylic acid residues, followed by the attachment of the PQQ cofactor units and the reconstitution of apo-glucose dehydrogenase in the PQQ sites led to an electrically-contacted enzyme-electrode for the bioelectrocatalyzed oxidation of glucose. The surface coverage of the PQQ units and the enzyme components on a 90 nm thick polymer film was estimated to be 1×10^{-11} mole cm^{-2} and 2×10^{-12} mole cm^{-2}, respectively. The bioelectrocatalyzed oxidation of glucose was followed electrochemically and by *in situ* electrochemically surface plasmon resonance measurements.[46] Figure 3-13 exemplifies the cyclic voltammograms that correspond to the bioelectrocatalyzed oxidation of glucose by the integrated electrode. The resulting calibration curve is depicted in Fig. 3-13 (inset). It was assumed that the redox-active polyaniline chains penetrate the protein matrix and reach sufficiently close distances in respect to the reconstituted PQQ units that facilitate the ET contact. Control experiments reveal that the direct covalent attachment of the glucose dehydrogenase to the polymer film does not yield an electrically contacted enzyme, implying that the alignment of the biocatalyst on the polymer is important to establish electrical communication between the PQQ redox-center and the electrode.

An interesting approach to electrically contact quinoproteins and electrodes involved the use of molecular wires to align the enzyme on the electrode and to conduct electrons between the quinone center and the electrode.[47] Amine oxidase catalyzes the oxidation of amines to aldehydes using copper ions and the topaquinone, TPQ, as redox center. The fact that the enzyme is inhibited by diethylaniline that binds to the redox-site pocket suggests that the diethylaniline-terminated oligophenylacetylene thiol (**13**) may act as a wire for the alignment of the enzyme on the electrode and the electrical contacting of its redox center by charge transport through the wire.

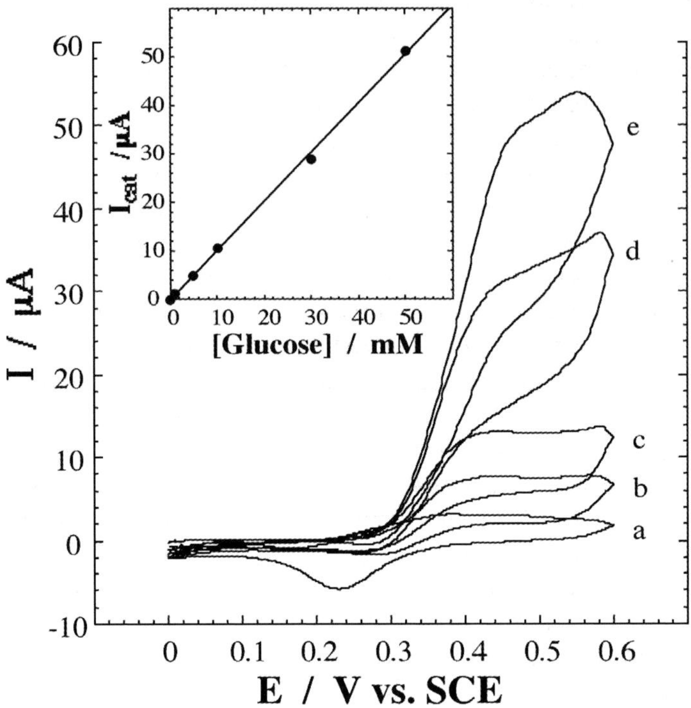

Figure 3-13. Cyclic voltasmmograms corresponding to the bioelectrocatalyzed oxidation of different glucose concentrations by the PQQ-reconstituted glucose dehydrogenase-polyaniline-functionalized electrode. Glucose concentrations: (a) 0 mM, (b) 5 mM, (c) 10 mM, (d) 20 mM, (e) 50 mM. Inset: Calibration curve corresponding to the electrocatalytic currents as a function of glucose concentration. Reproduced from ref. 46 by permission of the Royal Society of Chemistry (RSC).

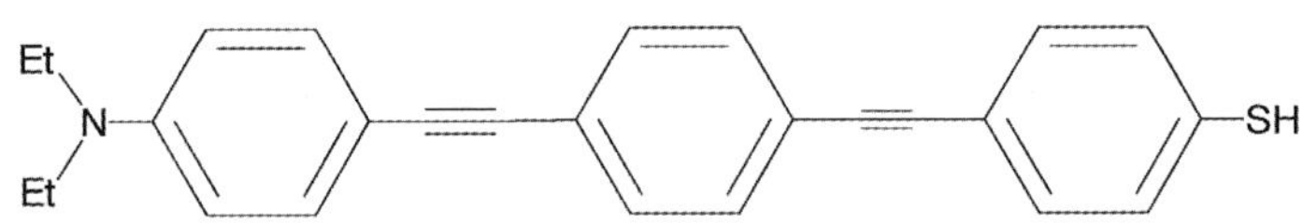

(13)

Indeed, the assembly of the thiol-functionalized molecular wire on an Au-electrode followed by the association of the protein to the inhibiting diethylaniline units, led to an electrically contacted protein where the TPQ moiety embedded in the protein revealed quasi-reversible electrochemical properties.

2.3 Reconstituted Electrically-Contacted Hemoproteins

Hemoproteins are a broad class of redox-proteins that act as cofactors, e.g. cytochrome *c*, or as biocatalysts, e.g. peroxidases. Direct ET between peroxidases such as horseradish peroxidase,[48] lactoperoxidase,[49] or chloroperoxidase[50] and electrode surfaces, mainly carbonaceous materials, were extensively studied. The mechanistic aspects related with the immobilized peroxidases on electrode surfaces and their utilization in developing biosensor devices were reviewed in detail.[51,52] The direct electrical contact of peroxidases with electrodes was attributed to the location of the heme site at the exterior of the protein that yields close contact with the electrode surface even though the biocatalyst is randomly deposited on the electrode. For example, it was reported[53,54] that non-oriented randomly deposited horseradish peroxidase on a graphite electrode resulted in 40–50% of the adsorbed biocatalyst in an electrically contacted configuration. For other hemoproteins such as cytochrome *c* it was found that the surface modification of the electrodes with promoter units[55,56] such as pyridine units induced the binding of the hemoproteins in an orientation that facilitated direct electron transfer. By this method, the promoter sites induce a binding-ET process-desorption mechanism at the modified electrode. Alternatively, the site-specific covalent attachment of hemoproteins such as cytochrome *c* resulted in the orientation of the protein on the electrode surfaces and direct ET communication.[57]

The reconstitution process was adopted for enhancing ET of peroxidase and for establishing direct electrical communication of horseradish peroxidase with electrodes. Apo-horseradish peroxidase (apo-HRP) was reconstituted[58] with a dyad consisting of ferrocene tethered to hemin, (**14**), Fig. 3-14. The resulting reconstituted HRP revealed enhanced ET features and the oxidation of 2,2'-azino-bis(3-ethylbenzothiazoline-6-sulfonic acid), ABTS, by H_2O_2 was ca. 2.5-fold faster in the presence of the (**14**)-reconstituted HRP as compared to the native HRP.

The surface reconstitution of apo-HRP on a hemin-functionalized monolayer was reported to yield a surface-oriented HRP, exhibiting direct ET with the electrode.[59] A mercaptobutanoic acid monolayer was assembled on an Au-electrode, followed by the covalent coupling of 1,12-diaminododecane to the base monolayer. Hemin was then covalently linked to the surface, and apo-HRP was reconstituted on the hemin sites, Fig. 3-15. The resulting surface-reconstituted HRP revealed direct ET with the electrode, and the bioelectrocatalyzed reduction of H_2O_2 was activated by the reconstituted HRP.

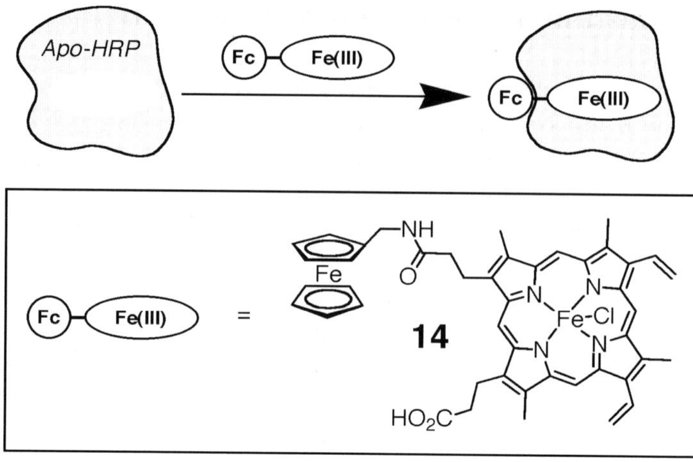

Figure 3-14. The reconstitution of apo-horseradish peroxidase, apo-HRP, with the ferrocene-tethered hemin (**14**).

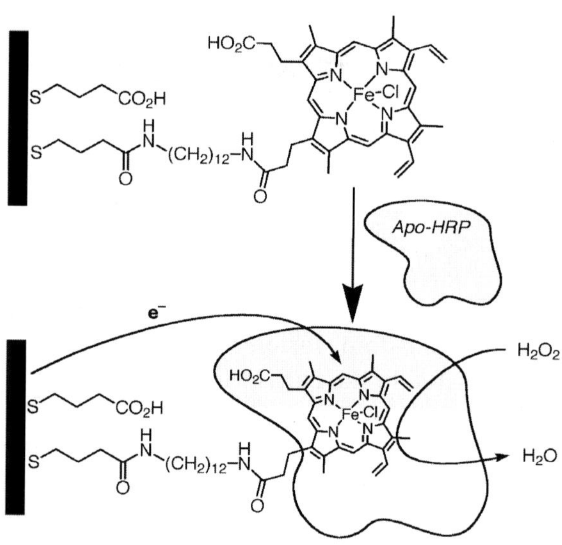

Figure 3-15. The surface-reconstitution of apo-horseradish peroxidase on a hemin monolayer assembled on an Au-electrode.

2.4 Reconstituted *de novo* Hemoproteins on Electrodes

Extensive research efforts were directed in recent years towards the synthesis of *de novo* proteins.[60,61] Specifically, the reconstitution of *de novo* proteins with metal prophyrin attracted interest as a biomimetic approach to tailor synthetic hemoproteins.[62,63]

The reconstitution of a tailored *de novo* protein with Fe(III)-protoporphyrin IX units, (**15**), on an electrode surface generates a hemoprotein-type assembly that mimics ET functions of cytochrome *c*.[64,65] A four-helix bundle *de novo* protein, (**16**), consisting of 128 amino acids and a mass of 14728 was synthesized.[66] It included two pairs of identical helices A and B. The B helices are terminated with Gly-Gly-Cys units for the assembly of the structures on surfaces, whereas the helices A include each two histidines acting as ligand. The position of the histidine ligands on the opposite helices A permits, in principle, the axial ligation of two Fe(III)-heme sites into the protein structure. The four helix-bundle *de novo* protein was assembled as a monolayer on an Au-electrode as depicted in Fig. 3-16.

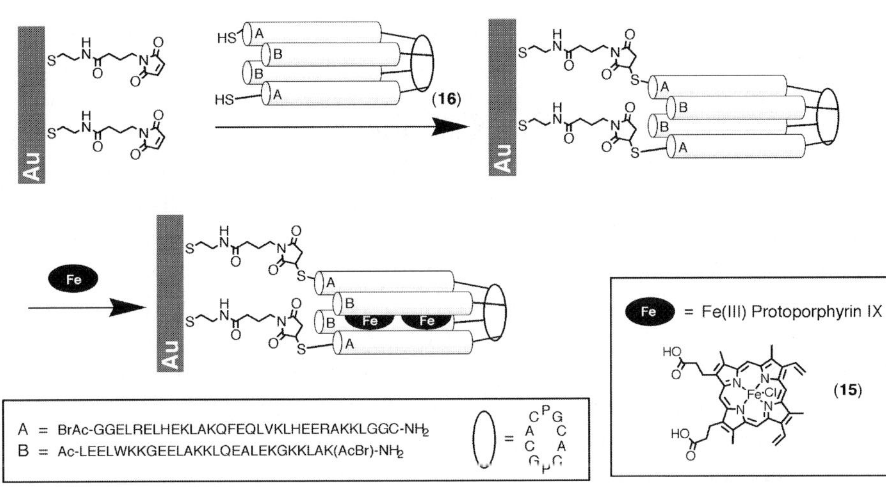

Figure 3-16. The reconstitution of Fe(III)-protoporphyrin IX, (**15**), in the four-helix bundle *de novo* protein, (**16**), bound to a functionalized Au-electrode. Reproduced with permission from ref. 64. Copyright Wiley–VCH, 1998.

Interaction of the functionalized-monolayer electrode with Fe(III)-protoporphyrin IX, (**15**), resulted in the reconstitution of the protein with the two heme units. The surface coverage of the reconstituted *de novo* hemoprotein was found to be 2.5×10^{-11} mole cm^{-2}. Assuming that the protein footprint area is ca. 25 Å^2, the saturated surface coverage of the *de novo* protein corresponds to 3.5×10^{-11} mole cm^{-2}, and thus the experimental value corresponds to ca. 70% of a saturated densely-packed monolayer. Detailed electrochemical characterization of the heme-reconstituted *de novo* protein indicated that the two heme units integrated in the protein exhibited slightly different redox potentials and different kinetics of reconstitution. The heme-site close to the electrode surface exhibited a more negative redox-potential that corresponded to –0.43 V *vs.* SCE, and it was reconstituted into the protein slowly, whereas the heme center that was reconstituted in the remote position was incorporated into the protein in a faster process and revealed a slightly more positive redox potential that corresponded to –0.36 V *vs.* SCE. The ordering of the redox potentials of the two heme sites resulted in a vectorial directional electron transfer from the electrode to the first heme site that acted as a relay unit for ET to the second heme center.[65]

The heme-reconstituted *de novo* protein associated with the electrode was found to act as an effective ET mediator to other hemoproteins. Also it acted as an artificial heme-cofactor that substituted cytochrome *c* in activating cytochrome *c* dependent enzymes. The cytochrome *c* dependent enzyme nitrate reductase, NR (E.C. 1.9.6.1 from *Escherichia coli*) was found to form an affinity complex with the surface-reconstituted hemo-*de novo* protein. The crosslinking of the affinity complex on the electrode surface with glutaric dialdehyde led to the formation of an integrated bioelectrocatalytic electrode, Fig. 3-17. The reduction of the *de novo* hemoprotein activated an ET cascade to NR that induced the bioelectrocatalyzed reduction of nitrate (NO_3^-) to nitrite (NO_2^-). Figure 3-18A shows the derived calibration curve. This system indicates that the heme-reconstituted *de novo* protein acts as a synthetic cofactor analog that mimics the functions of cytochrome *c* in mediating ET to the biocatalyst. Interestingly, this concept was further broadened by the biomimetic assembly of reconstituted bioelectrocatalytic protein structures that stimulate electrocatalytic transformations that do not proceed in nature. Apo-myoglobin was reconstituted with Co(II)-protoporphyin. While the Co(II)-reconstituted myoglobin lacked direct electrical communication with the electrode, it was found that the heme-reconstituted *de novo* hemoprotein activated the Co(II)-reconstituted myoglobin towards the bioelectrocatalytic hydrogenation of acetylene dicarboxylic acid to malic acid. It was found[65] that crosslinking of the affinity complex generated between Co(II)-reconstituted myoglobin and the hemo-*de novo* protein monolayer associated with the electrode yielded an integrated bioelectrocatalytic electrode for the hydrogenation of acetylene dicarboxylic acid. Figure 3-18B shows the cyclic voltammograms generated by the integrated electrode at different concentrations of the acetylenic substrate. Mechanistic analysis of the reaction has revealed that the hemo-*de novo* protein mediates ET to the Co(II)-reconstituted myoglobin. The resulting Co(I)-hydride myoglobin acted as the reactive species in the hydrogenation of acetylene dicarboxylic acid.

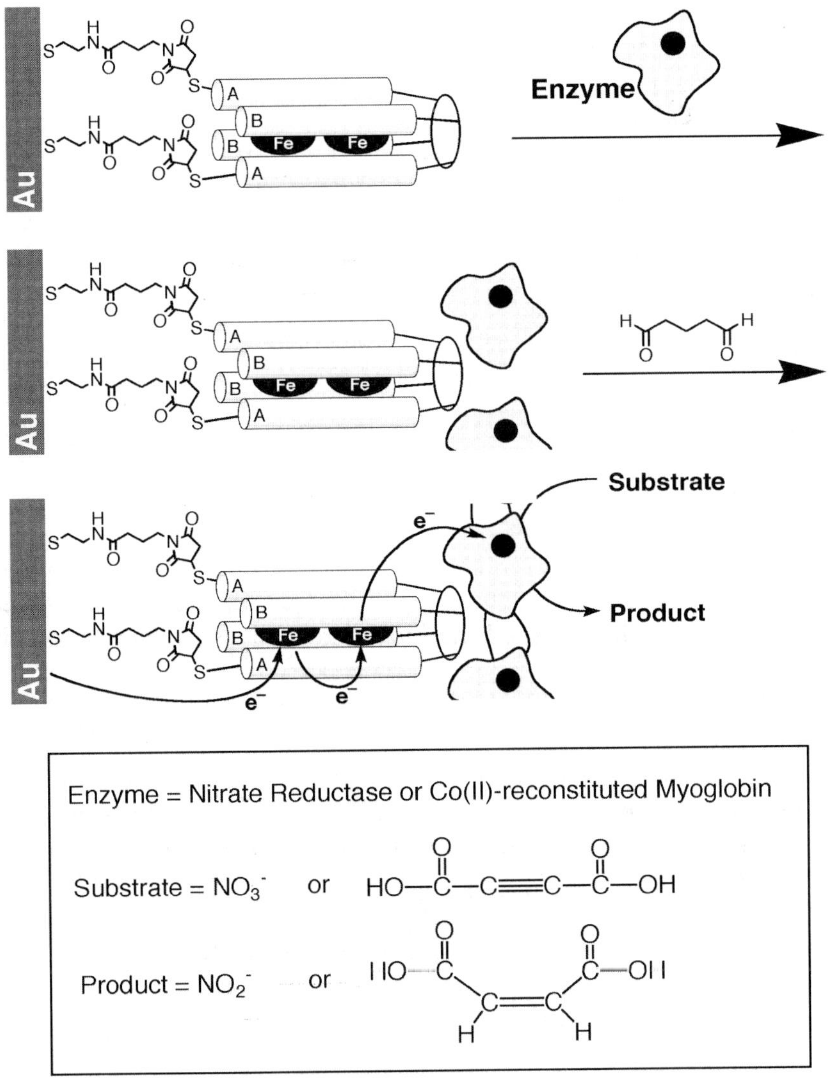

Figure 3-17. The assembly and bioelectrocatalytic functions of integrated, electrically-contacted, electrode consisting of the Fe(III)-protoporphyrin IX- reconstituted *de novo* protein and surface-crosslinked redox proteins. Reproduced with permission from ref. 65. Copyright 1999 American Chemical Society.

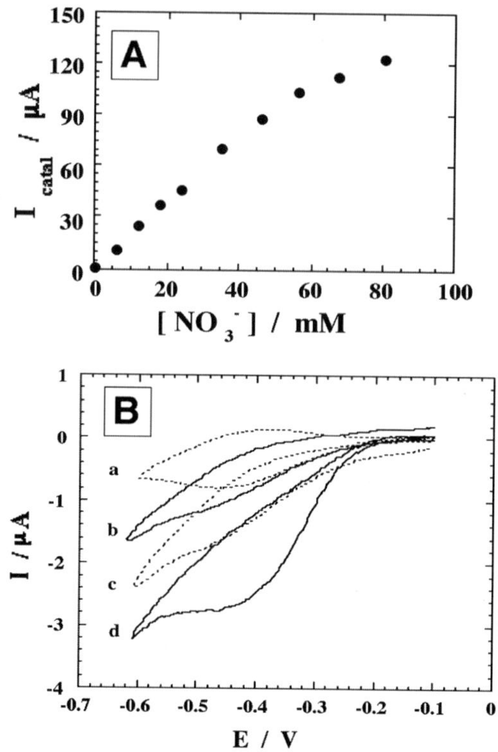

Figure 3-18. (A) Calibration curve corresponding to the bioelectrocatalytic currents generated by the Fe(III)-protoporphyrin-reconstituted *de novo* protein/nitrate reductase electrode in the presence of different concentrations of nitrate (NO_3^-). (B) Cyclic voltammograms corresponding to the bioelectrocatalyzed hydrogenation of different concentrations of acetylene dicarboxylic acid in the presence of Fe(III)-protoporphyrin IX-reconstituted *de novo* protein/Co(II)-reconstituted electrode. The concentration of acetylene dicarboxylic acid corresponds to: (a) 0 M, (b) 5×10^{-6} M, (c) 1×10^{-5} M, (d) 2×10^{-5} M. Data recorded in 0.1 M phosphate buffer, pH 7.0, scan rate 5 mV s^{-1}. Reproduced with permission from ref. 65. Copyright 1999 American Chemical Society.

3. ELECTRICAL CONTACTING OF REDOX PROTEINS BY CROSSLINKING OF COFACTOR-ENZYME AFFINITY COMPLEXES ON SURFACES

The previous section addressed the electrical contacting of redox-proteins by means of the reconstitution process. The concept was based on the exclusion of the firmly associated cofactor from the protein, and the reconstitution of the resulting apo-protein on cofactor units tethered to the electrode surface. The reconstitution process regenerated a tight cofactor-protein complex on the surfaces, and this facilitated the mediated ET by alignment of the protein on the surface. ET cofactors, such as nicotinamide dinucleotide, NAD^+, nicotinamide dinucleotide phosphate, $NADP^+$, or cytochrome *c*, act as diffusional ET mediators that form a labile equilibrium complex with the respective redox enzymes. The orientation of the enzyme's redox center in respect to the cofactor in the complex results in mediated ET and the biocatalytic activation of the enzyme. A method related to the reconstitution process for the electrical contacting of redox enzymes that are dependent on diffusional cofactors that form labile complexes with the enzymes, was developed, Fig. 3-19. The cofactor linked to the electrode surface is directly electrically communicated with the conductive support or by the use of an intermediary mediator. The enzyme generates the labile complex with the cofactor interface, and in order to generate an integrated rigid bioelectrocatalytic interface, the complexed proteins are crosslinked with a molecular crosslinker. Since the primary cofactor-enzyme complex exhibits the appropriate orientation for ET, the resulting array is anticipated to reveal bioelectrocatalytic functions.

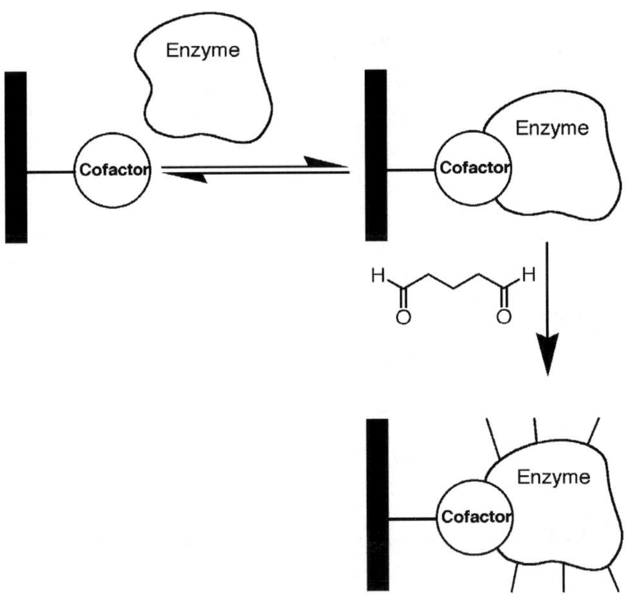

Figure 3-19. Organization of electrically-contacted redox enzymes on electrodes by the surface-crosslinking of cofactor-enzyme affinity complexes on the electrode surface.

3.1 Integrated NAD(P)$^+$-Dependent Enzyme Electrodes

This method was applied to assemble integrated electrically-contacted NAD(P)$^+$-dependent enzyme electrodes. The direct electrochemical reduction of NAD(P)$^+$ cofactors or the electrochemical oxidation of NAD(P)H cofactors are kinetically unfavored.[67] Different diffusional redox mediators such as quinones, phenazine, phenoxazine, ferrocene or Os-complexes were employed as electrocatalysts for the oxidation of NAD(P)H cofactors.[68,69] An effective electrocatalyst for the oxidation of the NAD(P)H is pyrroloquinoline quinone, PQQ, (**7**), and its immobilization on electrode surfaces led to efficient electrocatalytic interfaces (particularly in the presence of Ca^{2+} ions) for the oxidation of the NAD(P)H cofactors.[70,71] This observation led to the organization of integrated electrically contacted enzyme-electrodes as depicted in Fig. 3-20 for the organization of a lactate dehydrogenase electrode.[72]

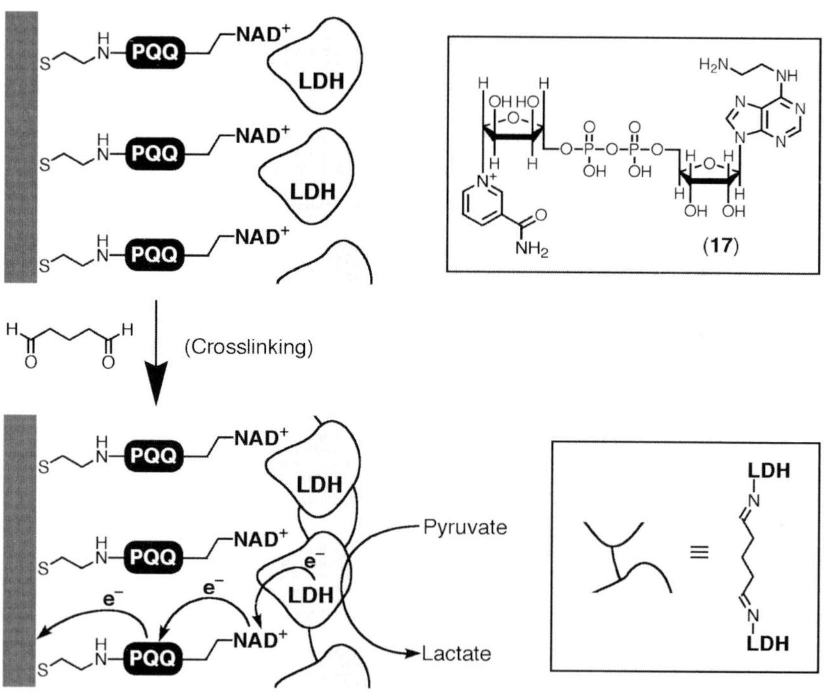

Figure 3-20. Assembly of an integrated lactate dehydrogenase, LDH, electrode for the bioelectrocatalyzed oxidation of lactate by the surface crosslinking of an affinity complex formed between LDH and a PQQ-NAD$^+$ monolayer assembled on an Au-electrode. Reproduced with permission from ref. 72. Copyright 1997 American Chemical Society.

The pyrroloquinoline quinone, PQQ, (**7**), cofactor was covalently linked to a cystamine monolayer associated with an Au-electrode. The modified cofactor N^6-(2-aminoethyl)-NAD^+, (**17**), was then linked to the PQQ monolayer. The NAD^+-dependent enzyme lactate dehydrogenase, LDH, was then interacted with the modified electrode to yield the NAD^+-LDH affinity complex on the electrode support. Crosslinking of the labile surface-associated complex with glutaric dialdehyde led to the formation of firm integrated electrically contacted LDH electrodes.[72] The integrated electrode revealed bioelectrocatalytic activities towards the oxidation of lactate to pyruvic acid, and lactate could be amperometrically analyzed by the electrode in the concentration range of $1\times10^{-3} - 1\times10^{-2}$ M. A similar approach was also applied to assemble an electrically contacted NAD^+/alcohol dehydrogenase electrode for the electrocatalyzed oxidation of ethanol. The biocatalyzed reduction of the NAD^+ cofactors by the lactate or ethanol substrates in the systems is followed by the electrocatalyzed oxidation of NADH by PQQ.

A related method for the preparation of integrated $NAD(P)^+$-dependent enzyme electrodes employed[32] 3-aminophenyl boronic acid (**9**) as coupling reagent of the native $NAD(P)^+$ cofactors to the PQQ monolayer, Fig. 3-21. This is exemplified by the preparation of the electrically-contacted $NADP^+$-dependent malate dehydrogenase, MalD, electrode, Fig. 3-21A, and of the integrated, electrically contacted NAD^+-dependent lactate dehydrogenase, LDH, electrode, Fig. 3-21B. Aminophenyl boronic acid (**9**) was covalently linked to the PQQ monolayer associated with the electrode. The ribose units of $NADP^+$ or NAD^+ that include vicinal hydroxyl functions were bound to the boronic acid ligands to form the respective boronate complexes. The MalD and LDH proteins were then interacted with the PQQ-$NADP^+$ and the PQQ-NAD^+ interfaces to form the respective labile complexes. Crosslinking of the couples with glutaric dialdehyde generated the firm electrically contacted enzyme electrodes. The detailed surface characterization of the systems revealed that the surface coverage of the PQQ and $NADP^+$ units corresponded to 1.4×10^{-10} mole cm^{-2} and 1.8×10^{-10} mole cm^{-2}, respectively, and the surface coverage of MalD was 1.0×10^{-12} mole cm^{-2}, that corresponded to 70% of a random densely-packed enzyme monolayer. Similarly, for the integrated LDH electrode, the surface coverage of the PQQ and NAD^+ units was 1.9×10^{-10} mole cm^{-2} and 2.0×10^{-10} mole cm^{-2}, respectively, and the surface coverage of LDH was 7.0×10^{-12} mole cm^{-2}. Figure 3-22A shows the electrocatalytic anodic currents observed in the presence of the MalD integrated electrode at variable concentrations of malate, and the respective calibration curve (inset). Figure 3-22B depicts the cyclic voltammograms corresponding to the oxidation of variable concentrations of lactate by the integrated LDH-electrode, and the respective calibration curve (inset).

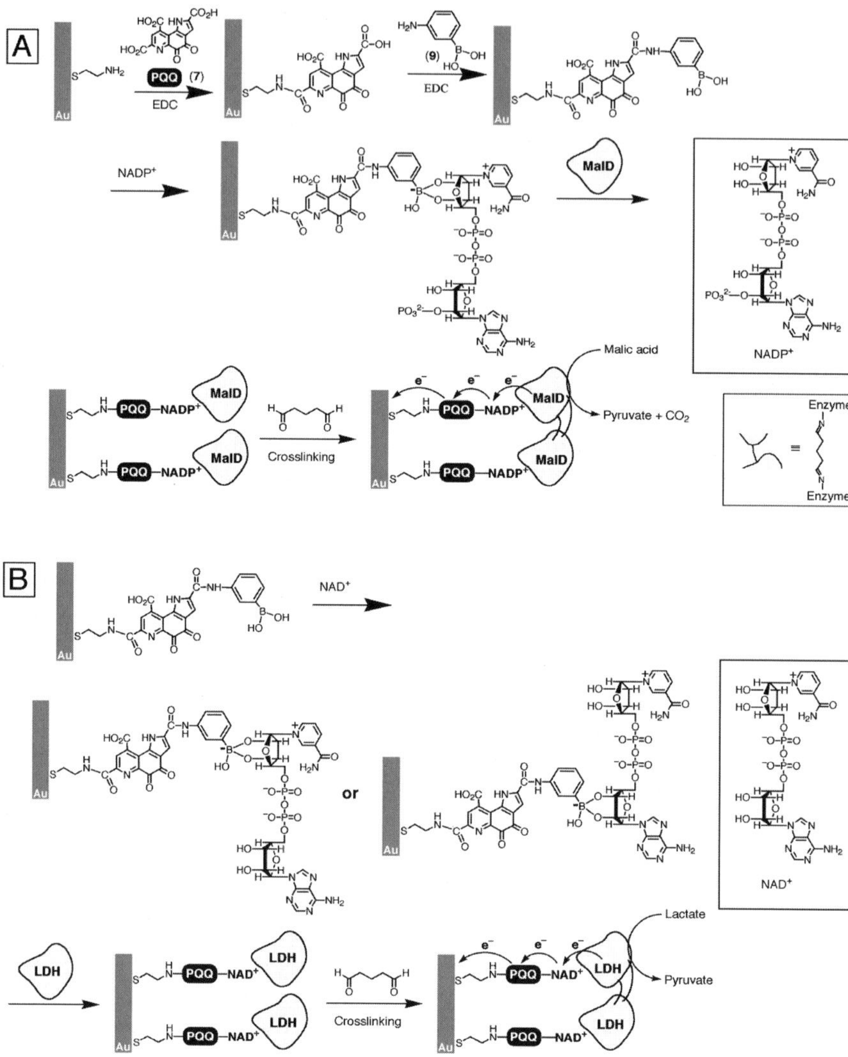

Figure 3-21. (A) Assembly of an integrated malate dehydrogenase, MalD, electrode for the bioelectrocatalyzed oxidation of malate by the surface crosslinking of an affinity complex formed between MalD and a boronate-linked PQQ-$NADP^+$ monolayer. (B) Assembly of an integrated lactate dehydrogenase, LDH, electrode for the bioelectrocatalyzed oxidation of lactate by the surface crosslinking of an affinity complex formed between LDH and different structures of a boronate-linked PQQ-NAD^+ monolayer. Reproduced with permission from ref. 72. Copyright 1997, American Chemical Society.

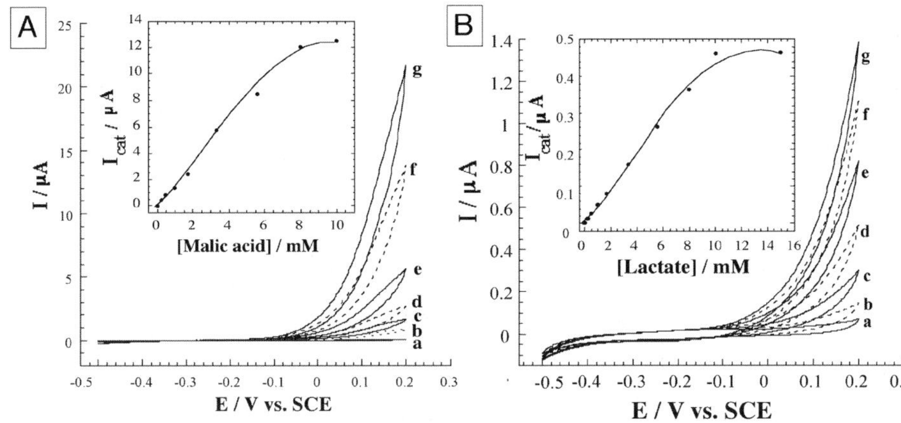

Figure 3-22. (A) cyclic voltammograms corresponding to the bioelectrocatalyzed oxidation of different concentrations of malate by the electrode configuration shown in Fig. 3-21(A). Malate concentrations are: (a) 0 mM, (b) 0.25 mM, (c) 0.5 mM, (d) 1 mM, (e) 1.7 mM, (f) 3.3 mM. (g) 5.6 mM. Inset: Calibration curve corresponding to the electrocatalytic anodic currents as a function of malate concentration. (B) Cyclic voltammograms corresponding to the bioelectrocatalyzed oxidation of different concentrations of lactate by the electrode configuration shown in Fig. 3-21(B). Lactate concentrations are: (a) 0 mM, (b) 0.5 mM, (c) 1.7 mM, (d) 3.3 mM, (e) 5.6 mM, (f) 8.0 mM, (g) 10 mM. Inset: Calibration curve corresponding to the electrocatalytic anodic currents as a function of lactate concentration. Reproduced with permission from ref. 72. Copyright 1997 American Chemical Society.

These systems reveal that the alignment of the $NAD(P)^+$-dependent enzymes on the $NAD(P)^+$ cofactors and their lateral crosslinking leads to the electrically contacted bioelectrocatalytic assemblies. The biocatalytic reduction of the $NAD(P)^+$ cofactors by the respective substrates yields the reduced NAD(P)H cofactors. The electrocatalytic oxidation of NAD(P)H by PQQ then activates the continuous bioelectrocatalytic functions of the enzymes. Impedance measurements performed on the electrically contacted MalD electrode enabled the extraction of the heterogeneous electron transfer rate-constant $k_{et} \approx 3.8\times10^{-6}$ cm s^{-1}. Knowing the surface coverage of MalD, the overall rate constant reflecting the reaction of malate with the enzyme electrode was estimated to be $k_{overall} = 8.5\times10^5\ M^{-1}\ s^{-1}$.

The alignment and electrical contacting of NAD^+-dependent enzymes on electrodes was also accomplished by the generation of the NAD^+-enzyme complex and its crosslinking on a conductive, redox-active, polymer that electrically contacts the cofactor-enzyme assembly with the electrode,[34] Fig. 3-23.

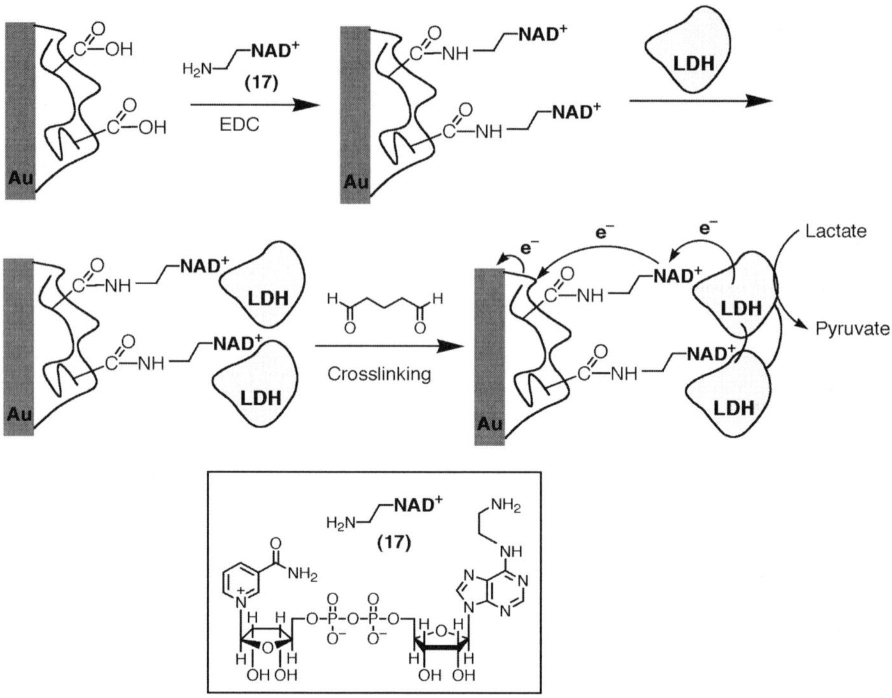

Figure 3-23. The assembly of an integrated lactate dehydrogenase, LDH, electrode for the bioelectrocatalyzed oxidation of lactate by the crosslinking of an affinity complex between LDH and PQQ on a polyacrylic acid-polyaniline film composite associated with the electrode. Reproduced with permission from ref. 34. Copyright 2002 American Chemical Society.

A polyaniline/polyacrylic acid composite that exhibits redox activities in neutral pH media, was assembled on an Au-electrode, and the N^6-aminoethyl-NAD^+ cofactor, (**17**), was covalently linked to the acrylic acid residues. The labile complex generated between lactate dehydrogenase, LDH, and NAD^+ was crosslinked on the polymer-modified surface. Figure 3-24 shows the cyclic voltammograms of the electrode in the presence of different concentrations of lactate, and the respective calibration curve is depicted in the figure. From the highest value of the experimental current density, 0.27 mA cm^{-2}, and knowing the enzyme content on the electrode, 4×10^{-2} mole cm^{-2}, the electron transfer turnover rate between the enzyme's redox site and the electrode was estimated to be ca. 350 s^{-1}. This value is substantially higher than the electron transfer turnover rate that was observed[32] for the crosslinked PQQ/NAD^+/LDH array, Fig. 3-20, ca. 2.5 s^{-1}.

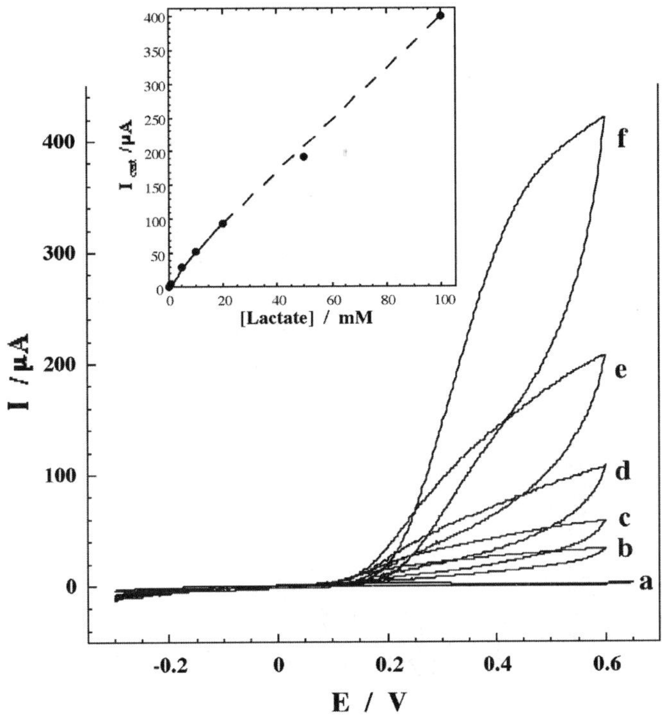

Figure 3-24. Cyclic voltammograms corresponding to the bioelectrocatalyzed oxidation of different concentrations of lactate by the electrode configuration shown in Fig. 3-23. Lactate concentrations are: (a) 0 mM, (b) 5 mM, (c) 10 mM, (d) 20 mM, (e) 50 mM, (f) 100 mM. Inset: Calibration curve corresponding to the electrocatalytic currents as a function of lactate concentration. Reproduced with permission from ref. 34. Copyright 2002 American Chemical Society.

3.2 Integrated Electrically-Contacted Hemoprotein Electrodes

Cytochrome *c*, cyt *c*, represents a further diffusional cofactor that activates many ET processes with proteins. While cytochrome c lacks direct electrical contact with electrodes, the modification of electrodes with promoter molecules such as pyridine proved as an effective means to establish electrical communication between cyt *c* and the electrode.[7] It was suggested that the interaction of cyt *c* with the promoter units via a diffusional equilibrium process aligns the hemoprotein on the electrode surface in a configuration that facilitates ET between the heme center and the electrode. A different approach to establish direct ET between cyt *c* and the electrode included the use of the wild-type cyt *c* (from *Saccharomyces cerevisiae*), that includes a single cysteine residue, or cyt *c* mutants with specific cysteine residues introduced by genetic engineering, for the directional immobilization of the hemoprotein on Au-electrodes.[57] The electrical contacting of cyt *c* by the covalent linkage of the hemoprotein to the electrode enables the organization of integrated, electrically-contacted cyt *c*-enzyme electrodes.[73] For example, a cyt *c*-cytochrome oxidase, COx, integrated electrode for the bioelectrocatalyzed reduction of O_2 to water was developed,[73] Fig. 3-25.

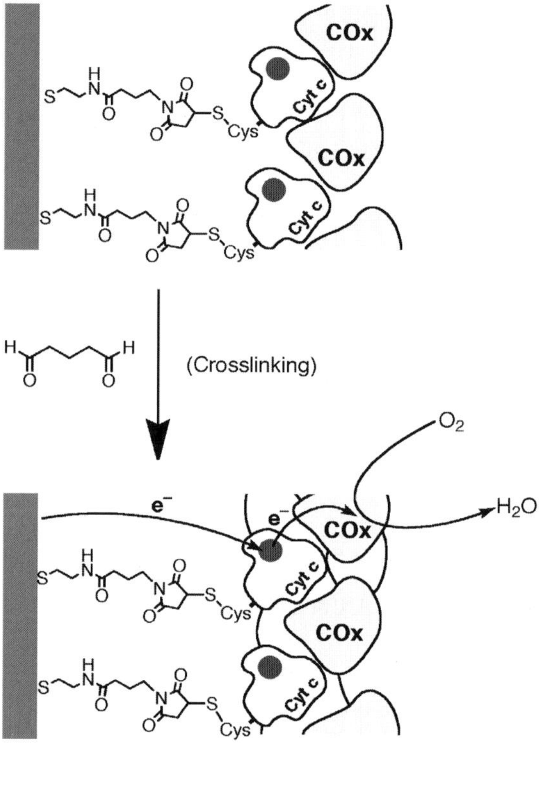

Figure 3-25. The assembly of an integrated electrically contacted electrode for the reduction of O_2 to water generated by the crosslinking of a cytochrome *c*/cytochrome oxidase (cyt *c*/COx) monolayer associated with the electrode. Reproduced with permission from ref. 73. Copyright Elsevier, 1999.

The wild-type cyt *c* was covalently tethered to a maleimide-functionalized monolayer assembled on an Au-electrode. The protein-protein complex generated between cyt *c* associated with the surface and COx was crosslinked with glutaric dialdehyde to yield the cyt *c*-COx electrically contacted electrode that acted as an effective surface for the bioelectrocatalyzed reduction of O_2 to H_2O.

Microperoxidase-11, MP-11, is a heme-undecapeptide that is prepared by the digestion of cytochrome *c* and it includes the active surrounding of cytochrome *c*.[74] MP-11 was immobilized on electrode surfaces[75] and its electrochemistry was characterized.[76] The MP-11-modified electrodes were reported[77] to act as effective electron mediator interfaces for the reduction of cytochrome *c*, hemoglobin, myoglobin and nitrate reductase (cytochrome *c*-dependent). The MP-11-mediated activation of nitrate reductase, NR, was employed to assemble an integrated MP-11/NR electrode for the bioelectrocatalyzed reduction of nitrate (NO_3^-) to nitrite (NO_2^-). An affinity complex between a MP-11-functionalized electrode and NR ($K_a = 3.7\times10^3$ M^{-1}) was crosslinked with glutaric dialdehyde to yield the electrically contacted electrode for the bioelectrocatalyzed reduction of nitrate to nitrite. In this system the reduced MP-11 mediates ET to NR and activates the enzyme towards the reduction of NO_3^-.

A related study employed[78] a MP-11 monolayer-functionalized electrode for the electrochemical activation of Co(II)-protoporphyrin IX towards the hydrogenation of acetylene dicarboxylic acid, (**18**), in an integrated bioelectrocatalytic electrode configuration, Fig. 3-26. A MP-11 monolayer was linked to an Au-electrode. It was found[76] that the association of MP-11 to the electrode surface occurs by two binding modes that include the covalent linking of the terminal carboxylic acid functionality of the peptide, or the binding of carboxylic residues that substitute the heme site. Co(II)-protoporphyrin IX-reconstituted myoglobin, Co(II)-Mb, was then interacted with the MP-11 interface, and the resulting complex was crosslinked with glutaric dialdehyde. The modified electrode revealed bioelectrocatalytic activities towards the hydrogenation of acetylene dicarboxylic acid (**18**) to maleic acid (**19**). Mechanistic studies revealed that the reduced MP-11 mediated the reduction of Co(II)-Mb to a Co(I)-hydride-Mb species that was the active intermediate in the hydrogenation process. The electrocatalyzed hydrogenation proceeded with a current yield that corresponded to ca. 80%.

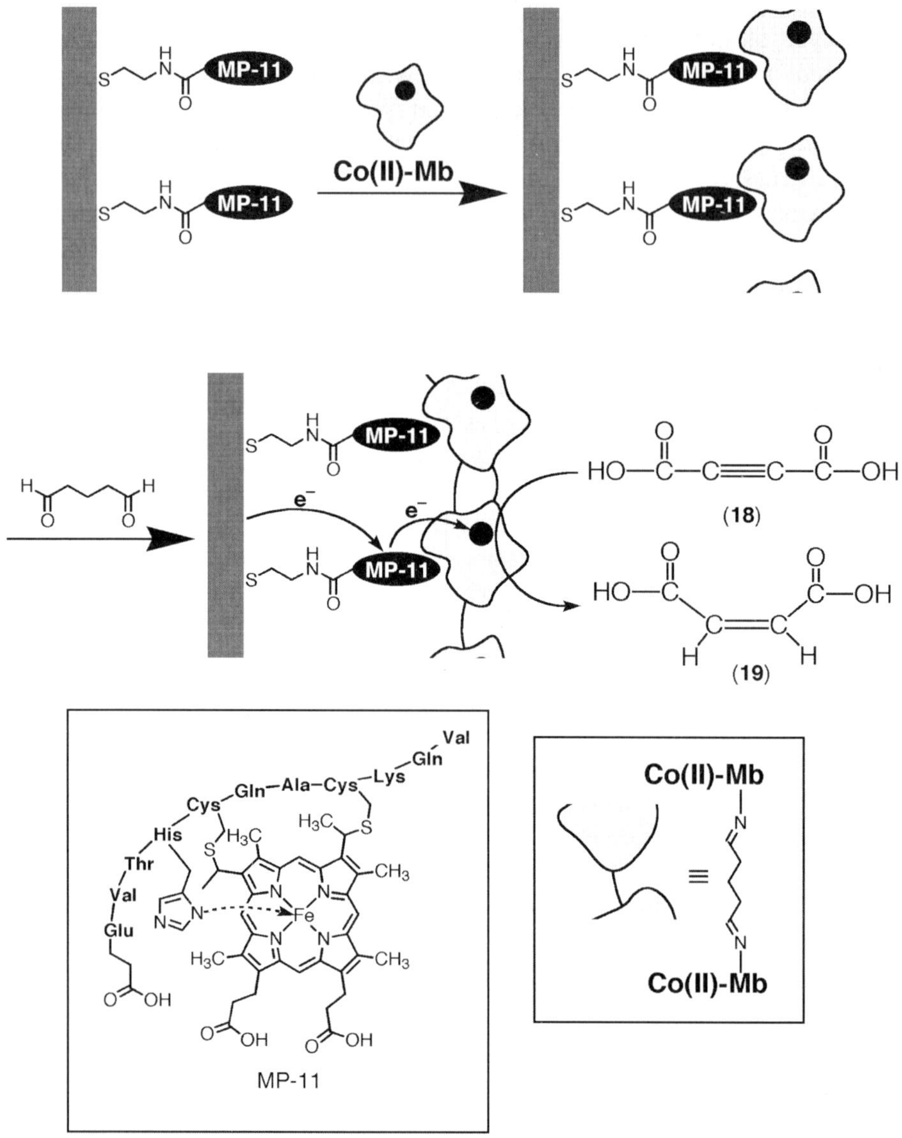

Figure 3-26. The organization of a functional electrode for the bioelectrocatalyzed hydrogenation of acetylene dicarboxylic acid, (**18**), to maleic acid, (**19**), by the surface crosslinking of an affinity complex formed between Co(II)-reconstituted myoglobin and a MP-11 monolayer associated with an electrode. Reproduced from ref. 78 by permission of the Royal Society of Chemistry (RSC).

4. RECONSTITUTED ENZYME-ELECTRODES FOR BIOFUEL CELL DESIGN

The development of biofuel cells that catalyze the conversion of chemical energy stored in abundant organic raw materials, or biomass, to electrical energy is a continuing challenging effort in bioelectronics.[17] The efficient electrical communication between surface-reconstituted redox-enzymes and electrode surfaces resulting in fast ET turnover rates, provides means to tailor bioelectrocatalytic electrodes for biofuel cell applications. Specifically, the high ET turnover rates of the glucose oxidase, GOx, reconstituted electrodes led to oxygen insensitive biocatalytic electrode, a feature that enabled the design of biofuel cells with oxygen (O_2) as oxidizer without the application of a separating membrane. Nanoengineered bioelectrocatalytic electrodes that apply the reconstitution principle were used to assemble a non-compartmentalized glucose-O_2 biofuel cell.[73] Figure 3-27 depicts the schematic configuration of the biofuel cell that includes the reconstituted glucose oxidase (GOx) on the PQQ-FAD monolayer electrode, Fig. 3-4A, as anode and the crosslinked cyt *c*/COx affinity complex on the Au-electrode, Fig. 3-25, as cathode.

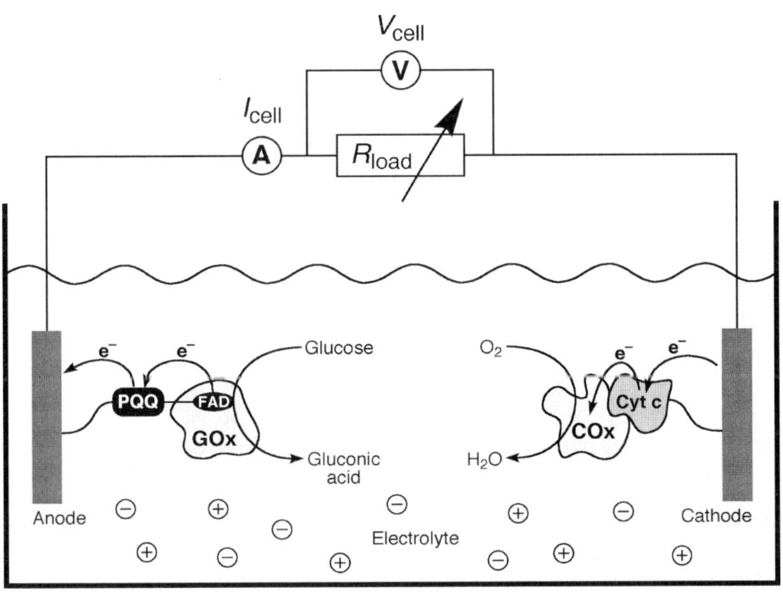

Figure 3-27. Schematic configuration of a biofuel cell element that utilizes glucose as a fuel and O_2 as the oxidizer. Reproduced with permission from ref. 73. Copyright Elsevier, 1999.

Glucose is used as the fuel substrate in the system, whereas O_2 acts as the oxidizer. The biocatalyzed oxidation of glucose to gluconic acid injects electrons into the electrode (anode), whereas electrons from the cathode are transferred to O_2 via the cyt *c*/COx biocatalyst. The current flow through the external circuit, or the formation of a potential difference across the external resistor, provide the route for the conversion of the chemical energy released upon the biocatalytic oxidation of glucose into electrical energy. Figure 3-28 shows the current-voltage behavior of the biofuel cell at different external loads. The power of the cell at different external loads is shown in Fig. 3-28, inset. The maximum power output is 4 μW at an external load of 0.9 kΩ.

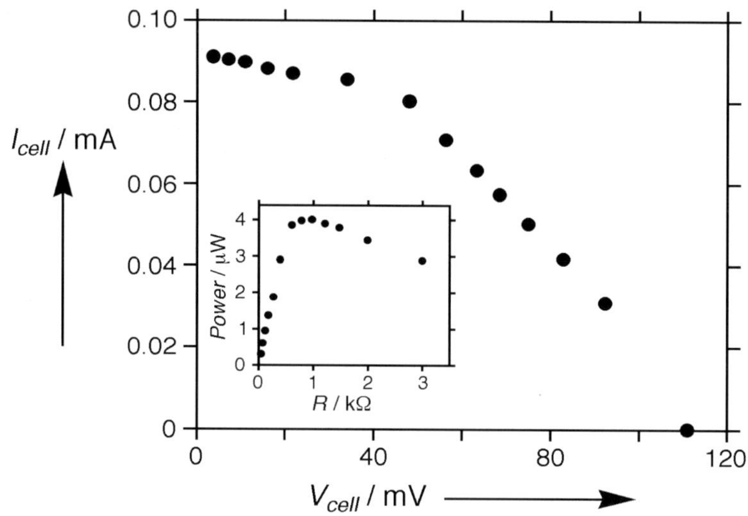

Figure 3-28. I-V curves at different external resistances of the biofuel cell configuration shown in Fig. 3-27. Glucose concentration corresponds to 1 mM. Inset: Extracted electrical power from the biofuel cell at different external resistances. Reproduced with permission from ref. 73. Copyright Elsevier, 1999.

The ideal voltage-current relationship for an electrochemical generator is rectangular. The experimental voltage-current curve deviates from the rectangular shape and the fill factor, $f = P_{max} \cdot I_{sc}^{-1} \cdot V_{oc}^{-1}$, corresponds to ca. 40%. The power output of the cell is low, but the fundamental understanding of the features of the system may stimulate routes to improve the system. In the configuration of the biofuel cell presented in Fig. 3-27, the anode and cathode consisted of enzymes in monolayer assemblies. Increasing the enzyme content by roughening of the electrode surface and/or the reconstitution of the biocatalyst in a thin film configuration rather than a monolayer, may enhance the power output. Also, increasing the thermodynamic redox-potential difference between the relay units associated with the anode and cathode are expected to improve the extracted power from the system. Most importantly, the charge transport of the electrode controls the generated current. For the system described in Fig. 3-27, the ET turnover rate at the anode is ca. 650 s^{-1}, while the ET turnover rate at the cyt *c*/COx cathode is only ca. 20 s^{-1}. Thus, the electrocatalytic reduction of O_2 is the rate limiting step in the current generation process. The substitution of the biomaterial on the electrode, *i.e.* with laccase, or the introduction of new mediators into the cyt *c*/COx system, may thus increase the power output from the cell. Such biofuel cells have been suggested as potential implantable devices that may use the sugar in the blood as the fuel substrate for the generation of electrical power for the activation of implanted pacemakers, insulin pumps or prosthetic units.

Recently, biofuel cell systems were suggested as self-powered biosensor devices.[79] The fact that the voltage output of the cell is controlled by the fuel concentrations suggests that the open-circuit voltage of the cell may be used as a signal for the sensing of the fuel product. Indeed, the biofuel cell consisting of the PQQ/reconstituted GOx as anode, and the cyt *c*/COx as cathode, Fig. 3-27, was used as a self-powered glucose sensing system. Figure 3-29A shows the open-circuit voltages of the cell in the presence of variable concentrations of glucose. Similarly, a biofuel cell configuration consisting of an integrated, electrically contacted, lactate dehydrogenase, LDH, electrode acting as the anode for the oxidation of lactate, and of the cyt *c*/COx cathode, was employed for the self-powered detection of lactate.[79] Figure 3-29B shows the open-circuit voltages of the biofuel cell at variable concentrations of lactate. The derived calibration curves corresponding to the sensing of glucose and lactate by the two biofuel cell structures are depicted in Fig. 3-29 (inset). Thus, the two electrodes that are the active components of the biofuel cells may be implanted into body fluids for the continuous monitoring of glucose or lactate levels.

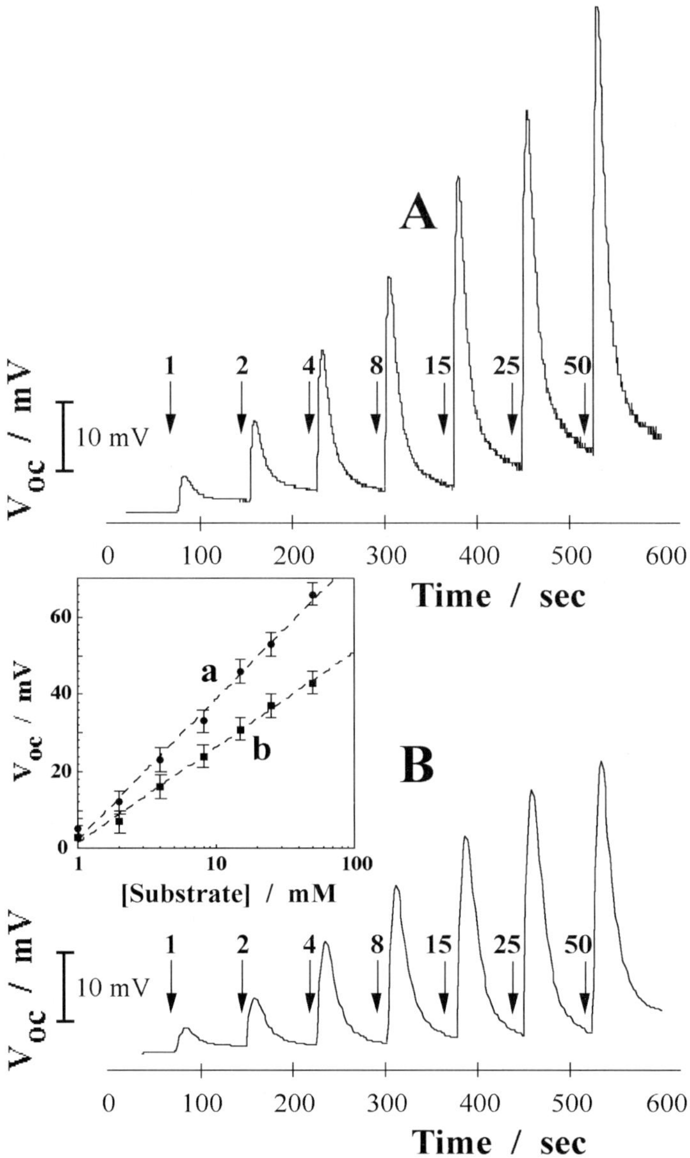

Figure 3-29. (A) Open-circuit voltages of a glucose-based biofuel cell consisting of the anode and cathode shown in Fig. 3-4 and Fig. 3-27, respectively, in the presence of 1 mM to 50 mM concentrations of glucose. (B) Open-circuit voltages of a lactate-based biofuel cell consisting of the anode and cathode configurations shown in Fig. 3-20 and Fig. 3-27 in the presence of 1 mM to 50 mM concentrations of lactate. Inset: Calibration curves corresponding to the analysis of: (a) glucose and (b) lactate by the biofuel cell based systems. Reproduced with permission from ref. 79. Copyright 2001 American Chemical Society.

5. SIGNAL-TRIGGERED SWITCHABLE BIOELECTROCATALYSIS BY RECONSTITUTION OF REDOX-PROTEINS

Tailoring of photoswitchable enzymes that are triggered by light signals to "ON" and "OFF" states has been a subject of extensive research efforts.[80,81] Different possible applications for photochemically triggered enzymes were suggested, including their use as therapeutic materials by their localized activation by light, their use as light signals amplifiers or actinometers, and their future application in biomaterial-based computers. Different methods for the photochemical switching of enzymes were reported, including the tethering of photoisomerizable units to the protein,[82] the application of photoswitchable inhibitors,[83] and the immobilization of enzymes in photosensitive matrices that switch their physical properties,[84] such as substrate permeability, by external light stimuli. The photoswitching of the activities of redox-enzymes is of particular interest since such systems enable the amplified amperometric transduction of the recorded photonic information. Thus, the integration of "ON"-"OFF" photoswitchable redox-enzymes with electrode surfaces may yield optobioelectronic "Write"-"Read" information storage and processing systems. Several methods to photoswitch the bioelectrocatalytic functions of enzymes were reported including the use of photoisomerizable relay units,[85] the use of "photo-command" interfaces that control the electrical contact between the redox-protein and the electrode,[86] and the tethering of photoisomerizable units to the protein.[87] The latter method that involves the covalent linkage of photoisomerizable units to the enzyme is based on the structural perturbation of the protein by one of the photoisomer states leading to the deactivation of the biocatalyst. This method, however, led only to the partial deactivation of the biocatalyst. This is attributed to the fact that the modification of the protein with the photoisomerizable units occurs randomly, and at positions remote from the active site environment, that do not lead to optimal distortion of the biocatalyst active site into a fully switched-off state.

The reconstitution method was suggested[88] as a means to introduce the photoisomerizable unit into the vicinity of the biocatalytic redox-center, thereby generating a light-switchable bioelectrocatalyst that operates between fully switched "ON" and "OFF" states. Apo-glucose oxidase, apo-GOx, was reconstituted with the nitrospiropyran-FAD cofactor unit, (**20**), Fig. 3-30.

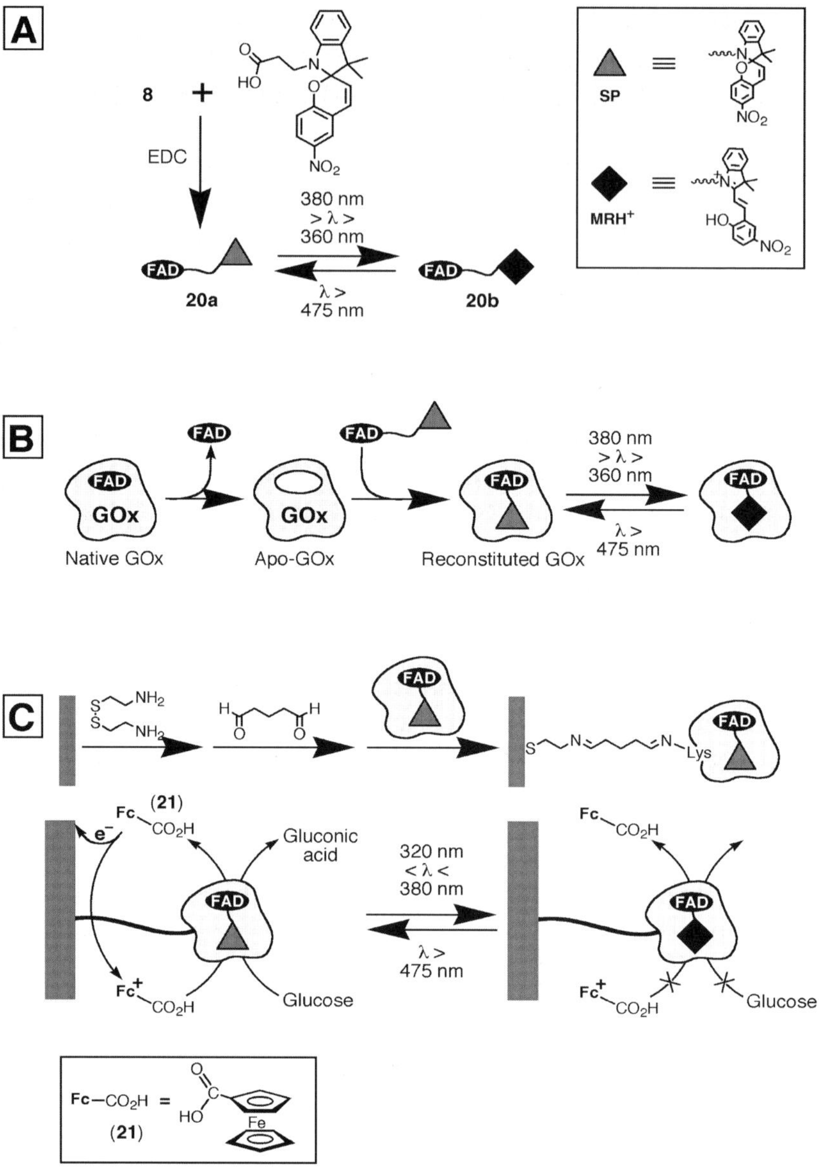

Figure 3-30. Organization of a photoswitchable glucose oxidase electrode for the bioelectrocatalyzed oxidation of glucose: (A) The synthesis of the photoisomerizable nitrospiropyran-FAD cofactor. (B) The reconstitution of apo-glucose oxidase, apo-GOx, with the photoisomerizable FAD-cofactor (**20a**). (C) The assembly of the reconstituted photoisomerizable GOx on an electrode surface and the photoswitching of the bioelectrocatalytic function of the enzyme electrode in the presence of ferrocene carboxylic acid (**21**) as mediator.

The resulting reconstituted GOx undergoes reversible photoisomerization between the neutral nitrospiropyran state, SP-GOx, (**20a**), and the positively charged protonated merocyanine, MRH$^+$-GOx (**20b**). Irradiation of the SP-GOx assembly with light in the region of 360 nm < λ < 338 nm yielded the MRH$^+$-GOx, whereas the illumination of the MRH$^+$-GOx with visible light, λ = 475 nm, generated the SP-GOx state. The photoisomerizable GOx was then immobilized on the Au-electrode in a monolayer configuration, Fig. 3-30. The photoisomerizable enzyme monolayer electrode revealed cyclic and reversible photoswitchable bioelectrocatalytic functions. Figure 3-31 shows the cyclic voltammograms of the photoactive enzyme-monolayer-electrode in the photoisomer states of the biocatalyst and in the presence of glucose and ferrocene carboxylic acid (**21**) that acted as a diffusional electron transfer mediator. In the presence of the SP-GOx monolayer electrode, the bioelectrocatalyzed oxidation proceeded and this was reflected by the formation of electrocatalytic anodic currents, Fig. 3-31, curves (a) and (c).

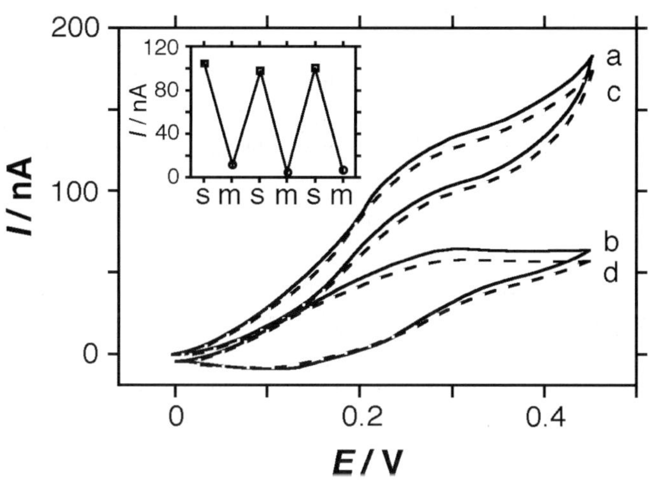

Figure 3-31. Cyclic voltammograms corresponding to the photoswitchable bioelectrocatalyzed oxidation of glucose, 50 mM, in the presence of ferrocene carboxylic acid, (**21**), 5×10^{-5} M, as diffusional electron mediator: (a) and (c): In the presence of the SP-GOx monolayer electrode generated by the irradiation of the electrode λ > 475 nm. (b) and (d): In the presence of the MRH$^+$-GOx monolayer electrode generated by the illumination of the electrode with filtered light 320 nm < λ < 380 nm. Inset: cyclic photoswitchable "ON" and "OFF" amperometric responses of the functionalized enzyme monolayer upon the light-induced isomerization of the interface between the SP-GOx and MRH$^+$-GOx, respectively. Reproduced with permission from ref. 88. Copyright 1997 American Chemical Society.

The bioelectrocatalyzed oxidation of glucose in this system originates from the primary oxidation of the ferrocene carboxylic acid, (**21**), to the respective ferrocenylium cation that mediates the oxidation of the enzyme's redox center and its activation towards the oxidation of glucose. Photoisomerization of the enzyme monolayer to the MRH^+-GOx state switched-off the bioelectrocatalytic functions of the protein monolayer, and only the electrical response of the diffusional electron mediator was observed, Fig. 3-31, curves (b) and (d). By the cyclic photoisomerization of the enzyme-monolayer electrode between the SP-GOx and MRH^+-GOx states, the reversible photoswitching of the enzyme activity between "ON" and "OFF" states was demonstrated, Fig. 3-31 (inset).

The signal-triggered activation and deactivation of the bioelectrocatalytic functions of redox-enzymes can be accomplished not only by photochemical means, but also by electrochemical triggering stimuli. Electrochemical switching of bioelectrocatalytic transformations was achieved by the reconstitution of enzymes in a polyacrylic acid/Cu^{2+} polymer film associated with electrodes.[89] The electrochemically-induced generation of conductive Cu^+-clusters or their dissolution, provides the mechanism for the electrical contacting of the redox-enzymes with the electrodes, Fig. 3-32. A polyacrylic acid/Cu^{2+} film was coated by a protecting polyethyleneimine (PEI) layer, and the pyrroloquinoline quinone, PQQ, (**7**), and the amino-FAD cofactor, (**8**), were covalently-linked to the PEI layer. Apo-glucose oxidase, apo-GOx, was then reconstituted onto the FAD cofactor sites, Fig. 3-32A. Similarly, cyt *c* was covalently linked to the PEI coating and the affinity complex of the cyt *c*/cytochrome oxidase (COx) complex was crosslinked on the modified polymer film, Fig. 3-32B. Electrochemical reduction (E = -0.5 V) of the Cu^{2+} ions in the polymer films associated with the electrode assemblies generated Cu^0 metallic clusters that provided the electrical contact between reconstituted GOx or the crosslinked cyt *c*/COx complex and the respective electrodes. Upon the oxidation of the Cu^0 clusters (E = 0.5 V), to the Cu^{2+} ions complexed to the polyacrylic acid film, the electrical contact between the redox proteins was blocked. Figure 3-33A shows the "ON" and "OFF" short-circuit current developed in the biofuel cell in the presence of a 80 mM glucose solution as a fuel and upon the reduction of the film to the Cu^0-polyacrylic acid state and the oxidation to the Cu^{2+}-polyacrylic acid configuration, respectively. The switched-on state of the biofuel cell can be applied not only for the generation of electrical power, but also as a potentiometric biosensor for glucose. Figure 3-33(B) shows the open-circuit voltage of the biofuel cell in the presence of different glucose concentrations and the respective calibration curve for analyzing glucose. Such electroswitchable biofuel cells may find important future applications as implantable electrical power generator devices that use body fluid (blood) as the fuel source. That is, the consumption of the fuel of the body fluid (e.g. glucose in blood) will occur only upon the electrical activation of the biofuel cell. One may argue that the current flow could be switched-off by the application of a high external resistor. Nonetheless, this method would result in the development of voltage on the anode and cathode. This may activate non-specific electrochemical processes of other ingredients in the body fluid, a process that is eliminated by the electrical triggering of the tailored electrodes of the biofuel cell.

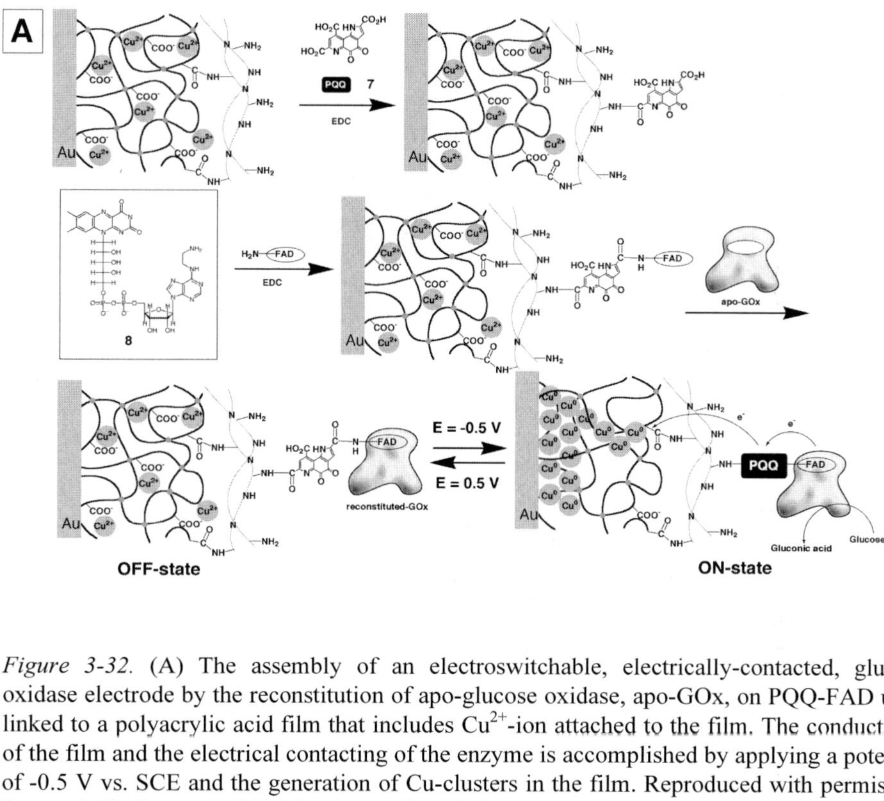

Figure 3-32. (A) The assembly of an electroswitchable, electrically-contacted, glucose oxidase electrode by the reconstitution of apo-glucose oxidase, apo-GOx, on PQQ-FAD units linked to a polyacrylic acid film that includes Cu^{2+}-ion attached to the film. The conductivity of the film and the electrical contacting of the enzyme is accomplished by applying a potential of -0.5 V vs. SCE and the generation of Cu-clusters in the film. Reproduced with permission from ref. 89. Copyright 2003 American Chemical Society.

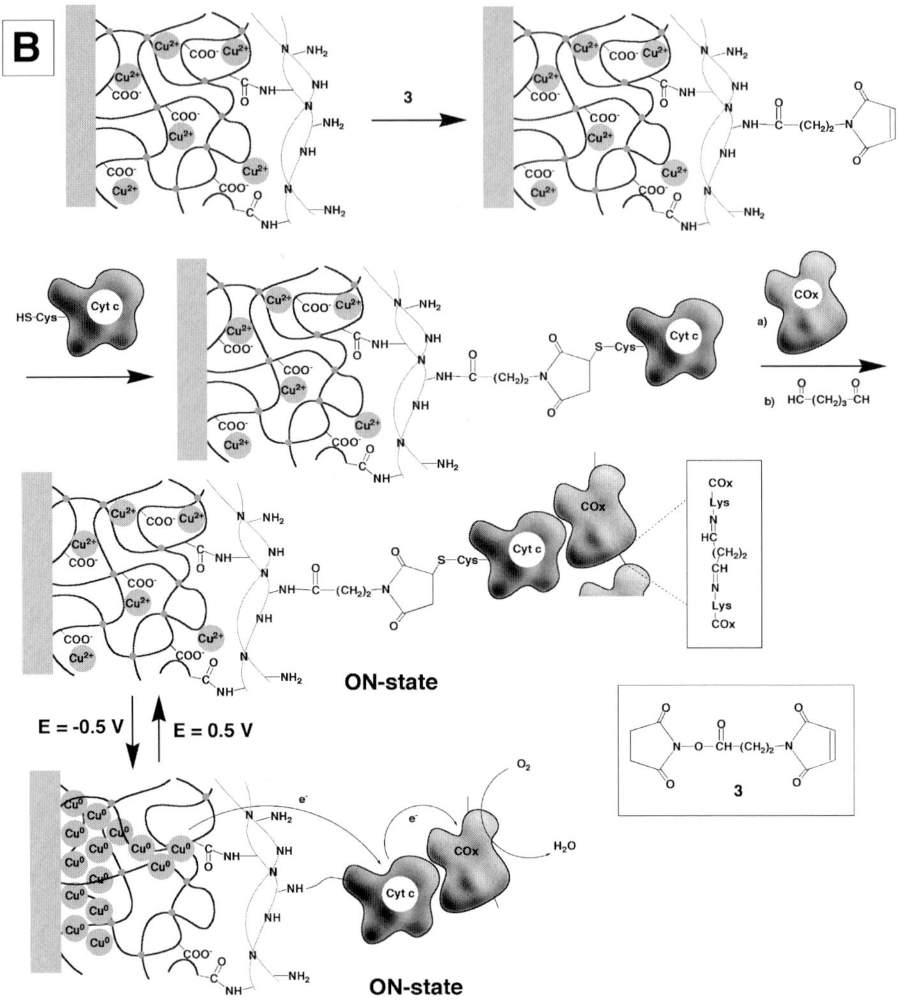

Figure 3-32. (B) The assembly of an electroswitchable, electrically-contacted electrode for the bioelectrocatalyzed reduction of O_2 consisting of a crosslinked affinity complex of cytochrome *c*/cytochrome oxidase (cyt *c*/COx) on a polyacrylic acid-polyethylene imine film that includes incorporated Cu^{2+}-ions. The conductivity of the film and the electrical contacting of the redox-proteins is accomplished by applying a potential of –0.5 V vs. SCE and the generation of Cu-clusters in the film. Reproduced with permission from ref. 89. Copyright 2003 American Chemical Society.

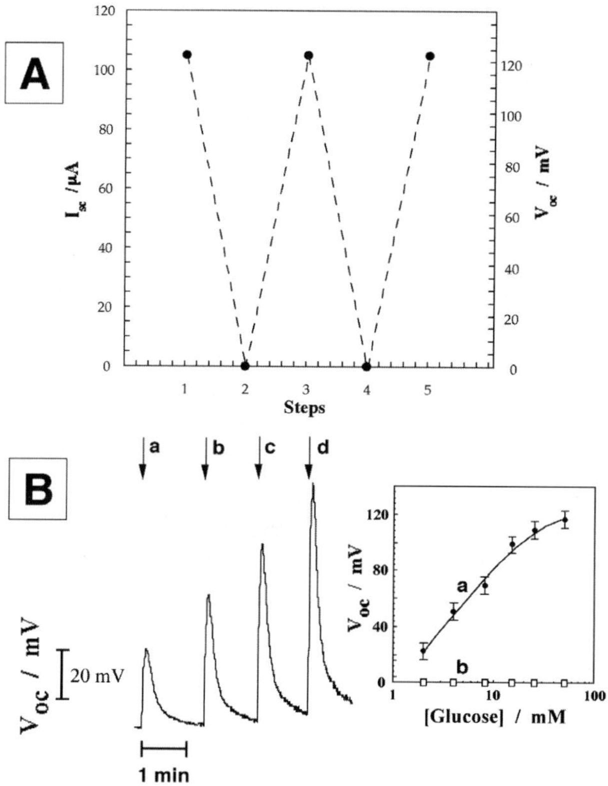

Figure 3-33. (A) Reversible "ON" and "OFF" switching of the short-circuit current, I_{sc}, and the open-circuit voltage, V_{oc}, generated by the biofuel cell consisting of the anode and cathode shown in Figs. 3-32A and 3-32B, respectively. The cell output is switched-on at steps 1, 3 and 5 by the application of a potential of –0.5 V vs. SCE on both electrodes and the generation of Cu-clusters in the respective polymer films. At points 2 and 4 the biofuel cell is switched-off by the application of a potential of +0.5 V vs. SCE on both electrodes that transforms the Cu-clusters into Cu^{2+}-ions. The biofuel cell operated in an air-saturated solution that included 80 mM glucose. (B) Open-circuit voltages (V_{oc}) at variable concentrations of glucose consisting of the Cu-clusters activated anode and cathode shown in Figs. 3-32A and 3-32B, respectively, upon the injection of different glucose concentrations: (a) 2 mM, (b) 3 mM, (c) 8 mM, (d) 40 mM. Inset: Calibration curve corresponding to the open-circuit voltage of the electrochemically-activated biofuel cell at variable glucose concentrations. (The system is always saturated with air). Reproduced with permission from ref. 89. Copyright 2003 American Chemical Society.

6. CONCLUSIONS AND PERSPECTIVES

The present chapter has described the protein reconstitution process on surfaces and the related crosslinking of cofactor-enzyme affinity complexes on surfaces as a means to structurally align redox-proteins on surfaces, and as a general method to establish electrical communication between the redox-enzymes and the electrode supports. Reconstitution of the redox-proteins on surfaces is basically an effort to nano-engineer tailored functionalities of redox-active proteins on electrode surfaces. The effectiveness of electrical communication between the redox-enzymes and the electrodes is far higher than the electrical contacting methods of the biocatalysts by any other chemical methodology. Besides the fundamental importance of the reconstitution process in the basic understanding of the structural factors that control electron transfer between the biocatalyst and the electrode, the method has important practical implications towards the future development of bioelectronic systems. The high electron transfer exchange rates between enzymes and electrodes yields high current outputs from the systems, and this will enable assembling sensitive biosensor devices, organizing miniaturized biosensors, and constructing efficient biofuel cell elements.

Recent developments in nanotechnology provide functionalized metal and semiconductor nanoparticles or nano-rods and other conductive nanowires such as carbon nanotubes. These nanomaterials could provide building blocks for the generation of hybrid structures with redox-proteins. The electrical contacting of redox-proteins with electrodes by the reconstitution of the respective apo-proteins on the respective cofactor-functionalized nanoelements could yield a new generation of bioelectronic systems where electrons are transported across long distances. The combination of nano-materials and redox-proteins by means of the reconstitution process has additional great promise in bioelectronics. The present chapter has addressed mainly the amperometric behavior of redox-proteins reconstituted on electrode surfaces. The effective electrical contacting within redox-proteins reconstituted on metal nanoparticle assemblies,[41] suggests that future bioelectronic devices based on such hybrid systems and voltammetric readout signals may be envisaged. Recent studies have demonstrated that shell-protected metal nanoparticles may be charged by electrons.[90] The assembly of organized hybrid structures on electrodes that consist of metal nanoparticles with reconstituted redox-enzymes may then lead to the charging of the particles by the biocatalyzed process. This charging process may then enable the voltammetric analysis of the respective substrates.

Another important broadening of the reconstitution process is in the area of *de novo* synthesized proteins. The present study has demonstrated the feasibility of organizing bioelectronic systems based on reconstituted *de novo* proteins (Cf. Section 2.4). These examples spark the future possibilities in the field. By the reconstitution of new electroactive synthetic cofactors into pre-designed *de novo* proteins, new man-made bioelectrocatalysts may be envisaged.

REFERENCES

1. Marcus, R.A.; Sutin, N. *Biochim. Biophys. Acta* **1985**, *811*, 265–322.
2. (a) Willner, I.; Katz, E. *Angew. Chem. Int. Ed.*, **2000**, *39*, 1180–1218. (b) Heller, A. *J. Phys. Chem.* **1992**, *96*, 3579–3587. (c) Heller, A. *Acc. Chem. Res.* **1990**, *23*, 128–134.
3. Bartlett, P.N.; Tebbutt, P.; Whitaker, R.G. *Prog. React. Kinetics* **1991**, *16*, 55–155.
4. (a) Williams, D.L.; Doig, A.P., Jr.; Korosi, A. *Anal. Chem.* **1970**, *42*, 118–121. (b) Janada, P.; Weber, J. *J. Electroanal. Chem.* **1991**, *300*, 119–124.
5. (a) Tarasevich, M.R. *Bioelectrochem. Bioenerg.* **1979**, *6*, 587–597. (b) Hill, H.A.O.; Higgins, I.J.; *Philos. Trans. R. Soc. London*, Ser. A. **1981**, *302,* 267–273.
6. Matsue, T.; Kasai, N.; Narumi, M.; Nishizawa, M.; Yamada, H.; Uchida, I. *J. Electroanal. Chem.* **1991**, *300*, 111–118.
7. (a) Armstrong, F.A.; Hill, H.A.O.; Walton, N.J. *Acc. Chem. Res.* **1988**, *21*, 407–413. (b) Taniguchi, I.; Toyosawa, K.; Yamaguchi, H.; Yasukouchi, K. *J. Chem. Soc., Chem. Commun.* **1982**, 1032–1033.
8. (a) Ye, J.; Baldwin, R.P. *Anal. Chem.* **1988**, *60*, 2263–2268. (b) Nassar, A.-E.F.; Willis, W.S.; Rusling, J.F. *Anal. Chem.* **1995**, *67*, 2386–2392.
9. (a) Schuhmann, W.; Ohara, T.J.; Schmidt, H.-L.; Heller, A. *J. Am. Chem. Soc.* **1991**, *113*, 1394–1397. (b) Degani, Y.; Heller, A. *J. Am. Chem. Soc.* **1988**, *110*, 2615–2620. (c) Schuhmann, W. *Biosens. Bioelectron.* **1995**, *10*, 181–192. (d) Willner, I.; Lapidot, N.; Riklin, A.; Kasher, R.; Zahavy, E.; Katz, E. *J. Am. Chem. Soc.* **1994**, *116*, 1428–1441.
10. Willner, I.; Riklin, A.; Shoham, B.; Rivenson, D.; Katz, E. *Adv. Mater.* **1993**, *5*, 912–915.
11. (a) Hale, P.D.; Inagaki, T.; Karan, H.I.; Okamoto, Y.; Skotheim, T.A. *J. Am. Chem. Soc.* **1989**, *111*, 3482–2484. (b) Kaku, T.; Karan, H.I.; Okamoto, Y. *Anal. Chem.* **1994**, *66*, 1231–1235.
12. Willner, I.; Katz, E.; Lapidot, N. *Bioelectrochem. Bioenerg.* **1992**, *29*, 29–45.
13. Maidan, R.; Heller, A. *Anal. Chem.* **1992**, *64*, 2889–2896.
14. Gregg, B.A.; Heller, A. *J. Phys. Chem.* **1991**, *95*, 5970–5975.
15. Badia, A.; Carlini, R.; Fernandez, A.; Battaglini, F.; Mikkelsen, S.R.; English, A.M. *J. Am. Chem. Soc.* **1993**, *115*, 7053–7060.
16. Eisenwiener, H.G.; Schultz, G.V. *Naturwissenschaften* **1969**, *56*, 563–564.
17. Katz, E.; Shipway, A.N.; Willner, I. In *Handbook of Fuel Cell Technology*, Vielstich, W.; Lamm, S.; Gasteiger, H.A.; Eds.; John Wiley & Sons: Chichester, **2003**, Vol. 1, Part 4, pp. 355–381.
18. (a) Winkler, J.R.; Di Bilio, A.J.; Farrow, N.A.; Richards, J.H.; Gray, H.B. *Pure Appl. Chem.* **1999**, *71*, 1753–1764. (b) Gray, H.B.; Winkler, J.R. *Annu. Rev. Biochem.* **1996**, *65*, 537–561.
19. Fox, L.S.; Kozik, M.; Winkler, J.R., Gray, H.B. *Science* **1990**, *247*, 1069–1071.

20. Winkler, J.R.; Gray, H.B. *Chem. Rev.* **1992**, *92*, 369–379.
21. McLendon, G. *Acc. Chem. Res.* **1988**, *21*, 160–167.
22. Hamachi, I.; Shinkai, S. *Eur. J. Org. Chem.* **1999**, 539–549.
23. (a) Willner, I.; Zahavy, E.; Heleg-Shabtai, V. *J. Am. Chem. Soc.* **1995**, *117,* 542–543. (b) Zahavy, E.; Willner, I. *J. Am. Chem. Soc.* **1996**, *118,* 12499–12514.
24. Casimirio, D.R.; Wong, L.L.; Colon, J.L.; Zewert, T.E.; Richards, J.H.; Chang, I.-J.; Winkler, J.R.; Gray, H.B. *J. Am. Chem. Soc.* **1993,** *115*, 1485–1489.
25. Hamachi, I.; Tanaka, S.; Shinkai, S. *J. Am. Chem. Soc.* **1993**, *115*, 10458–10459.
26. Hayashi, T.; Hisaeda, Y. *Acc. Chem. Res.* **2002**, *35*, 35–43.
27. Heleg-Shabtai, V.; Gabriel, T.; Willner, I. *J. Am. Chem. Soc.* **1999**, *121*, 3220–3221.
28. (a) Hamachi, I.; Tajiri, Y.; Shinkai, S. *J. Am. Chem. Soc.* **1994**, *116*, 7437–7438. (b) Hamachi, I.; Nagase, T.; Tajiri, Y.; Shinkai, S. *J. Chem. Soc., Chem. Commun.* **1996**, 2205–2206. (c) Hamachi, I.; Tajiri, Y.; Nagase, T.; Shinkai, S. *Chem. Eur. J.* **1997**, *3*, 1025–1031.
29. Hamachi, I.; Matsugi, T.; Wakigawa, K.; Shinkai, S. *Inorg. Chem.* **1998**, *37*, 1592–1597.
30. Riklin, A.; Katz, E.; Willner, I.; Stocker, A.; Bückmann, A.F. *Nature* **1995**, *376*, 672–675.
31. Willner, I.; Heleg-Shabtai, V.; Blonder, R.; Katz, E.; Tao, G.; Bückmann, A.F.; Heller, A. *J. Am. Chem. Soc.* **1996**, *118*, 10321–10322.
32. Zayats, M.; Katz, E.; Willner, I. *J. Am. Chem. Soc.* **2002**, *124,* 14724–14735.
33. Nakamura, S.; Hayashi, S.; Koga, K. *Biochim. Biophys. Acta* **1976,** *445*, 294–308.
34. Raitman, O.A.; Katz, E.; Bückmann, A.F.; Willner, I. *J. Am. Chem. Soc.* **2002**, *124*, 6487–6496.
35. Bartlett, P.N.; Simon, E. *Phys. Chem. Chem. Phys.* **2002,** *2*, 2599–2606.
36. Knoll, W. *Annu. Rev. Phys. Chem.* **1998**, *49*, 569–638.
37. Frutos, A.G.; Corn, R.M. *Anal. Chem.* **1998**, *A70*, 449A–455A.
38. (a) Sasaki, S.; Nagata, R.; Hock, B.; Karube, I. *Anal. Chim. Acta* **1998**, *368*, 71–76. (b) McDonnell, J.M. *Curr. Opin. Chem. Biol.* **2001**, *5,* 572–577.
39. Disley, D.M.; Cullen, D.C.; You, H.Y.; Lower, L.R. *Biosens. Bioelectron.* **1998**, *13*, 1213–1225.
40. Chegel, V.; Raitman, O.; Katz, E.; Gabai, R.; Willner, I. *Chem. Commun.* **2001**, 883–884.
41. Xiao, Y.; Patolsky, F.; Katz, E.; Hainfeld, J.F.; Willner, I. *Science* **2003**, *299*, 1877–1881.
42. Davidson, V.L.; Jones, L.H. *Anal. Chim. Acta* **1991**, *249*, 235–240.
43. Davidson, V.L. *Principles and Applications of Quinoproteins*; Dekker: New York, **1993**.
44. Ikeda, T. *Bull. Electrochem.* **1992**, *8,* 145–159.
45. Katz, E.; Schlereth, D.D.; Schmidt, H.-L.; Olsthoorn, A.A.J. *J. Electroanal. Chem.* **1994**, *368*, 165–171.
46. Raitman, O.A.; Patolsky, F.; Katz, E.; Willner, I. *Chem. Commun.* **2002**, 1936–1937.
47. Hess, C.R.; Juda, G.A.; Dooley, D.M.; Amii, R.N.; Hill, M.G.; Winkler, J.R.; Gray, H.B. *J. Am. Chem. Soc.* **2003**, *125*, 7156–7157.
48. Yaropolov, A.I.; Malovic, V.; Varfolomeev, S.D.; Berezin, I.V. *Dokl. Akad. Nauk SSSR* **1979**, *249*, 13999–1401.
49. Csöregi, E.; Jönsson, G.; Gorton, L. *J. Biotechnol.* **1993**, *30*, 315–317.
50. Ruzgas, T.; Gorton, L.; Emnéus, J.; Csöregi, E.; Marko-Varga, G. *Anal. Proc.* **1995**, *32*, 207–208.
51. Ghindilis, A.L.; Atanasov, P.; Wilkins, E. *Electroanalysis* **1997**, *9*, 661–675.
52. Ruzgas, T.; Csöregi, E.; Emnéus, J.; Gorton, L.; Marko-Vargas, G. *Anal. Chim. Acta*, **1996**, *330*. 123–138.

53. Ruzgas, T.; Gorten, L.; Emnús, J.; Marko-Varga, G. *J. Electroanal. Chem.* **1995**, *391*, 41–49.
54. Lindgren, A.; Munteau, F.-D.; Gazaryan, I.G.; Ruzgas, T.; Gorton, L. *J. Electroanal. Chem.* **1998**, *458*, 113–120.
55. Armstrong, F.A.; Hill, H.A.O.; Walton, N.J. *Q. Rev. Biophys.* **1986**, *18*, 261–322.
56. Allon, P.M.; Hill, H.A.O.; Walton, N.J. *J. Electroanal. Chem.* **1984**, *178*, 69–86.
57. Pardo-Yissar, V.; Katz, E.; Willner, I.; Kotlyar, A.B.; Sanders, C.; Lill, H. *Faraday Discussions* **2000**, *116*, 119–134.
58. Ryabov, A.D.; Goral, V.N.; Gorton, L.; Csöregi, E. *Chem. Eur. J.* **1995**, *5*, 961–967.
59. Zimmermann, H.; Lindgren, A.; Schuhmann, W.; Gorton, L. *Chem. Eur. J.* **2000**, *6*, 592–599.
60. (a) DeGrado, W.F.; Wassermann, Z.R.; Lear, J.D. *Science* **1989**, *243*, 622–628. (b) Hill, R.B.; DeGrado, W.F. *J. Am. Chem. Soc.* **1998**, *120*, 1138–1145.
61. Rabanal, F.; DeGrado, W.F.; Dutton, P.L. *J. Am. Chem. Soc.* **1996**, *118*, 473–474.
62. Arai, T.; Kobata, K.; Mihara, H.; Fujimoto, T.; Nishino, N. *Bull. Chem. Soc. Jpn.* **1995**, *68*, 1989–1998.
63. Gibney, B.R.; Mulholland, S.E.; Rabanal, F.; Dutton, P.L. *Proc. Natl. Acad. Sci. U.S.A.* **1996**, *93*, 15041–15046.
64. Katz, E.; Heleg-Shabtai, V.; Willner, I.; Rau, H.K.; Haehnel, W. *Angew. Chem. Int. Ed. Engl.* **1998**, *37*, 3253–3256.
65. Willner, I.; Heleg-Shabtai, V.; Katz, E.; Rau, H.K.; Haehnel, W. *J. Am. Chem. Soc.* **1999**, *121*, 6455–6468.
66. Rau, H.K.; Haehnel, W. *J. Am. Chem. Soc.* **1998**, *120*, 468–476.
67. (a) Moiroux, J.; Elving, P.J. *J. Am. Chem. Soc.* **1980**, *102*, 6533–6538. (b) Schmidt, H.-L.; Schuhmann, W. *Biosens. Bioelectron.* **1996**, *11*, 127–135.
68. (a) Kulys, J.; Gleixner, G.; Schuhmann, W.; Schmidt, H.-L. *Electroanalysis* **1993**, *5*, 201–207. (b) Schlereth, D.D.; Katz, E.; Schmidt, H.-L. *Electroanalysis* **1995**, *7*, 46–54.
69. Ohtani, M.; Kuwabata, S.; Yoneyama, H. *J. Electroanal. Chem.* **1997**, *422*, 45–54.
70. Katz, E.; Lötzbeyer, T.; Schlereth, D.D.; Schuhmann, W.; Schmidt, H.-L. *J. Electroanal. Chem.* **1994**, *373*, 189–200.
71. Willner, I.; Riklin, A. *Anal. Chem.* **1994**, *66*, 1535–1539.
72. Bardea, A.; Katz, E.; Bückmann, A.F.; Willner, I. *J. Am. Chem. Soc.* **1997**, *119*, 9114–9119.
73. Katz, E.; Willner, I.; Kotlyar, A.B. *J. Electroanal. Chem.* **1999**, *479*, 64–68.
74. Adams, P.A. In *Peroxidases in Chemistry and Biology*, Everse, J.; Everse, K.E., Eds.; CRC Press: Boston, 1991, Vol. 2, chapter 7, pp. 171–200.
75. Lötzbeyer, T.; Schuhmann, W.; Katz, E.; Falter, J.; Schmidt, H.-L. *J. Electroanal. Chem.* **1994**, *377*, 291–294.
76. Katz, E.; Willner, I. *Langmuir* **1997**, *13*, 3364–3373.
77. Narvaez, A.; Dominguez, E.; Katakis, I.; Katz, E.; Ranjit, K.T., Ben-Dov, I.; Willner, I. *J. Electroanal. Chem.* **1997**, *430*, 227–233.
78. Heleg-Shabtai, V.; Katz, E.; Levi, S.; Willner, I. *J. Chem. Soc., Perkin Trans. 2* **1997**, 2645–3652.
79. Katz, E.; Bückmann, A.F.; Willner, I. *J. Am. Chem. Soc.* **2001**, *123*, 10752–10753.
80. Willner, I. *Acc. Chem. Res.* **1997**, *30*, 347–356.
81. Willner, I.; Rubin, S. *Angew. Chem., Int. Ed. engl.* **1996**, *35*, 367–385.
82. Willner, I.; Rubin, S.; Riklin, A. *J. Am. Chem. Soc.* **1991**, *113*, 3321–3325.
83. Delcour, A.; Hess, G. P. *Biochemistry* **1986**, *25*, 1793–1798.
84. Willner, I.; Rubin, S.; Zor, T. *J. Am. Chem. Soc.* **1981**, *113*, 4013–4014.

85. Lion-Dagan, M.; Marx-Tibbon, S.; Katz, E.; Willner, I. *Angew. Chem., Int. Ed. Engl.* **1985**, *34*, 1604–1606.
86. Lion-Dagan, M.; Katz, E.; Willner, I. *J. Chem. Soc., Chem. Commun.* **1994**, 2741–2742.
87. Lion-Dagan, M.; Katz, E.; Willner, I. *J. Am. Chem. Soc.* **1994**, *116*, 7913–7914.
88. Blonder, R.; Katz, E.; Willner, I.; Wray, V.; Bückmann, A.F. *J. Am. Chem. Soc.* **1997**, *119*, 11747–11757.
89. Katz, E.; Willner, I. *J. Am. Chem. Soc.* **2003**, *125*, 6803–6813.
90. Tempelton, A.C.; Wuelfing, W.P.; Murray, R.W. *Acc. Chem. Res.* **2000**, *33*, 27–36.

Chapter 4

VOLTAMMETRY OF ADSORBED REDOX ENZYMES: MECHANISMS IN THE POTENTIAL DIMENSION

JULEA N. BUTT[1] AND FRASER A. ARMSTRONG[2]
[1]*Centre for Metalloprotein Spectroscopy and Biology, School of Chemical Sciences and Pharmacy, University of East Anglia, Norwich, NR4 7TJ, UK;* [2]*Department of Chemistry, Inorganic Chemistry Laboratory, University of Oxford, Oxford, OX1 3QR, UK.*

1. INTRODUCTION

Over 30% of all enzymes are oxidoreductases, *i.e.* they catalyse redox reactions. Many carry out long-range electron transfer (ET) and *couple* this process to discrete transformations of a substrate at the catalytic site and to conformational changes and transmembrane ion/proton pumping. This coupling means that the electrochemical potential (which controls oxidation states and the rate at which they are interconverted) should be a powerful and direct determinant of enzyme activity and ought to be an important experimental variable: however, until recently these predictions have remained largely unexplored. Traditional experimental strategies have resolved enzyme mechanisms in the time and concentration domains, with little control, or knowledge, of sample oxidation state once the catalytic reaction has been initiated. By contrast, voltammetric methods and other dynamic electrochemical techniques have long been used to study the reactions of small molecules, yielding kinetic information not only as functions of concentration and time but also revealing the dependence of rates on electrochemical potential. It was only when groups led by Hill, Kuwana and Niki demonstrated direct (*i.e.* unmediated) and reversible voltammetry of cytochromes at solid electrodes in the late 1970's that the powers of dynamic electrochemical methods (as opposed to potentiometry,

O. Hammerich and J. Ulstrup (eds.), Bioinorganic Electrochemistry, 91–128.

which measures systems at equilibrium) began to be exploited to resolve electron transfer and coupled chemistries within proteins.[1–3]

The utility of direct protein voltammetry is enhanced when the molecules under investigation are adsorbed at up to monolayer coverage on a suitable electrode – an approach known as *protein film voltammetry* (PFV), Fig. 4-1.

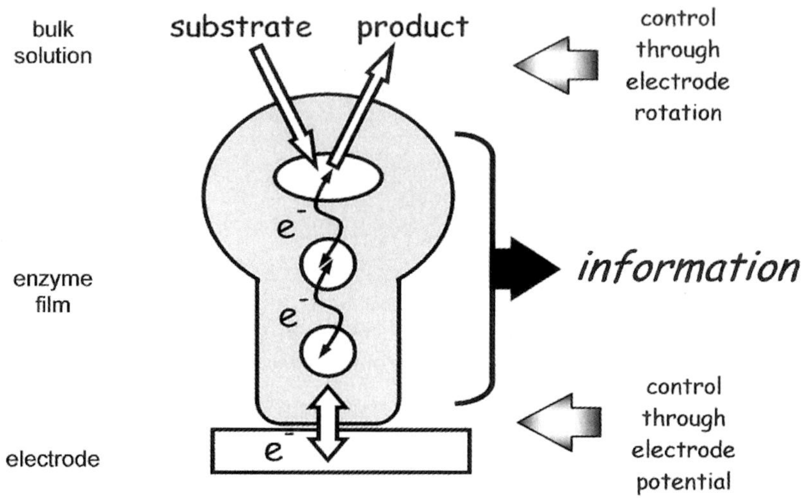

Figure 4-1. Protein film voltammetry as a technique for studying redox enzyme mechanisms. The catalytic current-potential profile provides information on the rate-defining catalytic processes occurring within the enzyme. It is important that interfacial electron transfer is facile and information is not masked by limitations due to the transport of substrate and product: for this reason the rotating disc electrode is an important tool in these studies.

Unlike having the reacting protein molecules free in solution, the voltammetry is no longer limited by the relatively sluggish process of protein diffusion and the redox driven chemistry of an active site is brought into sharp focus.[4–7] From a practical viewpoint, the electrode assumes the role of the physiological redox partner, an advantage being that activity (current) is resolved as a function of driving force (electrode potential) with the latter precisely defined and variable over a broad and continuous window. When control of substrate/product mass transport is combined with a model for interfacial electron transfer, the catalytic current-potential profile displayed by the film can be interpreted in terms of rate-defining events that are intrinsic to the enzyme's mechanism of catalysis, and will ideally reflect redox-linked chemistries relevant to cellular function.

As a tool for mechanistic enzymology, protein film voltammetry benefits from its sensitivity, the quantitative nature of the thermodynamic and kinetic information gained, and the extremely small sample requirements: indeed, monolayer coverage of a protein of 50,000 Da requires less than 10 picomoles cm^{-2} (note that a typical electrode surface area is just a few square millimetres).[4,5,7] In addition, real-time investigations of reactivity over a wide range of conditions are facilitated by the simple ability to transfer an electrode with its adsorbed film quickly and easily from one medium to another. As we will illustrate later, these various facets of protein film voltammetry combine to afford fresh perspectives on redox enzyme catalysis. In the next section we will set the scene by describing the electrode surfaces that are necessary to observe protein film voltammetry; then we will outline the techniques that are commonly used to quantify enzyme activity and discuss the factors that contribute to the voltammetric response.

2. ELECTROCHEMICAL STUDIES OF PROTEIN MOLECULES ADSORBED ON ELECTRODES

2.1 Electrode Surfaces for Protein Film Voltammetry

The pyrolytic graphite edge (PGE) electrode has been successful in affording PFV of a wide range of proteins in our laboratories and others. Prepared simply by polishing an 'edge-plane' exposed graphite electrode with an aqueous alumina slurry, followed quickly by ultrasonic cleaning, the resulting surface is rough, hydrophilic and rich in acidic oxides, Fig. 4-2A.[8–10] Unlike a metal electrode, the surface atoms and groups have satisfied valencies, so that the atomic-scale interactions with the protein are weak and non-covalent, more like the interactions that occur between proteins in solution or on a membrane. Both electrostatic and hydrophobic interactions contribute to the formation of reasonably stable films, and protein adsorption and stability can be optimised by choice of electrolyte conditions, e.g. ionic strength, pH, and certain co-adsorbates such as the polyamines, neomycin and polymyxin. These polyamines are protonated at neutral pH and bind to negatively charged patches on a protein, altering its charge and forming salt bridges between like charged regions of the graphite and protein surfaces, thus enhancing both protein-electrode and lateral protein-protein interactions within the film. Experience has shown that the quality of the protein sample is a crucial factor for successful film formation, as trace impurities, particularly apo-protein, compete favourably for adsorption at the electrode surface. Low temperatures are also helpful, although film stability is enhanced at the expense of catalytic activity.

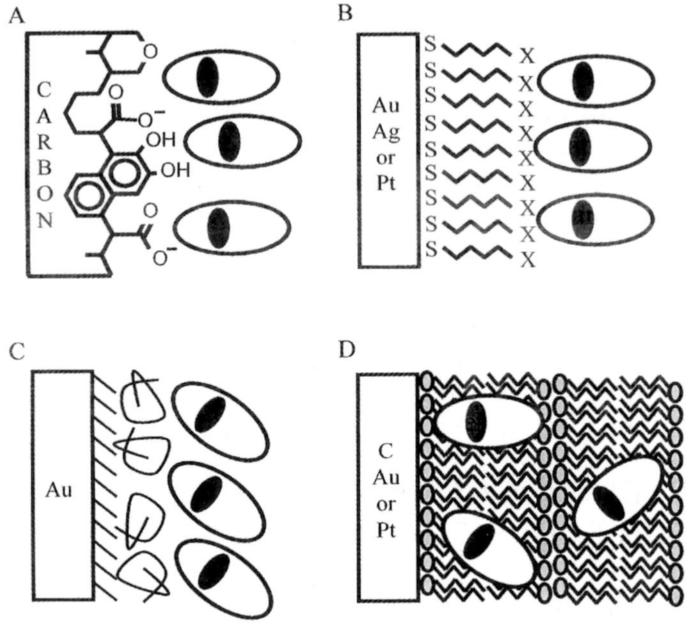

Figure 4-2. Cartoons of the ways in which proteins are attached to electrodes for protein film voltammetry. (A) Pyrolytic graphite edge (PGE) electrode. (B) SAM modified metal electrode. (C) Layer-by-layer polyion adsorption. (D) Electrode confined bilayers.

The PGE electrode presents a topographically and chemically heterogeneous surface for protein adsorption. Surface heterogeneity may be a key factor in the utility of PGE electrodes because it probably provides a 'combinatorial advantage' - *i.e.* at least some sites are likely to be favourable for binding a particular enzyme and engaging it for fast interfacial electron transfer. From a practical standpoint pyrolytic graphite is a robust material that provides a protein-friendly surface through a relatively simple procedure. Importantly it offers access to a wide range of potentials in aqueous solution (ca. –1V to +1V), a feature that is very desirable if highly oxidising or highly reducing states are to be studied. A potential disadvantage of PGE is that the surface groups include redox-active functionalities such as aldehydes and quinones that themselves undergo sluggish electron transfer reactions: these may complicate the interactions between the protein and electrode in addition to contributing signals to the voltammetry.[8,9,11–13]

Chemical modification of metal electrodes allows for more rationally designed surfaces for protein adsorption. Spontaneous adsorption of functionalised alkyl thiols onto Au, Ag or Pt generates a *self-assembled monolayer* (SAM) with the functional group(s) projected away from the electrode to engage with protein molecules, Fig. 4-2B. The redox chemistry of SAM-modified electrodes is less complex than PGE and generally limited to reductive desorption of the alkyl thiol, a process that is useful in characterising the SAM.[14,15] The large range of functionalised thiols possible makes these electrodes an important direction for research.[16–18] For example, by varying the chain length, the effects of distance and electric field on interfacial electron transfer rates can be examined.[19–21] It is also possible to attach proteins covalently to metal electrodes by way of the thiol/disulphide functionalities projected from surface-exposed cysteine or derivatised lysine residues. This produces robust films that facilitate studies in chemical or physical environments where non-covalent attachment is ineffective or produces an unproductive orientation.[22,23]

Protein molecules can also be confined within bilayer structures, or adsorbed along with polyion surfactants or DNA in a layer-by-layer assembly approach, as depicted in Figs. 4-2C and 4-2D.[24–28] The fact that in these assemblies many of the protein molecules contributing to the voltammetric signal are not in intimate contact with the electrode suggests the importance of entrapment and dynamics in eliciting the electrochemical response of proteins. Such approaches and considerations are particularly relevant for studying enzymes normally associated with biological membranes.

2.2 Electrochemical Techniques and the Information they Provide

Two of the electrochemical techniques used in protein film voltammetry are shown in Fig. 4-3. In cyclic voltammetry the electrode potential is swept in a linear manner back and forth between two limits. The rate at which the potential is scanned defines the time scale of the experiment and this can be varied from < 1 mV s^{-1} to > 1000 V s^{-1}. This is a very large dynamic range, and it is possible to carry out both steady-state and transient experiments on the same sample of enzyme.[29,30] Cyclic voltammetry is important because it provides the 'big picture' and produces a signal that links the reaction or active site of interest to a particular potential. In chronoamperometry, the current is monitored at a constant potential following a perturbation such as a step to this potential or addition of a substrate. This experiment is important because it separates the potential and time dependencies of a response.[31] In both types of experiment, it is usually important to be able to rotate the electrode in order to control transport of the substrate and product to and from the enzyme film.

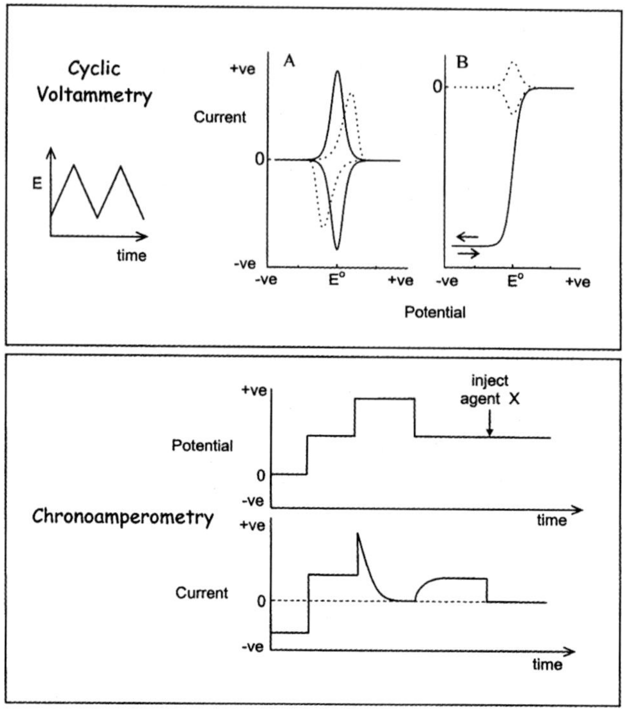

Figure 4-3. Electrochemical techniques and the redox-linked chemistries of an enzyme film on an electrode. Cyclic voltammetry provides an intuitive 'map' of enzyme activities. A. The non-turnover signal at 'low' scan rates (solid lines) provides thermodynamic information, while raising the scan rate leads to a peak separation (broken lines) the analysis of which gives the rate of interfacial electron exchange and additional information on how this is coupled to chemical reactions. B. Catalysis leads to a continual flow of electrons that amplifies the response and correlates activity with driving force under steady-state conditions: here the catalytic current reports on the reduction of an enzyme substrate (solid line). Chronoamperometry allows deconvolution of the potential and time domains: here an oxidoreductase is reversibly inactivated by application of the most positive potential, an example is [NiFe]–hydrogenase, and inhibition by agent X is shown to be essentially instantaneous.

First, we consider cyclic voltammetry and focus on the 'non-turnover' experiment in which there is no substrate in solution. As the potential is scanned, electrons transfer back and forth between the electrode and the redox-active site(s), producing a current peak in each direction. These two peaks constitute the 'signal'. Provided the scan rate is slow enough to equilibrate all the processes required in the redox reaction, the signal obtained will be as predicted by the Nernst equation that holds for a reversible electron-transfer process. The current peaks will be compact and symmetrical, as shown in Fig. 4-3A (solid line). The average of the peak positions gives the reduction potential, E°, and the area under the peaks is related *directly* to the number of sites involved in the electron exchange. For a one-electron reaction, the half-height peak width, $\Delta E_{p,1/2}$, at 25 °C should be close to 91 mV, although this is usually broadened by imperfections, the most probable of which is inhomogeneity among adsorbed molecules. Overlaid signals from enzymes with multiple redox centres can be deconvolved by calculating the defined shape of each contribution to the cyclic voltammetric response[48].

When an active site can transfer two (or more) electrons, an important term is n, which can be treated as the number of electrons transferred cooperatively, since the peak height i_p varies as n^2, eqn. 4-1A, while the peak width varies as $1/n$, eqn 4-1B.[32] In these equations, ν is the scan rate, A is the electrode surface area, Γ is the electroactive coverage, R is the gas constant, T is the absolute temperature and F is the faraday constant.

$$i_p = \frac{n^2 F^2}{RT} \frac{\upsilon A \Gamma \exp\{(nF/RT)(E - E^{\circ})\}}{\left(1 + \exp\{(nF/RT)(E - E^{\circ})\}\right)^2} \qquad \text{(4-1A)}$$

$$\Delta E_{p,1/2} = 3.53 \frac{RT}{nF} = \frac{90.6}{n} \text{ mV } (25\ ^{\circ}\text{C}) \qquad \text{(4-1B)}$$

This is best illustrated by considering a two-electron redox centre such as a flavin cofactor undergoing reduction: if the second electron is added at a much more favourable potential (in practice this means its reduction potential is more than 0.1 V more positive than for the first electron transfer) n approaches a value of 2 and the two electrons are considered to transfer *cooperatively*. Consequently the one-electron intermediate is inherently unstable. The resulting peaks are half as wide (approximately 45 mV) and four times as high as for the one-electron case. Although cooperativity is often lower than this, with n lying between 1 and 2, and inhomogeneity usually broadens the signal in any case, cooperative two-electron cofactors such as flavins are more likely to provide conspicuous features in the voltammograms of adsorbed enzymes.

The catalytic properties of the enzyme are observed by adding substrate to the electrolyte solution, whereupon the resulting flow of electrons between the electrode and molecules in solution is observed as a catalytic current. Catalysis amplifies the electrochemical response of an adsorbed enzyme to an extent ultimately determined by the turnover frequency k_{cat}, (an unfavourable K_M can be overcome by raising the substrate concentration) and the cyclic voltammogram typically converts to the form shown in Fig. 4-3B. The potential is usually cycled sufficiently slowly to ensure that the system is at steady state at all times, *i.e.*, the catalytic current does not vary with scan rate. The electrode is rotated at different speeds to investigate the role of substrate transport in determining the kinetics. If the catalytic reaction is controlled by the rate of supply of substrate, the current will depend on the square root of the rotation rate: this is given by the Levich equation, eqn. 4-2.[33] In this equation, D is the diffusion coefficient of the redox substrate in solution, ω is the rotation rate, ν is the kinematic viscosity of the solvent and C is the concentration of redox substrate in solution.

$$i_{\lim} = 0.62nFAD^{2/3}\omega^{1/2}\nu^{-1/6}C \tag{4-2}$$

The current will usually begin to level off at high rotation rate and the maximum value can be determined by plotting the reciprocal current against the reciprocal square root of the rotation rate, using the general Koutecky-Levich equation, eqn. 4-3A.

$$\frac{1}{i_{\lim}} = \frac{1}{i_{\max}} + \frac{1}{0.62nFAD^{2/3}\omega^{1/2}\nu^{-1/6}C} \tag{4-3A}$$

This procedure gives the current that would be observed at *infinitely high* rotation rate. From these values, measured for each substrate concentration, the K_M value is obtained, and provided we are able to determine the electroactive coverage from the size of the signals obtained before adding substrate, we can also obtain k_{cat}. The appropriate combination of Michaelis-Menten and Levich equations is then given by eqn. 4-3B.

$$\frac{1}{i_{\lim}} = \frac{(C + K_M)}{nFA\Gamma\, k_{cat}C} + \frac{1}{0.62nFAD^{2/3}\omega^{1/2}\nu^{-1/6}C} \tag{4-3B}$$

The electrochemically-derived kinetic constants can then be compared to those obtained by conventional methods, to judge how the activity of the enzyme immobilised on the electrode compares with that observed in solution. Even when the population of electroactive enzyme is too small to measure reliably (generally below 2 pmol cm^{-2} for one-electron signals on a graphite electrode) a catalytic wave can be observed whose current magnitude is directly related to turnover rate, and K_M (but not k_{cat}) can be measured.

While the Michaelis parameters are an important guide as to the integrity of the adsorbed sample, they are not of primary concern. By contrast, voltammetry reveals activity *as a function of potential* which can be varied and tuned to control and probe the enzyme in different ways. We will see that many enzymes display a 'potential optimum', *i.e.* they have maximum activity at a particular potential. Many enzymes operate well in both redox directions, and the 'bias' can be measured precisely by running the voltammetry in the presence of both oxidised and reduced substrates. The exact form of the catalytic wave provides important mechanistic details, and this is often more easily analysed by taking the derivative of the catalytic current (di/dE).

We now consider what happens when the scan rate is increased to determine the kinetics of electron transfer to and from an active site, and establish if this is limited by chemical events. For a simple isolated redox transition, *i.e.* one not coupled to chemical events (including proton transfer, conformational changes or catalysis), an increase in scan rate causes an increase in the peak separation when the rate of interfacial electron exchange starts to limit the voltammetric response, Fig. 4-3A (broken line). A good way to examine the electron-transfer kinetics is to plot the oxidation and reduction peak potentials as a function of log (scan rate): this yields a trumpet-shaped plot from which the electron-transfer exchange rate constant (k_0) can be determined using the Butler-Volmer (or Marcus) equation.[30] If this kind of experiment is carried out in the presence of substrate, the oxidation and reduction peaks appear once the scan rate is high enough to outrun the rate of turnover. This allows us to determine how a bound substrate changes the reduction potential, electron-transfer kinetics, or some other characteristic property of the active site.[34,35]

In chronoamperometry, the current is measured as a function of time at a constant potential. This can follow a potential step or injection of an agent that initiates a change in activity, Fig. 4-3. The advantage is that potential is no longer convoluted with time; instead the change in current due to a reaction that alters the catalytic activity is analysed to establish the order of the reaction and the rate constant. It is important to remember that the *rate of change* in current observed after the perturbation corresponds to the *rate of change* of catalytic rate. Figure 4-3 illustrates how the kinetics and potential dependence of enzyme activation and inactivation are readily resolved with the molecules adsorbed on an electrode. In this case the enzyme is inactivated at high potential, but reactivated by stepping to a less positive potential.

Another electrochemical technique being used in protein studies is square-wave voltammetry. The usefulness of this method is often ascribed to its ability to factor out the double-layer charging current; but for protein molecules confined to a film, the advantages are the increased sensitivity and additional kinetic information that can be obtained by varying the frequency and pulse amplitude. The data are more difficult to extract than cyclic voltammetry, and we will not attempt to elaborate on this aspect, although studies that have included square-wave voltammetry will be mentioned later in this chapter.

2.3 Properties of Enzymes Adsorbed on an Electrode

Our aim is to elucidate the redox-linked chemistry that is relevant to cellular function; thus the adsorbed enzyme will ideally retain the structural and functional characteristics displayed in the living cell. Adsorbed proteins which have strong and distinctive vibrational and electronic transitions can be probed by spectroscopic techniques such as surface enhanced resonance Raman and UV-visible reflectance spectroscopies. Otherwise the functionality of adsorbed enzymes is best assessed by comparing the activity observed on the electrode with that measured in solution assays of the purified proteins. The activities should be similar; but sometimes the electrode reveals a higher activity because electrons can be added or removed at any potential that optimises this rate (electrons are available 'on tap'). Indeed the film environment may provide a better resemblance to the cellular environment than dilute solution. In certain cases, the protein:electrode interface may mimic the interaction with the physiological redox partner in its manner of solvent exclusion and the formation of salt-bridges, while the high lateral concentration and the electric field experienced at the electrode interface may resemble that experienced by proteins bound at a membrane.

Ideally, PFV requires a film of active molecules (monolayer or sub-monolayer) that behave independently of each other and homogeneously with regard to their electrochemical and catalytic properties. Interactions between centres in neighbouring molecules are naturally minimised by the surrounding polypeptide and ensure that PFV is relatively free from complications induced by intermolecular interactions of the type that frequently distort the voltammetry of surface-confined small molecules. Usually, however, peak widths are larger than expected due to inhomogeneity (dispersion). The activity of enzyme molecules with dispersed and poor interfacial electron-transfer kinetics (low k_0) will distort the voltammogram and complicate analysis.

The factors determining interfacial electron-transfer rates to adsorbed proteins have been investigated by several groups. The subtlety of the problem is evident from studies carried out on horse heart and yeast cytochrome *c* adsorbed at different SAM-modified Au electrodes.[36] Yeast cytochrome *c* showed a 2500 fold increase in interfacial electron-transfer rate at the mixed SAM consisting of equimolar $HS(CH_2)_{10}COOH$ and $HS(CH_2)_8OH$ compared to the SAM composed only of $HS(CH_2)_{10}COOH$. By contrast, horse heart cytochrome *c* gave a low rate at the single-component SAM and the rate was only enhanced 5 fold at the mixed SAM. Since these cytochromes have similar electron-transfer reorganisation energies, the large differences in rate most likely reflect differences in the electronic coupling to the electrode. Experiments with horse heart cytochrome *c* and the 'blue' Cu protein azurin adsorbed at electrodes modified with $HS(CH_2)_nX$ SAMs have shown that for long chain lengths the interfacial electron-transfer rates decrease exponentially with distance, as expected for a tunneling mechanism.[19–21] However, as the chain length is decreased below a few $-CH_2-$ units, the rates approach a limiting value. In experiments designed to determine how the rate increases with driving force, square-wave experiments with azurin also suggest that a limiting rate is reached that is much smaller than expected from Marcus theory.[37] The implication from both types of experiment is that a reaction other than the elementary long-range electron-transfer process ultimately controls the rate. There is also increasing evidence from non-electrochemical studies that protein-protein electron-transfer reactions occur most rapidly when the precursor protein-protein complex is dynamic and able to explore, rapidly, a large number of different configurations; this enables the bimolecular complex to locate those configurations (perhaps just one) that allow an electron to tunnel at the optimal rate.[38] With electrochemical experiments, the electron-transfer rate may be scaled down by a factor reflecting the (low) probability of achieving a well-coupled orientation even once the protein is adsorbed on the electrode. Whatever the factors involved, it is clear that values of $k_o > 1000\ s^{-1}$ are easily achieved. Indeed, for most enzymes, an electron-transfer exchange rate constant (k_o) greater than 100 s^{-1} will be sufficient to resolve stages of catalytic turnover.[39]

The highest interfacial electron transfer rate constant yet reported (about 14,000 s^{-1}) is for a c-type cytochrome from *Aquifex aolicus*.[40] This protein has a 62-amino acid linker domain by which it is usually anchored to the periplasmic side of the inner membrane: this linker has a cysteine as the terminal residue before the signal region, and the sulfur atom provides an anchor point. The cytochrome adsorbs strongly onto a Au electrode that is already modified with a hexane-thiol SAM (note this requires that the molecules in the SAM move or vacate to allow this). The results are striking, with oxidation and reduction peaks remaining visible at scan rates exceeding 1000 V s^{-1}. As the scan rate is increased, starting from low values, the number of molecules exchanging electrons drops, so that at 1000 V s^{-1}, only a few percent of the total are involved in the reaction (a return to slow scans restores the normal intensity of peaks). Therefore the response observed reflects the number of cytochrome *c* molecules that are 'ready to go'.

3. INTERPRETING THE CATALYTIC VOLTAMMOGRAM

3.1 The Serial-Resistor Model of Enzyme Electrocatalysis

A good way to consider catalytic electron flow through an enzyme adsorbed on an electrode surface is to consider the different stages in terms of a series of resistors, each of which represents a step that could be rate limiting and therefore control the current. This is represented in Fig. 4-4A. In any series of resistors, the conductance of the total is given by a reciprocal law; this is why the rate-determining step is the one with the highest resistance. The three stages are (i) interfacial electron-transfer to and from the electron entry/exit centre of the enzyme, represented by $1/i_E$, (ii) the inherent catalytic properties of the enzyme, represented by $1/i_{cat}$, and (iii) substrate mass transport, represented by $1/i_{trans}$. The series resistor model describes the current flow due to electrocatalysis in terms of the contributions from intrinsic characteristics (most simply k_{cat}, K_M) and experimental variables (electrode potential, substrate concentration, electrode rotation rate). The current at any instant is determined mainly by the slowest event(s) in the process, so that our interpretation is simplified when $1/i_{cat}$ is larger than the other terms. For example, in the simplest analysis $1/i_E$ becomes negligible at large overpotentials and the current becomes independent of applied potential, $1/i_{lim} \sim 1/i_{cat} + 1/i_{trans}$. When the current is independent of rotation rate, or the rotation rate dependence allows the determination of i_{lim} from the intercept of a Koutecky-Levich (reciprocal) plot, the variation of i_{lim} with substrate concentration gives a direct route to i_{cat} and K_M (k_{cat} can then be deduced if the electrode surface coverage of electroactive enzyme is established).

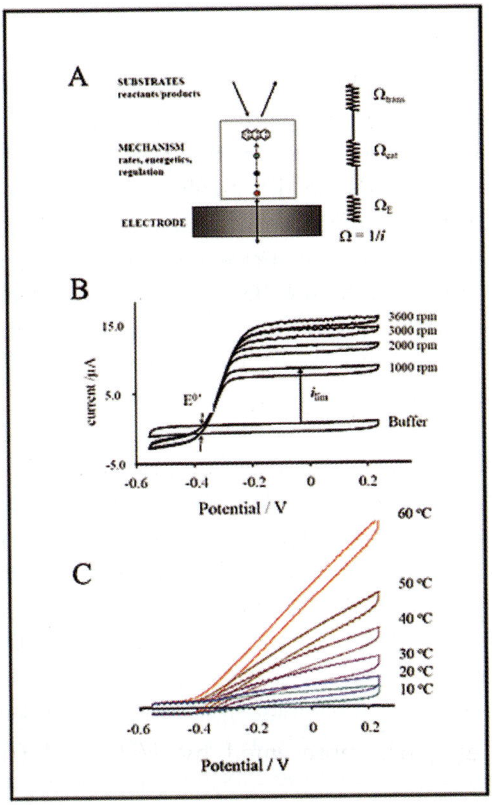

Figure 4-4. Representation of catalytic electron flow through an adsorbed enzyme in terms of a series of resistors. A. Cartoon showing the different processes that may control the catalytic current. B. Oxidation of H_2 by [NiFe]–hydrogenase adsorbed on PGE electrode (0.1 bar H_2, scan rate 0.2 V s^{-1}) illustrating control of catalytic rate by substrate mass transport (Ω_{trans} is largest). The current changes sign at the zero-current background at a potential corresponding to the ambient reduction potential for the H^+/H_2 couple. C. Oxidation of H_2 by a sparsely populated film of [NiFe]-hydrogenase (1 bar H_2, scan rate 1 V s^{-1}, rotation rate 2500 rpm) illustrating how the rate determining factor in catalysis shifts as the temperature is raised. At low temperature, the current is determined by enzyme turnover, while at high temperature it is controlled by interfacial electron-transfer ($1/i_E$ becomes dominant). Reproduced from refs. are 7, 41 and 43 with permission of the American Chemical Society.

It follows that if an enzyme has a very high turnover frequency, it may be difficult to determine its inherent catalytic properties. The [NiFe]–hydrogenase from *Allochromatium vinosum* is a good example.[41] A film of this enzyme formed on a PGE electrode displays very high activity for H_2 oxidation and, for an atmosphere of 10% H_2, the catalytic current varies with the electrode rotation rate as expected from the Levich equation. Figure 4-4B shows also that as the rotation rate is increased to increase the current, so does the potential at which the plateau for H_2 oxidation is reached, and this effect is seen also as the hydrogen partial pressure is raised. At relatively low rotation rates the determining factor is transport of H_2 (thus $i_{\mathrm{lim}} \rightarrow i_{\mathrm{trans}}$), but as we make greater demands on the catalytic flux through the enzyme by raising the rate of substrate transport, the current is eventually determined by the rate of interfacial electron-transfer (which has lower activation energy).

An alternative way to alleviate mass-transport control is to have less enzyme on the electrode: catalytic centres with sufficient separation are served by more efficient substrate transport through hemispherical diffusion than those in a more densely packed array where planar substrate diffusion predominates.[42] Figure 4-4C shows voltammograms measured for hydrogenase over a wide temperature range, the coverage in this case being much lower than for Fig. 4-4B, so that an increase in the electrode rotation rate does not cause an increase in catalytic current. At low temperatures the voltammogram has a sigmoidal shape, but as the temperature is raised it becomes linear, similar to that expected with Ohm's law. The basis for this observation is that the activation energy for enzyme catalysis is higher than for interfacial electron transfer. However, the linear relationship is not entirely expected, since if interfacial electron transfer is rate limiting, the current should increase more rapidly as the potential becomes more positive. The explanation is that the enzyme has not one unique rate for interfacial electron transfer, but several; *i.e.* there is a large dispersion among these rates.[43] This arises because of the dependence of interfacial electron transfer on the nature of the contact with the electrode surface, so that the process is influenced by small variations and fluctuations in orientation. The molecules of hydrogenase are adsorbed in many different ways, each with a different value of the interfacial electron transfer rate constant.

3.2 Visualising Rate Determining Events within the Enzyme

When interfacial electron exchange rate(s) are sufficiently high and the response is free from mass transport limitations, the catalytic current will be determined by the inherent activity of the enzyme. Variation of current (activity) with potential can be explained by an extension of the Michaelis-Menten description of enzyme kinetics that relates activity to oxidation state through incorporation of the Nernst equation.[44] The resulting expressions describe the catalytic cycle, and include rates of intramolecular electron exchange, chemical events, substrate binding and product release, together with the reduction potentials of centres in the enzyme, and the influence of their oxidation states on the catalytic rate. What we observe in an electrochemical experiment depends on the rate of the electron-transfer reactions *within* the enzyme, and it is useful to think of an 'electrochemical control centre': this is the redox centre up to which electron exchange with the electrode is fast, and whose properties are therefore being measured.

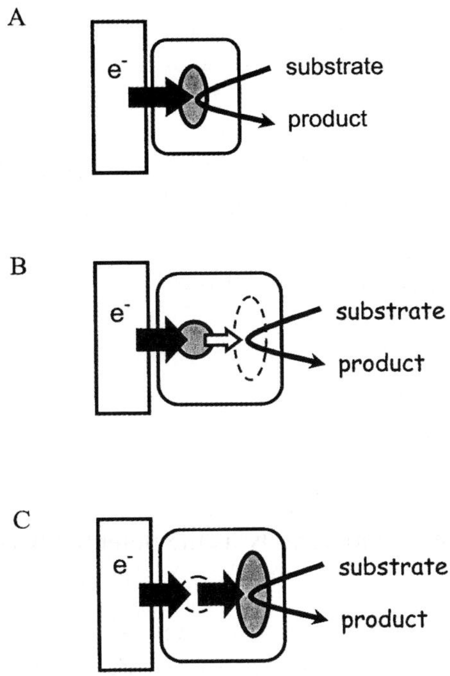

Figure 4-5. Points of electrochemical control in enzyme voltammetry. The voltammetry reports on the 'electrochemical control centre', the redox site (solid) up to which electron exchange with the electrode is fast. For single-centre redox enzymes (A), the electrochemical control centre is the active site. For multi-centered redox enzymes the electrochemical control centre may be a relay centre (B), or the active site redox group if catalysis is not impeded by the rate of intramolecular electron exchange (C). Further scenarios are also possible, as described in the text.

Figure 4-5 shows examples of different electrochemical control centres. In an enzyme with a single redox centre, Fig. 4-5A, this is both the catalytic centre and electrochemical control centre. When one or more relay centres lie between the catalytic centre and the electrode, it is possible either that electron exchange with the electrode is fast only up to one of the relay centres, Fig. 4-5B, or fast all the way to the catalytic centre, Fig. 4-5C. In the former case, the voltammetry we observe is that of the relay centre, not the active site.[45] In the latter case, operation of the relay is 'invisible' to the voltammetric experiment, which now conveys the action occurring at the active site. This is the scenario we seek; we can vary the substrate concentration and observe changes in the shape and position of the wave to obtain information on the chemistry occurring at the active site. An additional possibility is that one or more redox centres act as a device to control the catalytic rate according to their oxidation states. Both enhancements and attenuations in activity may be observed as the electrode potential is modulated across the reduction potentials of the centres concerned and we will describe examples of each of these effects below.

4. SPECIFIC EXAMPLES OF INFORMATION GAINED BY VOLTAMMETRY

4.1 Cytochrome *c* Peroxidase

Cytochrome *c* peroxidase (CcP) from bakers yeast is one of the best studied redox enzymes, and a paradigm for those heme proteins that utilise the ferryl group as a catalytic intermediate, including cytochromes P450. The active site is shown in Fig. 4-6A. Cytochrome *c* peroxidase catalyses the reduction of hydrogen peroxide (1.32 V at pH 7) by reduced cytochrome *c* (E^{o} = 260 mV). Two main intermediates have been characterised by spectroscopy: the first of these, 'Compound I' contains Fe(IV) and a cation radical located on a tryptophan residue; this is reduced to 'Compound II' (where the oxidising equivalent resides mainly as Fe(IV)), then a second electron returns the enzyme to the resting state Fe(III). Since one redox partner is a protein, cytochrome *c*, it would be expected that CcP is well suited for long-range electron transfer, and Fig. 4-6B shows this is the case.[46] At ice-bath temperature and with a low electrolyte concentration, CcP adsorbs at a PGE electrode and catalyses the reduction of hydrogen peroxide at quite high potentials (>700 mV). Without substrate in solution, a reversible non-turnover signal is observable, consisting of a pair of peaks centred at 700–800 mV depending on pH, over the range 7.0 to 4.5. Importantly, both peaks have a half-height width less than the one-electron value (84 mV at 0°C), thus $n > 1$, and therefore two electrons are transferred cooperatively. From the peak areas, the electroactive coverage of enzyme is approximately 3 pmol cm^{-2}. The conclusion is that the resting Fe(III) state is oxidised by two electrons to a state that is identical or equivalent to Compound I and the intermediate known as Compound II is inherently unstable.

A

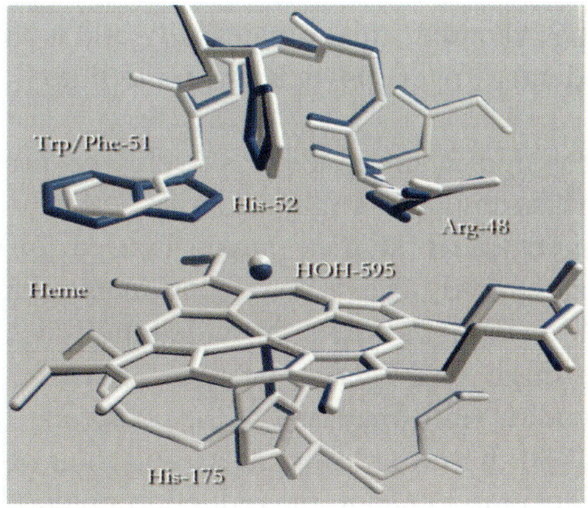

B

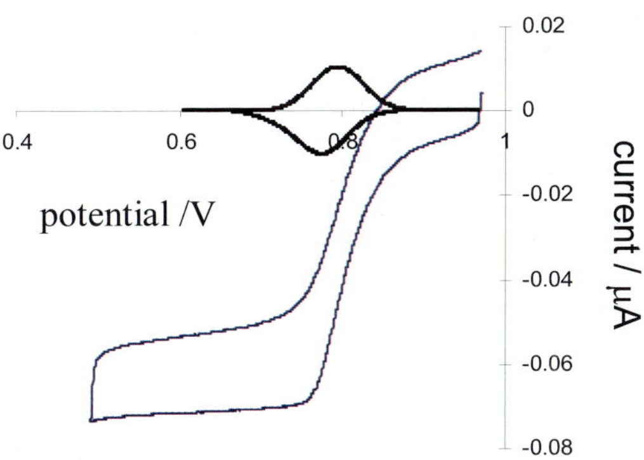

Figure 4-6. (A) A close-up view of the active site of yeast cytochrome *c* peroxidase showing the residues in the distal pocket at which hydrogen peroxide is reduced to water. Overlaid on the structure of the wild type enzyme are the positions of residues in the W51F mutant (tryptophan is replaced by phenylalanine). (B) Voltammograms of a film of wild type CcP on a PGE electrode, obtained in the absence and presence of H_2O_2 at ice temperature, pH 5.0. The electrode is rotating at 200 rpm, but the catalytic current in this case continues to increase as the rotation rate is increased; therefore under these conditions the electrocatalysis is diffusion controlled and few facts are revealed about the enzyme's chemistry. For the W51F mutant, the signal due to the reversible two-electron couple and the catalytic wave are both shifted >100 mV more positive in potential compared to the wild-type enzyme. Reproduced from ref. 46 and 47 with permission.

Although this at first appears to contradict the established fact that Compound II can be prepared stoichiometrically and isolated, the stability of Compound II evidently owes its existence more to kinetics rather than thermodynamics.

The results are significant since PFV clearly provides a route for observing and measuring the high-potential redox reactions involving Fe(IV), while conventional studies have focused on the less relevant Fe(III)/Fe(II) redox couple. The PGE electrode is suitable for measuring potentials of 1 Volt or higher at pH 7, provided the process is fast enough to allow a reasonably high scan rate to be used. A number of mutants have been studied, providing information on the different factors that control reduction potentials of these oxidising states and their activities. Mutants of the distal residues generally change the reduction potential but retain activity (with the exception of H52) while proximal mutants usually have zero or diminished catalytic activity together with poor electron-transfer kinetics. The W51F mutant, in which the distal tryptophan has been converted into a phenylalanine (as occurs in horseradish peroxidase) exhibits a much higher reduction potential (883 mV at pH 5.4, compared to 753 mV for wild type) and increased H_2O_2 reduction activity which also occurs at higher potential; but it is less stable.[47] These experiments therefore show how the physiologically relevant reduction potentials of peroxidases are controlled.

4.2 Fumarate Reductases

Fumarate reductases contain a FAD cofactor as active site, along with a 'wire' of Fe-S clusters or heme groups to relay electrons from the electron donor. The enzyme from *E. coli* has four subunits, two of which, C and D, are membrane-intrinsic and provide the interaction sites for the electron donor – a quinol. Subunits A and B are membrane-extrinsic and contain, respectively, a covalently bound FAD and three Fe-S clusters ([2Fe-2S], [4Fe-4S] and [3Fe-4S] in order of increasing distance from the FAD). The *E. coli* enzyme is closely related to succinate dehydrogenase, which is also expressed in *E. coli*, but more importantly occurs in mitochondria as Complex II of the electron-transport respiratory chain. A soluble subcomplex, FrdAB, can be obtained by chemical treatment of the holoenzyme or by genetic engineering. It adsorbs to high coverage on a PGE electrode, giving non-turnover voltammograms revealing the FAD as a prominent signal that overlays weaker one-electron signals due to the [2Fe-2S] and [3Fe-4S] clusters, along with a signal at more negative potential due to the [4Fe-4S] cluster.[35,48,49] This is shown in Fig. 4-7A.

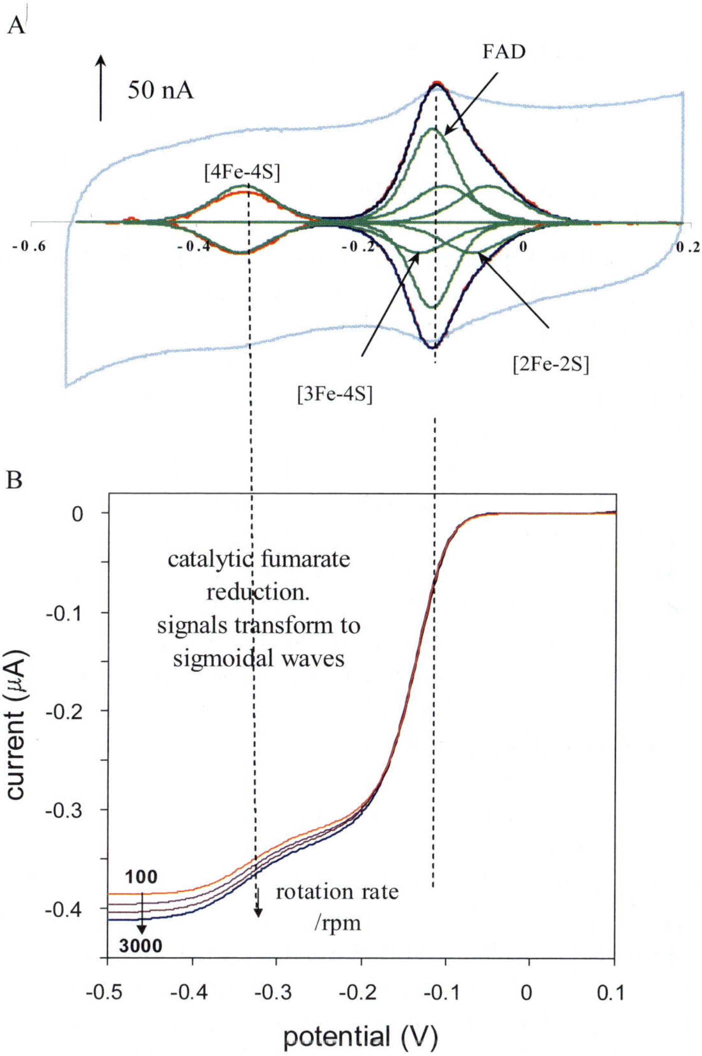

Figure 4-7. (A) Voltammogram of a film of *E. coli* fumarate reductase (FrdAB) adsorbed on a PGE electrode. This is a non-turnover experiment (no fumarate is present) and the voltammogram reveals discrete signals due to the redox active sites: these signals are deconvolved to show the components due to the three Fe-S clusters and the FAD (prominent because it undergoes a cooperative two-electron exchange). (B) Catalytic voltammogram measured in the presence of 0.2 mM fumarate, rotation rate varied from 100 to 3000 rpm, 20 °C. The catalytic activity is boosted at potentials below that of the [4Fe-4S] cluster. Reproduced from ref. 7 and 49 with permission.

Unlike the *E. coli* fumarate reductase, flavocytochrome c_3 (Fcc_3) from the marine bacterium *Shewanella frigidimarina* is a soluble enzyme. It has four heme groups instead of Fe-S clusters, and the FAD is not covalently bound, but otherwise the active site and the mechanism are similar to the *E. coli* enzyme. Conventional methods for studying the active site are not useful here, as the electronic spectrum of the flavin is masked by the intense transitions from the four heme groups, and the one-electron radical (which should be EPR active) is inherently unstable. As discussed above, the instability of the radical means that two electrons transfer cooperatively and this makes it easy to observe the FAD by voltammetry.[34,50]

For both of these fumarate reductases, the pH dependence of the FAD is easily studied by placing the enzyme-coated electrode into solutions of different pH and measuring the position of the prominent signal. In the presence of fumarate, the peak-like signals convert to a catalytic wave that is sigmoidal-like and rotation-rate dependent. Under conditions of high rotation rate and high fumarate concentration, which remove mass transport control, both enzymes exhibit a more complex waveform; in particular, a second boost in current is observed at a lower electrode potential. Figure 4-7B shows a result obtained for the *E. coli* enzyme, and studies with mutations that shift the [4Fe-4S] reduction potential have shown that the boost in catalytic rate correlates closely with reduction of this relay centre.[51]

The FAD signal from FrdAB and Fcc_3 can be observed even up to quite high scan rates, and this allows studies to be carried out in which the catalytic cycle is 'intercepted'. Thus for Fcc_3, a scan rate of 80 V s^{-1} results in electrons being returned to the electrode before they are passed on to fumarate.[34] With a high concentration of fumarate, the signal is now that of the Michaelis complex, and close analysis shows (a) that it is still sharp, *i.e.* both electrons are delivered and retrieved in the few milliseconds allowed for passage across the peaks; (b) no additional 'capacity' has been added (if product dissociation were rate limiting, we might expect to see four electrons transferred on the first sweep); (c) the reduction potential has barely altered compared with the enzyme in the absence of fumarate; (d) the electron-transfer kinetics are enhanced compared to the enzyme without substrate.

Similar results are observed with FrdAB, although the lower coverages that are obtained make it more difficult to observe the FAD at high scan rates. However, whereas Fcc_3 has no detectable succinate oxidation activity, FrdAB produces good catalytic voltammetry that can be studied easily without rotating the electrode (the turnover rate is low enough that the results reflect only the properties of the enzyme). This means also that the catalytic cycle can be easily outrun even at low scan rates. A study of the scan rate dependence in the presence of succinate showed that the FAD reduction potential is lowered significantly.[35] Thus, succinate binding depends on the FAD oxidation state, whereas fumarate binding does not, and the relative stability of the complex between succinate and oxidised FAD helps to drive the transfer of hydride and proton from reduced FAD to fumarate.

4.3 Succinate Dehydrogenase

Succinate dehydrogenase is structurally and functionally very closely related to the fumarate reductase from *E. coli,* indeed the catalytic subcomplex SdhAB lacking the membrane anchors can be obtained from beef heart mitochondria and other sources. SdhAB can be adsorbed at a PGE electrode, although it has proven difficult to obtain films that have a sufficiently high coverage to observe non-turnover signals. Even so, some interesting catalytic activity is revealed, as shown in Fig. 4-8.

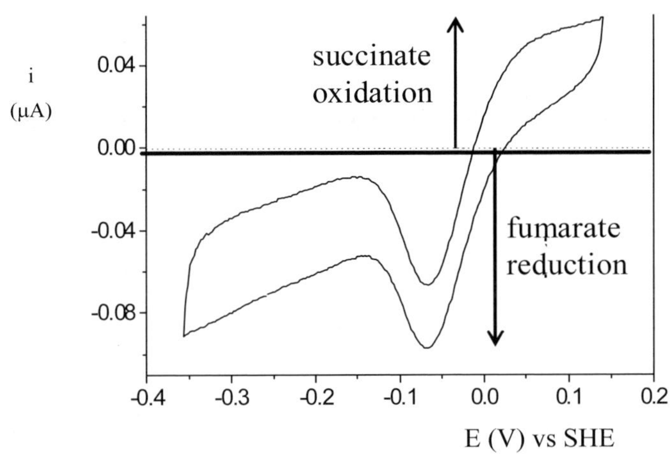

Figure 4-8. Voltammetry of succinate dehydrogenase adsorbed on a rotating PGE electrode in the presence of equal concentrations of succinate and fumarate.

Experiments carried out with equal concentrations of fumarate and succinate in solution reveal that at pH values below 7.5, succinate dehydrogenase is inherently biased to function better in the direction of fumarate reduction.[52] This is easily observed despite the instability of SdhAB (an interesting contrast with FrdAB) since successive cycles trace out isosbestic points for the scan in either direction. These points occur at zero Faradaic current and their average potential value corresponds to that provided by the fumarate/succinate ratio (for a 1:1 ratio, the average of the two isosbestic potentials gives the formal reduction potential). The exactness of the potential-activity relationship also provides a way to measure the origins of H/D isotope effects on the rates of catalysis in either direction, in particular making it possible to observe how the driving force is also affected.[53]

The reduction of fumarate is unusual because the activity is attenuated quite abruptly as the potential is lowered beyond a critical value; in other words the reaction rate decreases despite the increasing driving force.[54] Viewed in terms of analogies in electronics, the system displays *negative* resistance at potentials more negative than the peak, and in this respect it behaves as a device known as a tunnel diode. Several explanations can account for this, all requiring that the potential that is imposed by the electrode not only drives the catalytic reaction but also controls the enzyme's oxidation states, some of which may be inactive and unable to function even though the catalytic driving force is high. The rate of catalysis of fumarate reduction drops rapidly and dramatically below a certain potential because the active site is least active when the FAD is reduced. Raising the fumarate concentration does not remove this effect, suggesting that its origin is a drop in turnover frequency (k_{cat}) rather than substrate binding affinity (K_M).

Thus succinate dehydrogenase appears to be an excellent catalyst for fumarate reduction, but only within a narrow range of potential. This result is supported by conventional steady-state solution kinetics using benzyl viologen as an electron donor.[55] In these studies, carried out on a variety of enzymes in this family, those classed as fumarate reductases show normal time courses for viologen oxidation, while the succinate dehydrogenases show kinetics in which the rate of viologen oxidation accelerates with time, *i.e.* as the solution potential increases. The question remains as to whether this effect is useful or plays any role *in vivo*. Succinate dehydrogenase is evidently able to operate in either direction, so that this stage of the tricarboxylic acid cycle might be reversed if the quinone pool becomes too reduced. However, if the driving force in this direction became too large, the attenuation in activity would act as a ratchet.

4.4 NADH-dehydrogenase (Complex I)

NADH-ubiquinone oxidoreductase is a very large membrane-bound enzyme that catalyses the oxidation of NADH by ubiquinone. The mitochondrial enzyme, also known as Complex I, has more than 40 subunits and is established as a proton pump. It contains a flavin mononucleotide

(FMN) and eight Fe-S clusters, all of which are located in a membrane-extrinsic arm of the L-shaped molecule that projects out into the matrix for reaction with NADH. This can be resolved from the membrane-intrinsic domain to give a soluble fraction known as 1λ (MW 360 kDa) that is active in NADH oxidation. Hirst and co-workers have shown that the membrane-extrinsic subcomplex obtained from beef heart mitochondria is highly electroactive when adsorbed on a PGE electrode.[56,57] Figure 4-9 shows the voltammetry of a film of 1λ adsorbed on a rotating PGE disc electrode catalysing the interconversion of NAD^+ and NADH which are each present in the cell solution at 1 mM concentration.

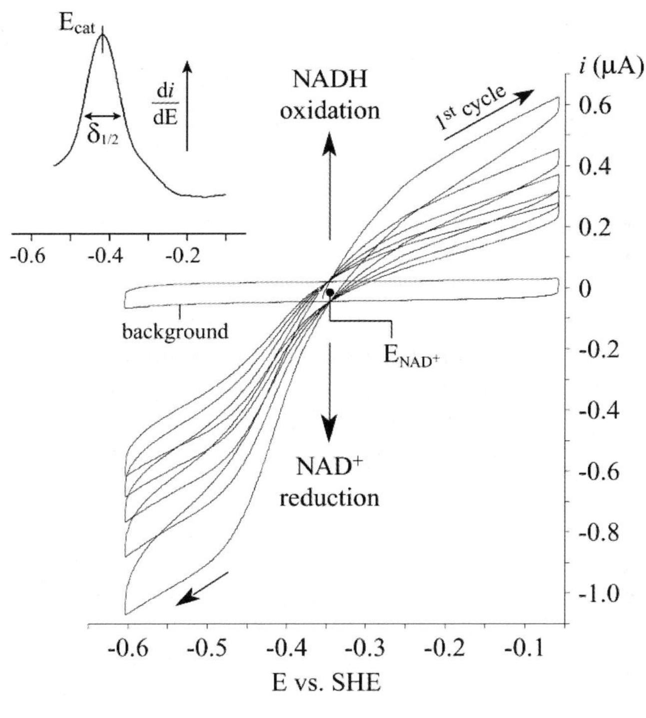

Figure 4-9. Catalytic voltammetry of the membrane-extrinsic subcomplex (1λ) of mitochondrial NADH-ubiquinone oxidoreductase adsorbed at a rotating PGE disc electrode in the presence of 1 mM NAD^+ and 1 mM NADH at pH 7.8, 20 °C. Reprinted from ref. 56 with permission.

The voltammetry is unstable (the enzyme's coverage is also too low to detect the redox centres directly) but the successive cycles that decrease in amplitude trace out an isosbestic potential which corresponds to the NAD^+/NADH couple at pH 7.8. The NAD^+/NADH reduction potential measured in this way shows a pH dependence of –29 mV over the pH range

5.5–10.5, exactly as expected for a $2e^-,1H^+$ stoichiometry. Several catalytic properties of the enzyme are revealed: first, at pH 7.8 as shown, there is actually a slight bias in the direction of NAD^+ reduction, with the catalytic potential E_{cat} (the potential at the steepest point of the slope – see derivative in inset) occurring at –420 mV. Second, a residual slope at high driving force is observed in each direction, suggesting that the enzyme molecules are inhomogeneously adsorbed on the electrode surface with some having interfacial electron transfer kinetics that are slow compared to the inherent turnover rate (see earlier discussion). Third, the potential dependence of the derivative yields n_{app} = 1 indicating that the 'control centre' is a species engaging in one-electron transfer. It was suggested that this step might be the transfer of electrons between the FMN (the FMN/FMNH radical potential is close to –420 mV) and the proximal [2Fe-2S] cluster.

Success with Complex I, like succinate dehydrogenase and fumarate reductase, is perhaps linked to the ability to obtain a membrane-extrinsic water-soluble subcomplex. The adsorption on carbon electrodes may reflect the tendency of such molecules to bind strongly through their exposed interfacial contact areas.

4.5 Hydrogenases

The active site of [NiFe]–hydrogenases contains a Ni atom and a Fe atom joined by two bridging cysteine thiolates. This is an unusual metal centre as the Fe is also bound to a CO and two CN^- ligands, and seems to remain as low-spin Fe(II) in all states of the enzyme that have been studied. The active site is buried well below the protein surface, and electron transfer requires a relay system of Fe-S clusters. The enzyme from the purple photosynthetic bacterium *Allochromatium vinosum* has a relay system comprising a [4Fe-4S], a [3Fe-4S] and a [4Fe-4S] cluster. It is extremely active when adsorbed on a PGE electrode, and at temperatures above 25 °C, it catalyses the oxidation of H_2 at diffusion-controlled rates. In fact, experiments carried out to compare the rotation rate dependence of catalytic current produced by adsorbed hydrogenase *vs.* the same PGE electrode or a Au electrode coated with Pt have suggested that the turnover rate at the active site is comparable to that of a Pt catalyst.[58] The amount of enzyme adsorbed is very low (less than 1 pmol cm^{-2} based on evaluation of the very weak non-turnover signal due to the single [3Fe-4S] cluster). To focus on the enzyme's properties, it is necessary to suppress the rotation rate dependence, and this is achieved easily by lightly repolishing the electrode after a film has been formed. We saw earlier (see Fig. 4-4C) how this high activity reveals a temperature-dependent competition between enzyme turnover (rate-determining at low temperature) and interfacial electron transfer (rate-determining at high temperature). Figure 4-10A compares two voltammograms for the same film of hydrogenase under 1 bar H_2 measured at pH 4.6 and pH 9.

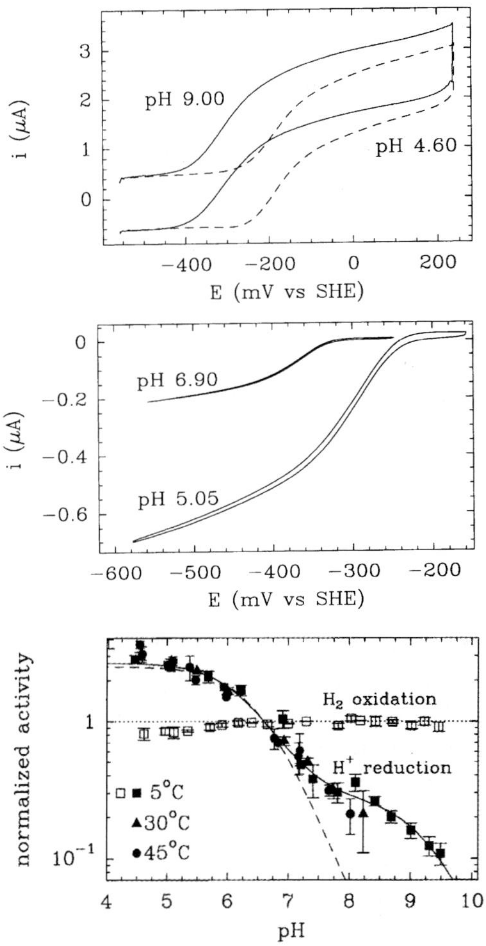

Figure 4-10. Voltammetric data for *Allochromatium vinosum* Ni-Fe hydrogenase adsorbed on a PGE rotating disc electrode. (A) Voltammograms for a sparsely populated film of hydrogenase catalysing the oxidation of H_2 (1 bar). The electrode can be transferred back and forth between solutions at pH 9.0 and pH 4.6, to reveal that similar limiting currents are obtained in each case: therefore, the limiting step in H_2 oxidation is not pH-dependent. (B) Voltammograms for H^+ reduction at two pH values, under 1 bar N_2 and rotation rate 2500 rpm. (C) Plots of the limiting currents for H_2 oxidation and H^+ reduction over a wide pH range. Reproduced from ref. 59 with permission.

These voltammograms were measured at 5 °C, with the electrode rotating at 2000 rpm, above which there is no further increase in catalytic current.[59] The result is independent of the order in which the buffer is exchanged. It shows that while there is a shift in catalytic potential, the turnover rate (given by the limiting current) is independent of pH. This is interesting, since without electrochemical control it is difficult to factor out the effect of driving force for different conditions. If for example, we studied this reaction using a redox dye such as a viologen as electron acceptor, the choice of one with a potential of –250 mV would give a high activity at pH 9 and very low activity at pH 4.6. By contrast, in the voltammetry experiment, it is easy to apply whatever electrode potential is required to reach the plateau at which the rate is maximised.

Proton reduction is also fast, but it is inhibited by H_2. By studying proton reduction under N_2 using a rotating disc electrode to disperse the H_2 product (the current increases with rotation rate because the product is swept away) we can examine proton reduction in some detail: using eqn. 4.3A that was given earlier, a plot of reciprocal current *vs.* reciprocal rotation rate gives an extrapolation to the limit of 'infinite rotation rate' where there is no inhibition by product. For each pH, these values can now be used to analyse 'uninhibited' proton reduction. Voltammograms at two pH values are shown in Fig. 4-10B, and the results of an extensive study of the optimal hydrogen oxidation and proton reduction rates as a function of pH are shown in Fig. 4-10C. By contrast with hydrogen oxidation, two p*K* values are observed in the proton reduction direction.

The catalytic activity is complicated by inactivation that occurs on exposure to oxygen or other oxidants. In this reaction another bridging species (believed to be OH^-) is incorporated and the Ni atom is oxidized to Ni(III). Reactivation is achieved by reduction. This interconversion, represented in Scheme 4-1, is easily studied with the enzyme adsorbed on an electrode.[60]

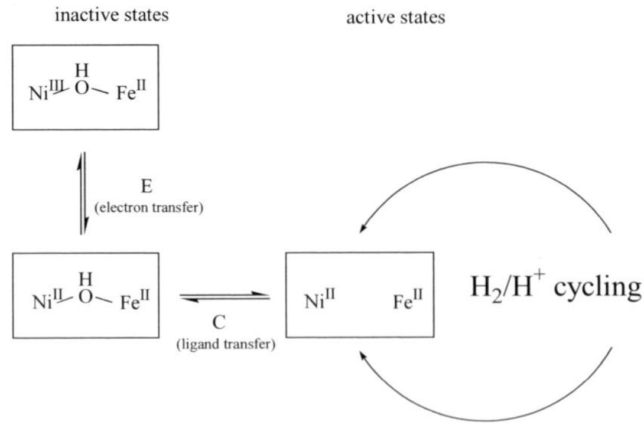

Scheme 4-1

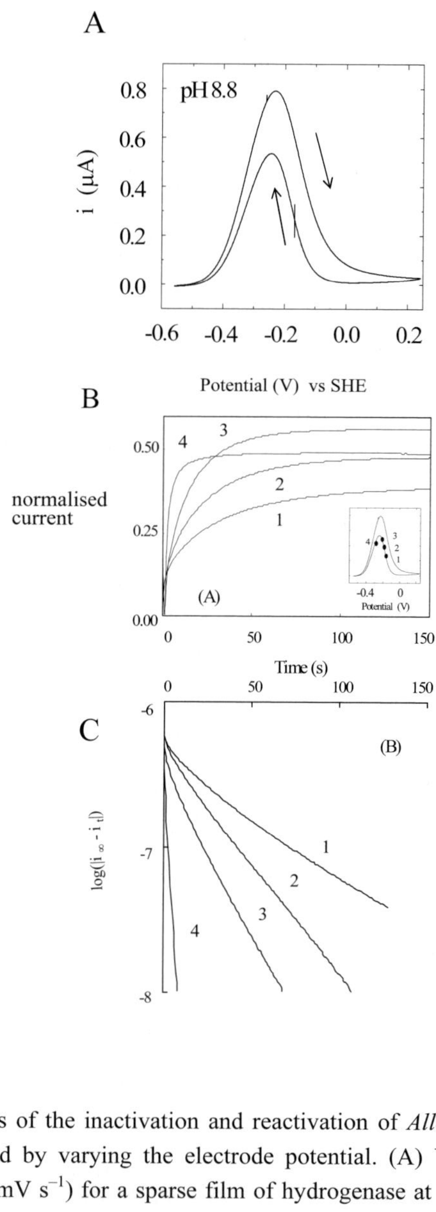

Figure 4-11. Studies of the inactivation and reactivation of *Allochromatium vinosum* Ni-Fe hydrogenase induced by varying the electrode potential. (A) Voltammogram measured at slow scan rate (0.3 mV s^{-1}) for a sparse film of hydrogenase at a PGE electrode, pH 8.8, 45 °C, 1 bar H_2. (B) Activation kinetics observed following different reductive potential steps to potentials indicated on the voltammogram shown in inset. (C) Semi-log plots for these kinetics, indicating nearly first-order kinetics and marked dependence on step potential. Reproduced from ref. 60 with permission.

The activity is related directly to current, so that the rate of activation or inactivation is observed as the rate of change of current, as a function of potential and time. Detailed studies have been made of the reaction occurring in the absence of O_2, which produces an inactive state known as 'ready' on the basis of its rapid reactivation (within minutes upon exposure to hydrogen and a reducing agent). Figure 4-11A shows a cyclic voltammogram measured at a very slow scan rate (0.3 mV s^{-1}) which shows the enzyme converting to the inactive state then reactivating as the potential is subsequently lowered. The inactivation process is slow, even at 45 °C. The kinetics of both inactivation and activation have been studied in detail, using potential steps to drive the reaction in either direction, as shown in Figs. 4-11B and 4-11C. Oxidative inactivation is faster at higher pH and is independent of potential, whereas the rate of reductive activation depends strongly on the electrode step potential over a wide range (see inset for potentials relative to the voltammogram shown in Fig. 4-11A). This is consistent with an 'EC' sequence (electron transfer followed by a chemical change) and the first step in the reaction is electron transfer to the Ni(III) state. The 'EC' reaction implicates an intermediate that is still inactive but in the Ni(II) state.

4.6 Nitrate Reductases

Membrane-anchored nitrate reductases, NarGHI, catalyse the reduction of nitrate to nitrite at a deeply buried active site in NarG that contains the $Mo[MGD]_2$ cofactor. Heme groups bound in the NarI subunit relay electrons from the quinol pool to membrane extrinsic subunits, NarG and NarH. Five Fe-S clusters complete an electron-transfer relay to the active site.[61] The soluble subunits are readily liberated from membranes of *Paracoccus pantotrophus* yielding a functional nitrate reductase, NarGH.[62] At pH 6, where NarGH displays maximum activity, the reduction potentials of the $Mo^{6+/5+}$ and $Mo^{5+/4+}$ couples are separated by ca. 500 mV making the semi-reduced active site inherently stable, in contrast to the flavin semiquinone forms of succinate dehydrogenase and the fumarate reductases.

NarGH adsorbs on PGE electrodes giving clear catalytic waves although non-turnover signals have yet to be visualised.[63] At a static electrode mass transport limitations preclude any attempts to extract mechanistic information but these are removed on rapid electrode rotation, Fig. 4-12A. The steady-state response, with catalysis occurring at potentials well below the reduction potential of the nitrate/nitrite couple (+480 mV at pH 6), reveals an enzyme biased heavily towards nitrate reduction over nitrite oxidation. Confirmation of the enzyme's functional integrity is provided by the reversible increase in catalytic current observed on transferring films to solutions of chlorate, in line with the enzyme's higher intrinsic reduction rate for chlorate than nitrate.[64] However, the increase of signal magnitude is achieved at the expense of waveform definition, Fig. 4-12B.

At most negative potentials the continual increase in catalytic current reflects limitation from a dispersion of k_o values that has greater impact with the higher rate of chlorate reduction compared to nitrate reduction. Thus, although smaller, the waves describing nitrate reduction provide a better framework for elucidation of NarGH activity. At lower nitrate concentrations the sigmoidal waveshape is lost to reveal 'tunnel-diode' characteristics, Fig. 4-13A, and the behaviour contrasts with that of succinate dehydrogenase for which these characteristics are observed regardless of substrate concentration.

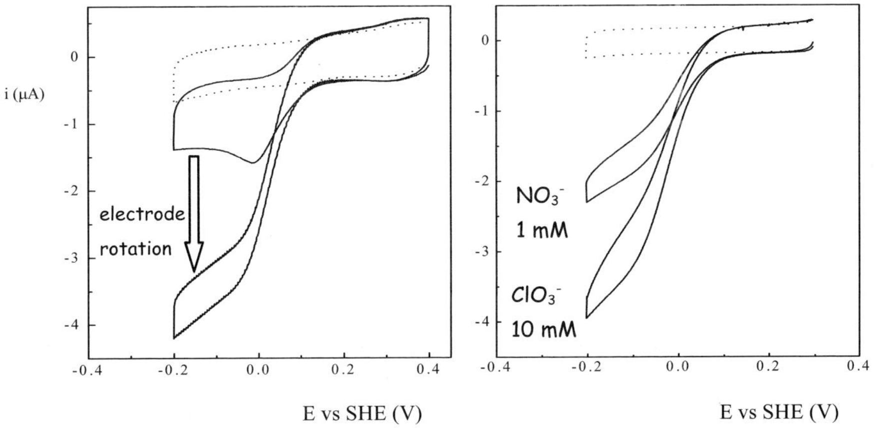

Figure 4-12. Catalytic voltammetry of *Paracoccus pantotrophus* nitrate reductase (NarGH) adsorbed as a film on a PGE electrode at pH 6. (A) Increasing the electrode rotation rate from 0 to 3000 rpm removes the mass transport limitation of the catalytic response in 50 μM NO_3^-. (B) The enzyme's greater rate of chlorate reduction compared to nitrate reduction is reflected in greater distortion of the waveform through dispersion of sluggish interfacial electron transfer rates (see also Fig. 4-4C). Scan rate 10 mV s^{-1}. Adapted from ref. 64, with permission.

For NarGH, rapid and reversible inter-conversion of kinetically (k_{cat}/K_M) distinct enzyme forms occurs on sweeping across the electrochemical potential domain. Raising the pH has two effects, both of which are fully reversed on returning the film to solutions of pH 6. First, there is an overall decrease in activity of the film at both high and low nitrate concentrations. Second, the peak in activity at low nitrate concentrations becomes more pronounced. Thus, variation of pH has distinct effects on the rate-defining events of enzyme- and nitrate-limited turnover.

It is significant that the reduction potentials of the $Mo^{5+/4+}$ couple and at least two of the Fe-S clusters occur in the potential window where reduction of NarGH triggers conversion to a state with lower nitrate-limited activity. Previous mechanisms for nitrate reduction have described nitrate binding to Mo^{4+} followed by simultaneous formation of nitrite and an Mo^{6+}=O species. If this is revised to assign nitrate a greater affinity for Mo^{5+} than Mo^{4+}, Fig. 4-13B, the key features of NarGH behaviour can be reproduced; at low nitrate concentrations catalysis proceeds via binding of NO_3^- to Mo^{5+} until the potential is made sufficiently negative to drive catalysis via Mo^{4+} and at a slower rate, while reduction of the nitrate bound $Mo^{5+/4+}$ centre defines the position and steepness of the sigmoidal catalytic response at high nitrate concentrations. Alternatively, an Fe-S cluster acts as a remote redox switch that triggers perturbations of the active site to modulate k_{cat} and K_M , Fig. 4-13C.

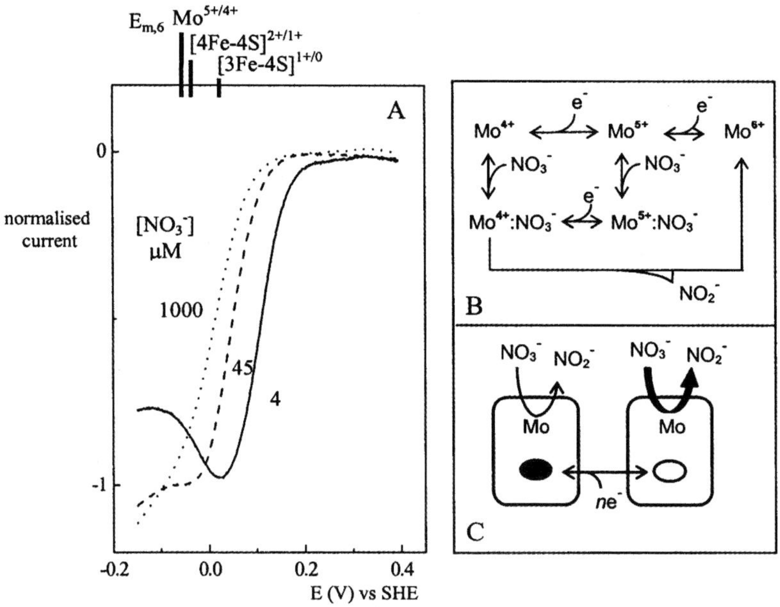

Figure 4-13. Changes in the rate-defining events of NarGH activity on variation of nitrate concentration are reflected in the steady-state catalytic waveforms at pH 6. (A) Reduction potentials of the $Mo^{5+/4+}$ couple and Fe-S clusters lie in the potential window where reduction of the enzyme triggers an attenuation in the rate of nitrate–limited turnover and suggest mechanisms, (B) and (C) respectively, that may account for NarGH behaviour. Scan rate 10 mV s^{-1}, rotation rate 3000 rpm. Adapted from ref. 64, with permission.

Additional observations are significant in distinguishing between these mechanistic possibilities, and indeed whether the behaviour of the enzyme is induced by immobilisation on a graphite electrode. First, NarGH displays similar catalytic behaviour at gold and PGE electrodes, materials that present quite different surfaces for protein adsorption. Second, *E. coli* NarGHI, an enzyme with similar reduction potentials and spectroscopic properties to the *P. pantotrophus* enzyme also displays an optimum potential for activity that is lost as nitrate concentration is increased.[65] The amino acid similarity of these enzymes is high but distinct in surface regions. Thus it seems most likely that the catalytic behaviour of the respiratory nitrate reductases is defined by redox-linked chemistries intrinsic to conserved features of these enzymes rather than the nature of the protein:electrode interface. Finally, experiments on *E. coli* NarGHI have shown that addition of azide (N_3^-) greatly attenuates the high-potential activity relative to the low-potential activity, removing the activity optimum.[66] Azide is a competitive inhibitor, so this observation links the potential optimum directly to events at the active site rather than a more remote or less specific target. It therefore appears most likely that optimum catalytic activity occurs when the enzyme is held at a potential favouring the Mo^{5+} form at steady state.

In addition to the membrane-anchored enzymes that support anaerobic respiration, bacteria express a number of functionally, and structurally, distinct nitrate reductases related by the presence of an $Mo[MGD]_2$ containing active site. PFV has demonstrated tunnel-diode behaviour from two of these; the periplasmic *Rhodobacter sphaeroides* NapAB and the assimilatory *Synechococcus elongatus* NarB.[67,68] In both cases preferential binding of nitrate to the Mo^{5+}, over Mo^{4+}, oxidation state provides an explanation for the catalytic voltammetry and this may prove to be a conserved feature of the catalytic cycle in these enzymes.

4.7 Cytochrome *c* Nitrite Reductase

E. coli cytochrome *c* nitrite reductase performs the six-electron reduction of nitrite to ammonium ion. Although catalysis proceeds without releasing any detectable intermediates a reaction pathway is suggested by the enzyme's ability to produce ammonium from nitric oxide and hydroxylamine (NH_2OH). Crystallised as a homodimer, the enzyme contains ten heme groups, $E_{m,7}$ = –37 to –320 mV, positioned in close proximity but with a bifurcated, rather than linear, arrangement of electron relay centres leading from the active site to the surface of each monomer.[69] Substrates bind to a heme that has axial ligands of water and, rather unusually, a lysine ($-NH_2$) side chain. In the absence of substrates this heme is spin-coupled to its nearest neighbour to give characteristic signals in electron paramagnetic resonance (EPR) spectroscopy. Potentiometric titrations suggested cooperative, two-electron reduction of these centres at a midpoint potential of -107 mV since no new signals appeared on reduction of the enzyme. However, the variation of signal intensity with potential was best described by an $n = 1$ Nernstian plot that was postulated to arise from a drift of the sample potentials on transfer to EPR tubes. The opportunity to study nitrite reductase at precisely defined potentials was provided by the formation of highly active films adsorbed onto a PGE electrode.[70] Although non-turnover signals have not been detected, the catalytic response again provides a route to deconvolve the enzyme's redox-driven chemistries. While the response is never entirely free of mass transport limitations, curvature of the Levich plots indicates that voltammetry performed with electrode rotation at 3000 rpm is defined predominantly by intrinsic properties of the enzyme. Figure 4-14A illustrates how the voltammogram at low nitrite concentrations is comprised of two features separated by a plateau of maximum activity that, for these nitrite-limited conditions, can be correlated with equilibrium reduction potentials of heme centres in the enzyme. The catalytic wave can thus be deconvolved showing that the enzyme is 'switched on' by reduction of an $n = 2$ centre with $E_m = -105$ mV, confirming the cooperative nature of electron transfer at the active site hemes. The activity is then attenuated by reduction of an $n = 1$ centre with $E_m = -320$ mV. These are properties ascribed to hemes located near the interface between monomers.[69] At more acidic pH the the waveshape becomes essentially sigmoidal as the attenuation becomes less distinct.[71] Thus the mechanism attenuating the catalytic rate is prevented when heme reduction is coupled to enzyme protonation. It is also interesting that hydroxylamine-limited turnover is described by a sigmoidal catalytic wave, Figure 4-14C. Thus, reduction of the low-potential hemes may slow down a step in the pathway for transformation of nitrite to hydroxylamine. Increasing the concentration of nitrite or hydroxylamine produces changes in the waveform. These are best summarised in overlaid derivative (di/dE) plots that highlight the features in each wave, Figs. 4-14B and 4-14D. Positive features reflect regions where activity increases on increasing the driving force for the catalysed reaction while the converse is true of negative features. For both substrates, as the reduction approaches maximum velocity the activity experiences a boost at low potentials, suggesting a common rate-defining event in product formation from either of the Michaelis complexes. Finite differences in the appearance of the waveforms can be explained if the rates of the defining steps of catalysis are distinct in each case and film transfers confirm that hydroxylamine reduction supports an electron flux at least 10 times that observed at the maximum rate of nitrite reduction. It is notable that assays in which dithionite reduced methyl viologen acts as the electron donor show no difference in the maximum electron flux achievable during reduction of either substrate. This suggests an artificial limitation to turnover in the solution phase assay, perhaps because dithionite or reaction products compete with substrate for access to active site while the electrode provides a better source of electrons (another example of electrons 'on tap'). A further observation is the small boost to hydroxylamine activity at ca. –500 mV in high concentrations of hydroxylamine. This is not readily accounted for by the equilibrium reduction potentials of the hemes but may reflect the reduction potential of a centre under turnover conditions; information that is clearly difficult to access by other methods.

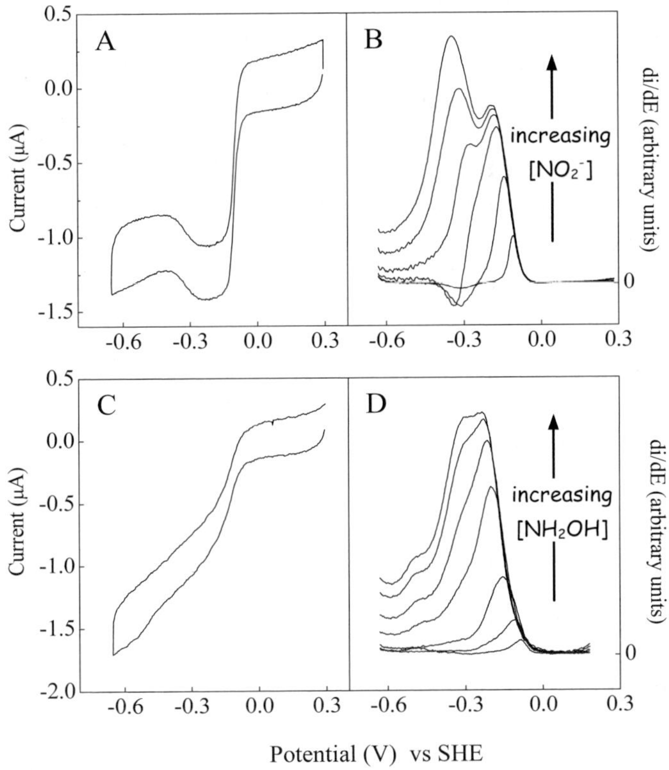

Figure 4-14. Catalytic voltammetry of nitrite reductase is critically dependent on the identity and concentration of the substrate at pH 7. (A) In 1 μM nitrite, reduction of the enzyme successively, turns on, but then attenuates activity. (B) Overlaid derivatives emphasise changes in the waveshape as the nitrite concentration is increased from 1.7 to 265 μM. At higher nitrite concentrations, activity increases upon application of a more negative potential, and this is reflected by the two positive features in the derivative plot. (C) At 1 mM hydroxylamine, reduction of the enzyme does not attenuate the rate of catalysis. (D) Overlaid derivatives show that as the hydroxylamine concentration is raised from 1.1 to 347 mM the waveshape develops two positive features similar to the waveshapes displayed at high rates of nitrite reduction. Reproduced from ref. 70 with permission of the American Society for Biochemistry and Molecular Biology.

Another interesting property of nitrite reductase is revealed when cyanide is introduced into the experiments.[29,72] Typical effects on the catalytic response during hydroxylamine reduction are illustrated in Fig. 4-15. Most obviously the activity is diminished, but this inhibitory effect is readily reversed on transfer to a cyanide free solution. More subtle is the appearance of a peak in the sweep towards more negative potentials that is reproduced during consecutive scans. The peak reflects cyanide binding and inhibition of reduced enzyme. That it is reproduced on consecutive scans indicates that cyanide is released from the oxidised enzyme at a rate that is slow compared to the timescale of the experiment as there is no recovery of activity observed during the sweep to positive potentials. Similar behaviour is observed during nitrite reduction.[29] Rapid scans to positive potentials trap cyanide on the enzyme and reveal a catalytic waveshape for nitrite reduction that is distinct from that displayed by the uninhibited enzyme. Thus, cyanide is clearly not a simple competitive inhibitor of nitrite reductase. Instead the cyanide bound enzyme retains activity but displays rate-defining steps distinct from those that operate in the uninhibited enzyme.

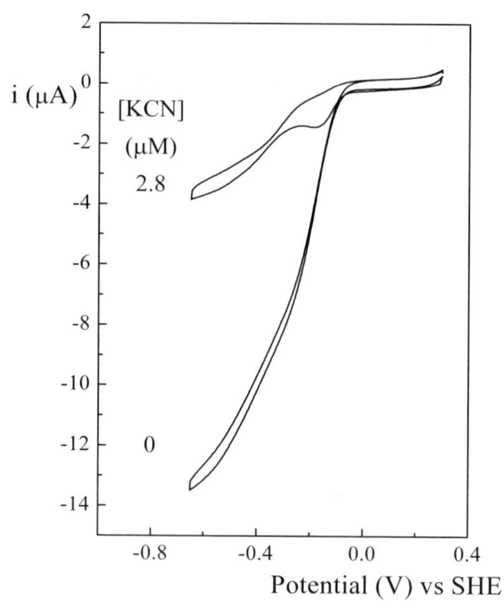

Figure 4-15. The influence of cyanide on PFV of nitrite reductase in 20 mM hydroxylamine, pH 6. Scan rate 30 mV s^{-1} with electrode rotation at 3000 rpm. Adapted from ref. 29 with permission of the American Chemical Society.

5. CLOSING PERSPECTIVES

We have drawn on a number of examples to illustrate the diversity of mechanistic insight afforded by PFV when mass transport and interfacial electron exchange are defined/controlled. It is apparent that protein size, long considered a barrier to the utility of the approach, is not an impediment in itself, as demonstrated profoundly by the catalytic voltammetry of Subcomplex 1λ of NADH-dehydrogenase (ca. 360 kDa).[56] Proteins with active sites 'wired' to their surfaces should all be amenable to voltammetric interrogation when the interface with the electrode accommodates the orientation, spacing and mobility required for facile electron exchange. Indeed the number of enzymes studied by PFV is increasing rapidly: properties previously unrevealed or convoluted because of their complexity have been brought under control by access to a defined potential domain, and in the majority of cases new facets of enzyme behaviour have been revealed. A major challenge for the future is to confirm the origin of the new activities at the molecular and electronic levels aided by *in situ* spectroscopies, scanning probe microscopies and protein engineering. There are also exciting possibilities to advance the composition of the films in order to probe macromolecular communication and build a more holistic picture of redox enzyme action; films comprised of cytochrome *c* complexed to cytochrome *c* oxidase have allowed intramolecular electron transfer to be built on to catalysis[73] and the ability to host proteins in electrode supported or suspended bilayers opens opportunities to study redox transformations of the quinol/quinone pool by integral membrane proteins or *vice versa*.[74–76] It remains to be seen how examples of the exquisite electrochemical regulation of enzyme activities revealed by PFV will be manifested in the cellular context.

ACKNOWLEDGEMENTS

We would like to thank all members of our research groups, past and present, for their enthusiastic contributions to the work presented here and the BBSRC, EPSRC and Wellcome Trust for their financial support.

REFERENCES

1. Eddowes, M.J.; Hill, H.A.O., *Chem. Commun.* **1977**, 771–772.
2. Yeh, P.; Kuwana, T., *Chem. Lett.* **1977**, 1145–1148.
3. Niki, K.; Yagi, T.; Inokuchi, H.; Kimura, K., *J. Am. Chem. Soc.* **1979**, *101*, 3335–3340.

4. Armstrong, F.A.; Heering, H.A.; Hirst, J., *Chem. Soc. Rev.* **1997,** *26*, 169–179.
5. Armstrong, F.A., *J. Chem. Soc., Dalton Trans.* **2002**, 661–671.
6. Butt, J.N., *Recent Res. Devel. Biochem.* **2003**, *4*, 159–180.
7. Léger, C.; Elliott, S.J.; Hoke, K.R.; Jeuken, L.J.C.; Jones, A.K.; Armstrong, F.A., *Biochemistry* **2003**, *42*, 8653–8662.
8. Engstrom, R.C.; Strasser, V.A., *Anal .Chem.* **1984**, *56*, 136–141.
9. Cabaniss, G.E.; Diamantis, A.A.; Murphy, W.R.; Linton, R.W.; Meyer, T.J., *J. Am.Chem. Soc.* **1985**, *107*, 1845–1853.
10. Armstrong, F.A.; Cox, P.F.; Hill, H.A.O.; Lowe, V.J.; Oliver, B.N., *J. Electroanal. Chem.* **1987**, *217*, 331–336.
11. Armstrong, F.A.; Camba, R.; Heering, H.A.; Hirst, J.; Jeuken, L.J.C.; Jones, A.K.; Léger, C.; McEvoy, J.P., *Faraday Disc.* **2000**, *116***,** 191–203.
12. Jeuken, L.J.C.; Armstrong, F.A., *J. Phys. Chem. B* **2001**, *105*, 5271–5282.
13. Spain, I.L. In *Chemistry and Physics of Carbon,* P.L. Thrower, Ed., Marcel Dekker: New York, 1980; Vol. 16, pp. 119–304.
14. Schneider, T.W.; Buttry, D.A., *J. Am. Chem. Soc.* **1993**, *115*, 12391–12397.
15. Weisshaar, D.E.; Lamp, B.D.; Porter, M.D., *J. Am. Chem. Soc.* **1992**, *114*, 5860–5862.
16. Petrović, J.; Clark R.A.; Yue, H.; Waldeck, D.H.; Bowden, E.F., *Langmuir*, **2005,** *21*, 6308–6316.
17. Yue, H.; Waldeck, D.H.; Petrović, J., Clark, R.A., *J. Phys. Chem. B.* **2006**, *110*, 5062–5072.
18. Chen, X.; Ferringo, R.; Yang, J.; Whitesides, G.M., *Langmuir* **2002**, *18*, 7009–7015.
19. Murgida, D.H.; Hildebrandt, P., *J. Phys. Chem. B* **2002**, *106***,** 12814–12819.
20. Chi, Q.; Zhang, J.; Anderson, J.E.T.; Ulstrup, J., *J. Phys. Chem. B* **2001**, *105*, 4669–4679.
21. Avila, A.; Gregory, B.W.; Niki, K.; Cotton, T.M., *J. Phys. Chem. B* **2000**, *104*, 2759–2766.
22. Chi, Q.; Zhang, J.; Nielsen, J.U.; Friis, E.P.; Chorkendorff, I.; Canters, G.W.; Anderson, J.E.T.; Ulstrup, J., *J. Am. Chem. Soc.* **2000***, 122*, 4047–4055.
23. Butt, J.N.; Thornton, J.; Richardson, D.J.; Dobbin, P.S., *Biochem. J.* **2000**, *78*, 1001–1009.
24. Kong, J.; Lu, Z.; Lvov, Y.M.; Desamero, R.Z.B.; Frank, H.A.; Rusling, J.F., *J. Am. Chem. Soc.* **1998**, *120*, 7371–7372.
25. Lvov, Y.M.; Lu, Z.; Schenkman, J.B.; Zu, X.; Rusling, J.F., *J. Am. Chem. Soc.* **1998**, *120*, 4073–4080.
26. Munge, B.; Das, S.K.; Ilagan, R.; Pendon, Z.; Yang, J.; Frank, H.A.; Rusling, J.F., *J. Am. Chem. Soc.* **2003**, *125*, 12457–12463.
27. Rusling, J.F., *Acc. Chem. Res.* **1998**, *31*, 363–369.
28. Bayachou, M.; Elkbir, L.; Farmer, P.J., *Inorg. Chem.* **2000**, *39*, 289–293.
29. Gwyer, J.D.; Richardson, D.J.; Butt, J.N., *Biochemistry*, **2004**, *43*, 15086–15094.
30. Hirst, J.; Armstrong, F.A., *Anal. Chem.* **1998**, *70*, 5062–5071.
31. Vincent, K.A.; Armstrong, F.A., *Inorg. Chem.* **2005**, *44*, 798–809.
32. Laviron, E. In *Electroanalytical Chemistry*, Bard, A.J. Ed., Marcel Dekker: New York, 1982; Vol. 12, pp. 53–157.
33. Bard, A.J.; Faulkner, L.R. *Electrochemical methods: Fundamentals and applications*, 2nd Edition. John Wiley and Sons, Inc.: New York, 2001, p. 339.
34. Jones, A,K.; Camba, R.; Reid, G.A.; Chapman, S.K.; Armstrong, F.A., *J. Am. Chem. Soc.* **2000**, *122*, 6494–6495.
35. Léger, C.; Heffron, K.; Pershad, H.R.; Maklashina, E.; Luna-Chavez, C.; Cecchini, G.; Ackrell, B.A.C.; Armstrong, F.A., *Biochemistry* **2001***, 40*, 11234–11245.

36. Kasmi, A.E.; Wallace, J.M.; Bowden, E.F.; Binet, S.M.; Linderman, R.J., *J. Am Chem. Soc.* **1998**, *120*, 225–226.
37. Jeuken, L.J.C.; McEvoy, J.P.; Armstrong, F.A., *J. Phys. Chem. B* **2002**, *106*, 2304–2313.
38. Leys, D.; Basran, J.; Talfournier, F.; Sutcliffe, M.J.; Scrutton, N.S., *Nat. Struct. Biol.* **2003**, *10*, 219–225.
39. Hirst, J.; Duff, J.L.C.; Jameson, G.N.L.; Kemper, M.A.; Burgess, B.K.; Armstrong, F.A., *J. Am. Chem. Soc.* **1998**, *120*, 7085–7094.
40. Baymann, F.; Barlow, N.L.; Aubert, C.; Schoepp-Cothenet, B.; Leroy, G.; Armstrong, F.A., *FEBS Lett.* **2003**, *539*, 91–94.
41. Pershad H.R.; Duff, J.L.; Heering, H.A.; Duin, E.C.; Albracht, S.P.; Armstrong, F.A., *Biochemistry* **1999**, *38*, 8992–8999.
42. Armstrong, F.A.; Bond, A.M.; Buchi, F.N.; Hamnett, A.; Hill, H.A.; Lannon, A.M.; Lettington, O.C.; Zoski, C.G., *Analyst* **1993**, *118*, 973–978.
43. Léger, C.; Jones, A.K.; Albracht, S.P.J.; Armstrong, F.A., *J. Phys. Chem. B.* **2002**, *106*, 13058–13063.
44. Heering, H.A.; Hirst, J.; Armstrong, F.A., *J. Phys. Chem. B* **1998**, *102*, 6889–6902.
45. Elliott, S.J.; McElhaney, A.E.; Feng, C.; Enemark, J.H.; Armstrong, F.A., *J. Am. Chem. Soc.* **2002,** *124*, 11612–11613.
46. Mondal M.S.; Fuller, H.A.; Armstrong, F.A., *J. Am. Chem. Soc.* **1996**, *118*, 263–264.
47. Mondal M.S.; Goodin, D.B.; Armstrong, F.A., *J. Am. Chem. Soc.* **1998**, *120*, 6270–6276.
48. Sucheta, A.; Cammack, R.; Weiner, J.; Armstrong, F.A., Biochemistry **1993**, *32*, 5455–5465.
49. Heering, H.A.; Weiner, J.H.; Armstrong, F.A., *J. Am. Chem. Soc.* **1997**, *119*, 11628–11638.
50. Turner, K.L.; Doherty, M.K.; Heering, H.A.; Armstrong, F.A.; Reid, G.A.; Chapman, S.K., *Biochemistry* **1999**, *38*, 3302–3309.
51. Hudson, J.M.; Heffron, K.; Kotlyar, V.; Sher, Y.; Maklasshina, E.; Cecchini, G.; Armstrong, F.A., *J. Am. Chem. Soc.* **2005**, *127*, 6977–6989.
52. Hirst, J.; Sucheta, A.; Ackrell, B.A.C.; Armstrong, F.A., *J Am. Chem. Soc.* **1996**, *118*, 5031–5038.
53. Hirst, J.; Ackrell, B.A.C.; Armstrong, F.A., *J. Am .Chem. Soc.* **1997**, *119,* 7434–7439.
54. Sucheta, A.; Ackrell, B.A.C.; Cochran, B.; Armstrong, F.A., *Nature* **1992**, *356*, 361–362.
55. Ackrell, B.A.C.; Armstrong, F.A.; Cochran, B.; Sucheta, A.; Yu, T., *FEBS Lett.* **1993**, *326*, 92–94.
56. Zu, Y.; Shannon, R. J.; Hirst, J., *J. Am. Chem. Soc.* **2003**, *125*, 6020–6021.
57. Reda, T.; Hirst, J., *J. Phys. Chem. B* **2006**, *110*, 1394–1404.
58. Jones, A.K.; Sillery, E.; Albracht, S.P.J.; Armstrong, F.A., *Chem. Commun.* **2002**, 866–867.
59. Léger, C.; Jones, A. K.; Roseboom, W.; Albracht, S.P.J.; Armstrong, F.A., *Biochemistry* **2002**, *41*, 15736–15746.
60. Jones, A.K.; Lamle, S.E.; Pershad, H.R.; Vincent, K.A.; Albracht, S.P.J.; Armstrong, F.A., *J. Am. Chem. Soc.* **2003**, *125*, 8505–8514.
61. Bertero M.G.; Rothery, R.A.; Palak, M.; Hou, C.; Lim, D.; Blasco, F.; Weiner, J.H.; Strynadka, N.C.J., *Nature Structural Biology* **2003**, *10*, 681–687.
62. Anderson, L.J.; Richardson, D.J.; Butt, J.N., *Biochemistry* **2001**, *40*, 11294–11307.
63. Anderson, L.J.; Richardson, D.J.; Butt, J.N., *Faraday Discussions* **2000,** *116*, 155–169.
64. Anderson, L.J., *Protein film voltammetry and spectroscopic studies of bacterial nitrate reductases*, Ph.D. Thesis, University of East Anglia, Norwich, U.K, 2002.

65. Elliott, S.J.; Léger, C.; Pershad, H.R.; Hirst, J.; Heffron, K.; Ginet, N.; Blasco, F.; Rothery, R.A.; Weiner, J.H.; Armstrong, F.A., *Biochim. Biophys. Acta* **2002**, *1555*, 54–59.
66. Elliott, S.J.; Hoke, K.R.; Heffron, K.; Palak, M.; Rothery, R.A.; Weiner, J.H.; Armstrong, F.A., *Biochemistry* **2004**, *43*, 799–807.
67. Frangioni, B.; Arnoux, P.; Sabaty, M.; Pignol, D.; Bertrand, P.; Guigliarelli, B.; Léger, C., *J. Am. Chem. Soc.* **2004**, *126*, 1328–1329.
68. Jepson, B.; Anderson, L.J.; Rubio, L.M.; Taylor, C.J.; Butler, C.S.; Flores, E., Herrero, A.; Butt, J.N.; Richardson, D.J., *J. Biol. Chem.* **2004**, *279*, 32212–32218.
69. Bamford, V.A.; Angove, H.C.; Seward, H.E.; Thomson, A.J.; Cole, J.A.; Butt, J.N.; Hemmings, A.M.; Richardson, D.J., *Biochemistry* **2002**, *41*, 2921–2931.
70. Angove H.C.; Cole, J.A.; Richardson, D.J.; Butt, J.N., *J. Biol. Chem.* **2002**, *277*, 23374–23381.
71. Gwyer, J.D.; Richardson, D.J.; Butt, J.N., *J. Am. Chem. Soc.* **2005**, *127*, 14964–14965.
72. Gwyer, J.D,; Angove, H.; Richardson, D.J.; Butt, J.N., *Bioelectrochemistry*, **2004**, *63*, 43–47.
73. Haas, A.S.; Pilloud, D.L.; Reddy, K.S.; Babcock, G.T.; Moser, C.C.; Blasie, J.K.; Dutton, P.L., *J. Phys. Chem. B* **2001**, *105*, 11351–11362.
74. Cornell, B.A.; Braach-Maksvytis, V.L.; King, L.G.; Osman, P.D.; Raguse, B.; Wieczorek, L.; Pace, R.J., *Nature*, **1997**, *387*, 580–583.
75. Heyse, S.; Ernst, O.P.; Dienes, Z.; Hofmann, K.P.; Vogel, H., *Biochemistry* **1998**, *37*, 507–522.
76. Jeuken, L.J.C.; Connell, S.D.; Henderson, P.J.F.; Gennis, R.B.; Evans, S.D.; Bushby, R.J., *J. Am. Chem. Soc.* **2006**, *128*, 1711–1716.

Chapter 5

ELECTROCHEMISTRY AT THE DNA/ELECTRODE INTERFACE:

New Approaches to Nucleic Acids Biosensing

MICHAEL G. HILL[1] AND SHANA O. KELLEY[2]
[1]*Occidental College, Department of Chemistry, Los Angeles, CA 90041, USA;* [2]*Boston College, Eugene F. Merkert Chemistry Center, Chestnut Hill, MA 02467, USA.*

1. INTRODUCTION

Although the first electrochemical reactions of DNA were reported more than forty years ago,[1] it was not until the last decade that the field of nucleic acid electrochemistry began to attract a large and diverse pool of researchers. Much of this recent interest has been driven by transformative advances in genomics, and the resulting need for new analytical techniques for routine DNA analysis. Transcriptional profiling, single-nucleotide polymorphism (SNP) discovery, and pathogen detection are all dependant on high-throughput, accurate methods for identifying nucleic acid sequences.

The DNA/electrode interface provides a unique platform for developing such analytical methods. The chemical structure of DNA is readily amenable to functionalization for biosensing applications: not only is DNA easily amplified and manipulated using standard biochemical methods, but nucleic acid base pairings are specific and robust, providing a strategy for capturing target sequences onto electrode surfaces. The negatively charged phosphate backbone can be used to recruit and concentrate redox-active cations or intercalators into DNA thin films for use as electrochemical reporter molecules,[2–13] and the DNA bases are themselves redox active, allowing direct electrochemical readout at an electrode surface.[14–20] Following the pioneering work of Palecek who first reported the detection of nucleic acid base residues using cathodic-stripping voltammetry at mercury electrodes, a number of innovative electrochemical assays have been developed to signal

O. Hammerich and J. Ulstrup (eds.), Bioinorganic Electrochemistry, 129–160.

the presence and composition of target DNA sequences immobilized onto electrode surfaces, Fig. 5-1.

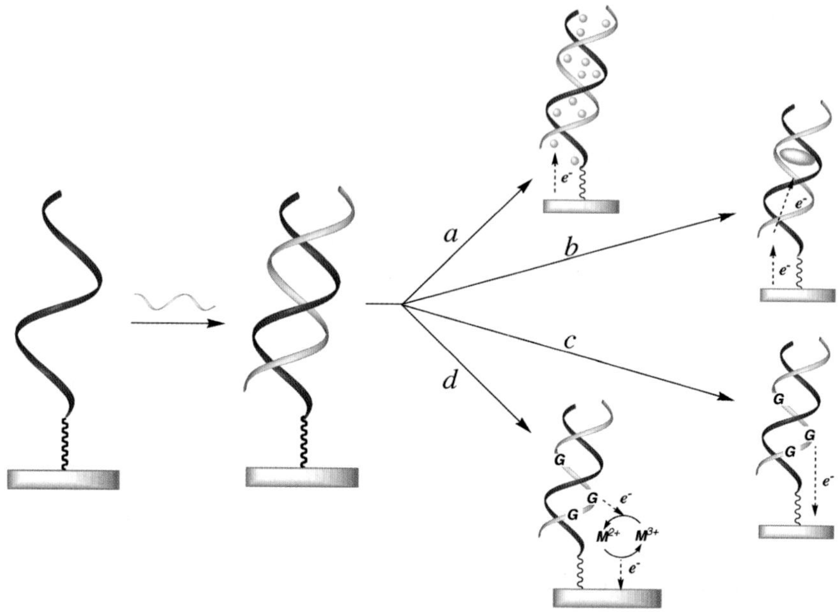

Figure 5-1. Schematic representation of various approaches to nucleic acids biosensing at the DNA/electrode interface. In scheme (a), an immobilized single-stranded probe sequence is used to capture a complementary target DNA from solution onto the electrode surface. Redox-active cations are then introduced, where they bind to the negatively charged phosphate backbone of DNA and function as electrochemical reporter groups for DNA hybridization. In (b), a similarly prepared DNA film is treated with a redox-active intercalator. Once inserted into the π-stack, the intercalator becomes electrochemically active through a DNA-mediated process. The presence of a single-base mismatch or other lesion, however, dramatically attenuates the electrochemical response, providing a straightforward assay for mutational analysis. Scheme (c) illustrates the direct electrochemical readout of DNA at a solid electrode surface using the inherent redox chemistry or the DNA bases (typically guanines), while (d) depicts the electrocatalytic oxidation of DNA bases using a transition-metal complex (e.g., $Ru(bpy)_3^{2+}$) as the redox mediator.

Many of these approaches rely on the differential hybridization of target DNAs that are perfectly matched with probe sequences to achieve highly specific and accurate target selection. Other assays use the sensitivity of DNA-mediated charge transport to duplex structure in order to signal the presence of a sequence of interest. Efforts also have been directed toward exploiting non-faradaic processes unique to DNA-modified surfaces as the basis for electrochemical readout. This chapter discusses each of these methods, and is intended to highlight how the DNA/electrode interface can be used to design and implement electrochemical assays for nucleic acids detection. Several comprehensive reviews on DNA films and nucleic acids biosensing have recently been published,[21–28] to which the reader is referred for additional information.

2. ELECTROCHEMISTRY OF CATIONIC REPORTER MOLECULES

The idea that electrochemical readout of DNA hybridization at a solid surface could be exploited for biosensing was first explored in the early 1990s.[2,3] Many of the original systems relied on electrochemical signals of cationic compounds to report the presence of single- *vs.* double-stranded DNA at an electrode surface.[2–4] As the number of backbone phosphate linkages increases upon duplex formation, more cations are recruited to the surface and an increase in the electrochemical response of the reporter molecule signals the hybridization event, Fig 5-1a. Small molecules that bind to DNA electrostatically are ideal choices for this assay, as their association is generally not sequence specific - an important design feature if the method is to have broad applicability to different DNA targets.

The first successful application of this methodology involved glassy carbon surfaces derivatized with single-stranded DNA.[2,3] Electrostatically-bound octahedral complexes of Co(III) were used to quantitiate immobilized DNA and to monitor the binding of a complementary oligonucleotide. This work verified that electrochemical detection of target sequences was feasible.

Over the last decade, many different cationic probes have been investigated for use in this assay. Transition-metal complexes with polypyridyl or other nitrogenous ligands are particularly common, as the reversible redox chemistry of these compounds provides excellent signal reproducibility. $Co(phen)_3^{2+}$,[2,3,4] $Co(bpy)_3^{2+}$,[2,3] $Ru(bpy)_3^{2+}$,[4] $Ru(phen)_3^{2+}$,[4] $Ru(NH_3)_6^{3+}$,[9–11] $Os(bpy)_2Cl_2$,[7] and $Os(bpy)_2dppz^{2+}$ [8] have all been used to detect DNA at an electrode surface. The use of protein-nucleic acid probes has also been explored as a means to increase the sensitivity of hybridization assays, and these materials impart enhanced sequence specificity and base-mismatch discrimination.[5] One important consideration in the selection of a redox probe is the stability of the DNA-electrode linkage. For example, many of the polypyridyl complexes undergo oxidation at potentials higher than those required to oxidize a gold-thiol linkage, so those surface/probe pairs are not compatible.

The cationic probe $Ru(NH_3)_6^{3+}$ has proven useful both for the analysis of DNA hybridization and for monitoring the behavior of immobilized oligonucleotides.[9–11,18,19,29] Tarlov and Steele have developed a versatile assay for determining the DNA density on an electrode based on chronocoulometric measurements of $Ru(NH_3)_6^{3+}$ bound electrostatically to the surface.[9] The single-stranded DNA probes used in these studies were derivatized with an alkylthiol tether, and covalently attached to gold electrodes through a gold-thiol bond. $Ru(NH_3)_6^{3+}$ binds readily to the DNA-modified surfaces at micromolar concentrations, and undergoes a reversible $1e^-$ reduction at potentials near ~ –0.35 V *vs.* SCE. Chronocoulometry[30] is then used to calculate the surface concentration of bound $Ru(NH_3)_6^{3+}$ (Γ_{Ru}, mol cm^{-2}) in the DNA film, according to the integrated Cottrell equation, eqn. 5-1,[31]

$$Q = \frac{2nFAD^{1/2}C_o^*}{\pi} t^{1/2} + Q_{dl} + nFA\Gamma_{Ru} \qquad (5\text{-}1)$$

where Q is the overall charge, t is time, n is the number of electrons transferred, F is the Faraday constant, A is the electrode area, D is the diffusion coefficient of $Ru(NH_3)_6^{3+}$, C_0* is the bulk concentration of $Ru(NH_3)_6^{3+}$, and Q_{dl} is the double-layer charging. If the binding interaction is purely electrostatic and the ionic strength of the supporting electrolyte is sufficiently low so that the film is saturated in $Ru(NH_3)_6^{3+}$, Γ_{Ru} can be used to calculate the surface concentration of DNA (Γ_{DNA}) according to eqn. 5-2.

$$\Gamma_{DNA} = \Gamma_{Ru}\left(\frac{z}{\mathrm{m}}\right) \qquad (5\text{-}2)$$

(Here z is the charge on the redox reporter molecule (3+) and m is the number of nucleotides in the DNA sequence). Owing to its simplicity, this assay has been used to study the effects of coadsorbents (*e.g.*, mercaptohexanol) as well as deposition time on the surface density of the DNA film.[10]

A recent study involving this same system used cyclic voltammetry instead of chronocoulometry to assay the electrostatic binding of $Ru(NH_3)_6^{3+}$ at both ss- and dsDNA-modified electrodes.[29] Langmuir[32] binding was observed at both surfaces, yielding association constants of 1.5×10^6 M^{-1} and 2×10^6 M^{-1}, respectively. The slightly higher affinity of $Ru(NH_3)_6^{3+}$ for duplex DNA was attributed to the higher negative charge density of the double-stranded DNA films. Notably, the value for Γ_{Ru} at saturation was nearly twice as large for double- *vs.* single-stranded films, consistent with the presence of twice the number of backbone phosphates in duplex DNA. Because the binding constant of $Ru(NH_3)_6^{3+}$ depends strongly on both ionic strength and the identity of the ions in the supporting electrolyte, this technique has been extended to evaluate the electrostatic binding constants of non-redox-active cations (e.g., Mg^{2+}, Ca^{2+}, K^{+}) *via* competitive-binding experiments.[33]

Indeed, because establishing the interfacial ionic equilibria required for maintaining charge neutrality within the DNA films is potentially so complicated (for example, $Ru(NH_3)^{3+}$ may itself participate in this equilibria during a faradaic process where mobile cations must cross the film/solution interface to balance charge within the monolayer) an alternative assay for hybridization based on long-range electrostatic repulsion of $Fe(CN)_6^{3-}$ from DNA-modified surfaces has been developed.[34] This latter technique utilizes electrochemical impedance spectroscopy (EIS), where a small oscillating voltage (~5 mV) is applied at the formal redox potential of $Fe(CN)_6^{3-}$ and the impedance is measured over a frequency spectrum, e.g., 0.1–10^5 Hz. A plot of the impedance in the complex plane (a Nyquist plot) shows a semi-arc at high frequencies yielding information about capacitive processes, and a 45° line at low frequencies resulting from slower events such as mass transfer to the electrode. Within the context of a Randles equivalent circuit, the diameter of the impedance arc is related to the electron-transfer resistance, R_{ET}, which in turn is related to the extent of electrode passivation. The degree to which $Fe(CN)_6^{3-}$ penetrates the film is reflected in R_{ET}, thus providing a quantitative measure of the relative DNA surface coverage. The EIS method has been applied to assay hybridization/dehybridization events of both pure DNA films as well as DNA/mercaptohexanol mixed monolayers (Figure 5-2). It has proven to be a simple and reliable technique to report on the composition of DNA films under a wide range of experimental conditions.

These electrochemical assays have shed light onto the probe/target DNA hybridization process.[10] Very low surface coverages of DNA are required for good hybridization yields; values for Γ_{Ru} larger than a few pmol cm^{-2} evidently impede the capture of target sequences, such that only a small percentage of immobilized probe sequences bind to the complement. The large negative charge density at the DNA-modified surface results in a significant electrostatic barrier to duplex formation, a factor not encountered in analogous solution-phase processes.

Complementary spectroscopic studies have also helped to elaborate the factors controlling hybridization at DNA-modified surfaces.[35–37] Surface plasmon resonance has been used to monitor DNA target capture in real time, and has confirmed that the efficiency of hybridization is maximized at surfaces sparsely covered with probe oligonucleotides.[36,37]

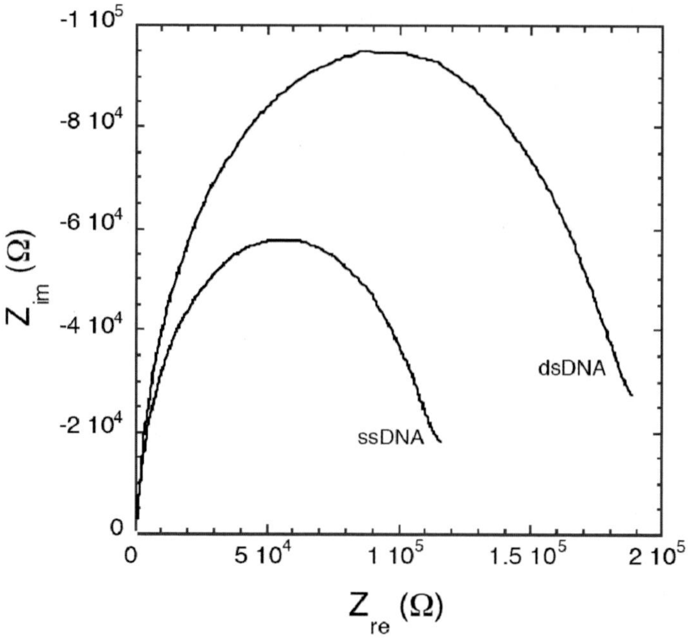

Figure 5-2. Nyquist plots for 1.0 mM $Fe(CN)_6^{3-}$ in 50 mM NaP_i pH 7 at a mercaptohexanol/DNA-modified gold electrode. The high-impedance arc represents fully hybridized DNA and the low-impedance arc represents ssDNA. Dehybridization was carried out by heating dsDNA films to 64°C in 10 mM Tris buffer pH 8 for 20 minutes. Films were rehybridized at room temperature in the presence of 1 μM complementary DNA in 50 mM NaP_i pH 7.

This technique was also used to monitor the effects of applied electric fields on hybridization and dehybridization;[35] not surprisingly, it was found that even small fields can significantly accelerate these processes. Mismatched sequences were particularly susceptible to potential-induced dehybridization, an effect that is potentially useful in discriminating between closely related sequences. As similar electric fields are involved in electrochemical assays, the effects of these fields on the DNA-film structure must be considered in the design and interpretation of DNA detection experiments.

2.1 Signal Amplification of Cation-Based Readout using Electrocatalysis

Despite the ability to quantify DNA densities using $Ru(NH_3)_6^{3+}$ in laboratory studies of interfacial hybridization, the finite surface concentrations of the redox reporter molecules (determined by the electrode area and DNA surface density) ultimately limit the inherent sensitivity of this assay and may hinder its practical application to the detection of pathogenic and human DNA. Recently, a method based on electrocatalytic amplification of the electrochemical signals was developed to improve the overall sensitivity of the hybridization assay.[11] Like the previous work, this system uses $Ru(NH_3)_6^{3+}$ to report DNA hybridization, but couples the direct $Ru(NH_3)_6^{3+}$ electrode reaction to a catalytic cycle involving $Fe(CN)_6^{3-}$ in solution, Fig. 5-3. Because homogeneous oxidation of electrochemically generated $Ru(NH_3)_6^{2+}$ by ferricyanide regenerates the Ru(III) probe for electrochemical recycling, the currents are greatly amplified.

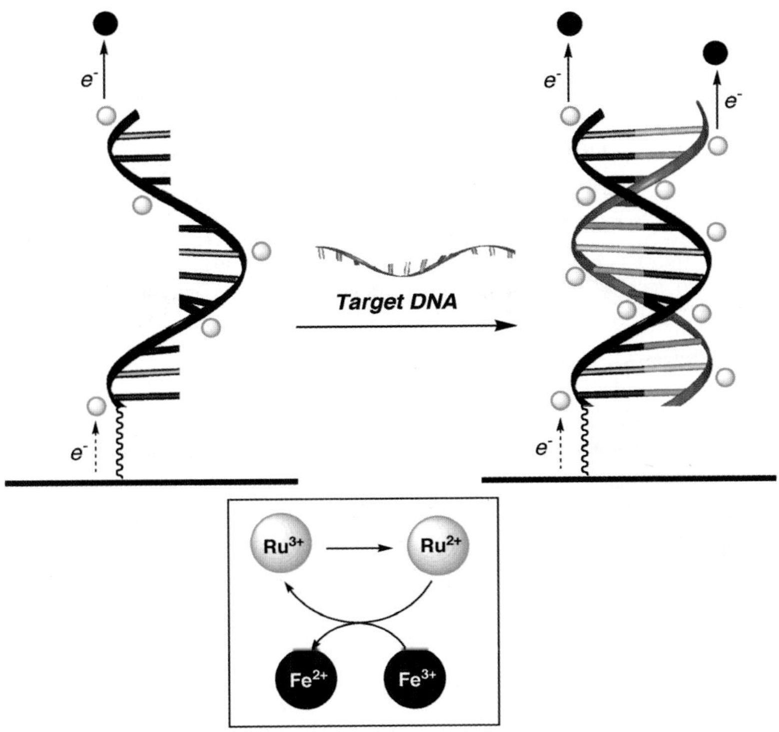

Figure 5-3. Electrocatalytic nucleic acids detection based on the reduction of $Fe(CN)_6^{3-}$ mediated by $Ru(NH_3)_6^{3+}$. The Ru(III) complex is electrostatically bound to the anionic backbone of an immobilized DNA oligonucleotide and upon the hybridization of a complementary target sequence is present at a higher concentration at the electrode surface. This increases the surface concentration of the redox mediator and leads to an enhanced electrochemical signal that can be used to detect specific DNA sequences. The presence of $Fe(CN)_6^{3-}$ chemically regenerates electrochemically reduced Ru(II), and thereby catalytically amplifies the signals. (Adapted from ref. 11).

The modified surfaces used in this study were gold electrodes treated with thiolated single-stranded DNA (generally 18–30 nucleotides long). The use of short deposition times (~60 minutes) and inclusion of a mercaptohexanol co-adsorbent allowed good control of probe density. In a typical experiment, the electrochemical response from a solution of $Ru(NH_3)_6^{3+}$ (27 μM) and $Fe(CN)_6^{3-}$ (2 mM) was measured before and after exposure of the electrode to a complementary target sequence. As shown in Fig. 5-4, the signals obtained in the presence of $Fe(CN)_6^{3-}$ are 10–20-times larger than those obtained with $Ru(NH_3)_6^{3+}$ alone; the irreversible electrode response is consistent with the chemical, rather than electrochemical oxidation of $Ru(NH_3)_6^{2+}$. Importantly, the differential increase in the electrochemical response is greater for those electrodes exposed to the target DNA, allowing a more definitive signaling of the hybridization event.

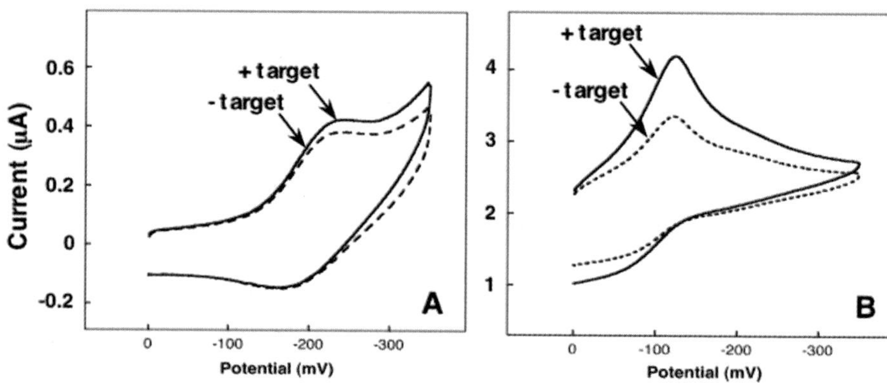

Figure 5-4. Electrocatalytic nucleic acids detection based on the reduction of $Fe(CN)_6^{3-}$ mediated by $Ru(NH_3)_6^{3+}$. Addition of $Fe(CN)_6^{3-}$ results in a 10–20-fold increase in the electrochemical response. Importantly, the differential response is much greater for double- vs. single-stranded surfaces. (Adapted from ref. 11.)

2.2 Detection of Pathogen-Related DNA Sequences using Electrocatalysis

To test the utility of the $Ru(NH_3)_6^{3+}/Fe(CN)_6^{3-}$ electrocatalytic assay, the detection of sequences corresponding to the *hpn* gene of *Helicobacter pylori* and a portion of the 23S rRNA from this organism was investigated. A simple diagnostic test for *H. pylori* would be of great value, as this pathogen is a leading cause of gastric ulcers and stomach cancer.[38] Moreover, strains resistant to clarithromycin — the antibiotic typically used to treat *H. pylori* — have also emerged, so the added ability to distinguish aberrant genotypes would also be important. In this case, the resistance has been linked to a single A2143C substitution within the 23S rRNA gene,[39] thus an effective assay would have to be able to resolve single-nucleotide sequence changes. Synthetic DNA oligonucleotides corresponding to the *hpn* and 23S rRNA sequences were thiolated and deposited onto gold. Hybridization of nanomolar concentrations of *hpn* and 23S rRNA targets to these electrodes was readily detected by the electrocatalytic assay. Because of the small volumes required for this analysis, detection limits in the femtomole range are possible using commercially available electrodes (A = 0.02 cm^2). Synthetic oligonucleotides, in addition to large (>200 nt) DNA and RNA targets generated enzymatically, were successfully hybridized and detected using electrocatalysis.[11] Short hybridization times (30–60 minutes) and mild hybridization conditions (40°C) were sufficient to obtain reproducible signals, making the assay straightforward and efficient. The detection of RNA products also indicates that gene-expression profiling might be possible using this approach.

Notably DNA targets possessing a single-base replacement gave much smaller readout signals, demonstrating that the electrocatalytic detection method is able to discriminate between closely related sequences. For example, the characteristic increase in the electrochemical signal that occurs upon treating the surface with the WT rRNA sequence failed to appear when an analogous sequence containing the A2143C substitution was used instead. Interestingly, however, as the electrode was incubated with the target sequence for longer periods of time, the catalytic response for the A2143C sequence steadily grew in, approaching the level of the WT rRNA signal after several hours, Fig. 5-5.

The finding that the sequence specificity of this assay is governed by hybridization dynamics rather than thermodynamics emphasizes the important differences between heterogeneous *vs.* solution-phase behavior. Surface-plasmon resonance studies of analogous interfacial DNA hybridization revealed not only slower binding kinetics for mismatched duplexes, but also fewer probe sites within the ssDNA film capable of capturing the incoming mutated sequence.[37] This latter finding may indicate that the hybridization process requires multiple probe sequences to capture mismatched targets, and/or the ensemble of immobilized probe-sequence conformations offers a comparatively small number of active capture sites for mutated targets.

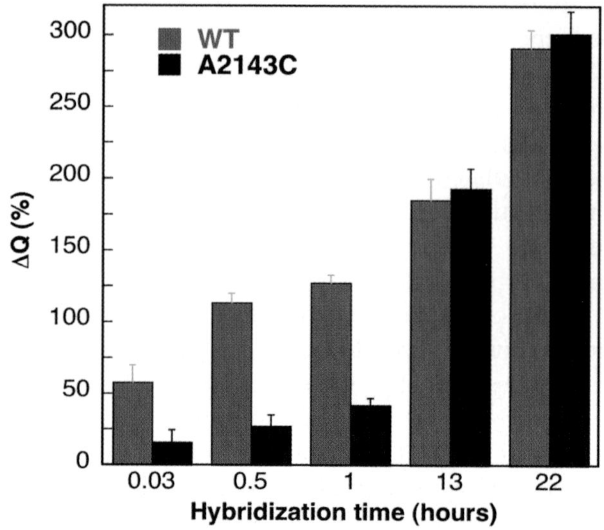

Figure 5-5. Dependence of electrocatalytic signal on hybridization time and the presence of a point mutation associated with antibiotic resistance. Electrochemical signals obtained as described in Fig. 5-4. (Adapted from ref. 11.)

Even with fully matched targets, very low hybridization yields were found for the surfaces used for electrochemical experiments. Fluorescently labeled DNA probes and targets were used to quantify the DNA surface coverages, as well as the yield of target-DNA capture. This analysis showed hybridization efficiencies on the order of only ~5–10%. The roughly 200% increase in the electrocatalytic currents after exposure to the target DNA - despite the low hybridization yields - highlights the enhanced sensitivity of this assay.

While not yet validated using clinical samples, the electrocatalytic assay appears to constitute a promising and sensitive method to detect nucleic acid sequences belonging to infectious pathogens. A particularly attractive feature of the method is the large signal enhancement that results from the formation of duplex DNA on the electrode surface. The unexpected finding that a single point mutation drastically attenuates the kinetics of duplex formation at the electrode surface - and therefore the magnitude of the electrocatalytic currents - indicates that hybridization-based bacterial genotyping is feasible, and that high-resolution sequence discrimination can be achieved with immobilized DNA probes.

Clearly, understanding the factors that govern DNA hybridization at disordered solid surfaces is of critical importance for the routine application of electrochemical assays based on the hybridization event. Extrapolation of models that describe the complexation of two complementary oligonucleotides in solution may not provide an accurate model of the analogous surface process. Because of the high local concentration of immobilized DNA, intermolecular complexes involving multiple probe sequences bound to a single target sequence may exist. Detailed studies of the mechanics of hybridization will provide useful information about the factors that modulate the efficiency of target capture. Questions involving the optimal length and densities of probe, as well as the target sequences also remain largely unexplored. Given the strong sequence dependence of non-specific DNA adsorption to gold surfaces,[44] it is also important to consider the sequence context of the probe and target. As biosensing applications are pushed to allow real-time pathogen detection of just a few copies of the sequence, having a detailed understanding of DNA-surface interactions will be critical in the development of assays requiring hybridization of extremely low concentrations of target DNA.

3. DNA SENSING WITH INTERCALATING PROBES

An alternative approach for utilizing DNA thin films for electrochemical sensors involves using redox-active probe molecules intercalated into the π-stack, Fig 5-1b. This approach has been successfully employed in the design of electrochemical assays for hybridization,[12,34] single-base mismatches,[12,13] and chemical modifications and lesions within duplex DNA.[13] It has also provided provocative experimental results regarding charge transport through the double helix.[40–42]

3.1 Formation and Structure of Double-Stranded DNA Films

Much of the work in this area has focused on the electrochemistry of intercalators bound to short (15 to 20-mer) duplexes self assembled onto gold surfaces. Figure 5-6 shows the basic approach for fabricating DNA monolayers.[43] After hybridizing a synthetic single-stranded sequence labeled with an alkylthiol linker (typically 16 σ-bonds between the terminal sulfur and the first base) in solution, the resulting duplex is self assembled onto gold by soaking the electrode for 10-12 hours in a 100 μM DNA/0.1 M

$MgCl_2$ solution at pH 7. The presence of Mg^{2+} is required to get good DNA coverages, presumably because the divalent cation minimizes electrostatic repulsion between adjacent duplexes during the self-assembly process. (These conditions may also help to limit the type of duplex dehybridization and associated non-specific adsorption of the resulting single strands recently observed for similar thiolated ds-oligonucleotides.[44]) Indeed, kinetics measurements on the film formation, followed by radioactive tagging as well as electrochemical blocking studies, show an initial fast phase in which ~1/4 of the duplexes adsorb to the gold within several minutes, followed by a much slower process during which the monolayer is filled in. Interestingly, it is only after this first phase that the film formation becomes especially sensitive to [Mg^{2+}], suggesting that the slow step in the monolayer formation may involve packing the duplexes tightly within the film. A similar two-dimensional film reconstruction has been proposed for the formation alkylthiol monolayers.[45–47]

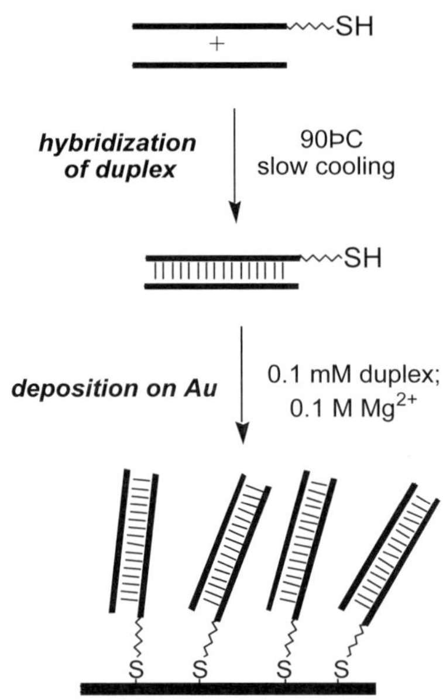

Figure 5-6. Schematic representation of the formation of thiol-terminated, double-stranded DNA films.

Duplex surface coverages (as determined by ^{32}P labeling, phosphate counting with $Ru(NH_3)_6^{3+}$, and the coulometry of intercalating redox probes covalently bound to the individual duplexes) generally range from 40–55 pmol cm^{-2}.[43] Using the dimensions of the individual duplexes to calculate possible surface morphologies, this corresponds to a close-packed structure where the individual helices line up at an angle of ~45° with respect to the electrode surface. Remarkably, inter-helical spacing is on the order of only ~6 Å under these conditions, suggesting an unusual degree of Debye screening of the phosphate-backbone charges within the 2D films. Not surprisingly, films prepared in the absence of Mg^{2+} generally have surface coverages that are considerably lower.[48]

To investigate the film morphology more directly, atomic-force microscopy (AFM) in fluid solution was used to visualize the surfaces.[49] Though demonstrating surface coverages that are apparently dense and uniform, these experiments failed to reveal local order in the monolayers. Subsequent use of an alkylthiol-modified AFM tip to minimize attractive forces between the SiN probe and the charged duplexes gave improved resolution, but still no clear picture of the surface topography.[50]

As a result, somewhat less direct methods were used to investigate the film morphology. Toward that end, a small patch of DNA was removed from the monolayer by applying a large vertical force with the AFM tip to mechanically scrape off the adsorbed molecules, Fig. 5-7.[49]

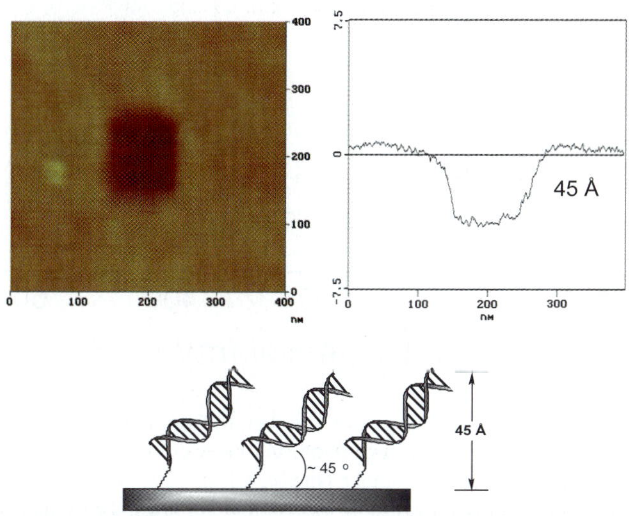

Figure 5-7. AFM image of DNA-modified gold (sequence: 5'-SH-AGT ACA GTC ATC GCG-3') after removal of a small patch (~100 nm^2) of DNA by mechanically scrapping the probe tip against the surface. (Adapted from ref. 49.)

Because the depth of the resulting hole is equal to the monolayer thickness, height-contrast measurements between the hole and the film surface gave a direct measurement of the film height.[51] Notably, the measured value of 45 Å suggests an average helical packing orientation of ~45° normal to the surface, based on the dimensions of the 15-base-pair duplex. Although the film is very stable under open-circuit conditions, application of either large positive or negative potentials results in rapid desorption of the monolayer, consistent with redox reactions that destroy the gold-thiol linkages.[49]

Applying less extreme potentials also affects the DNA film morphology. Figure 5-8 shows the effect of electric fields on a 15-base-pair thiol-modified duplex film, recorded by tapping-mode AFM under electrochemical control.

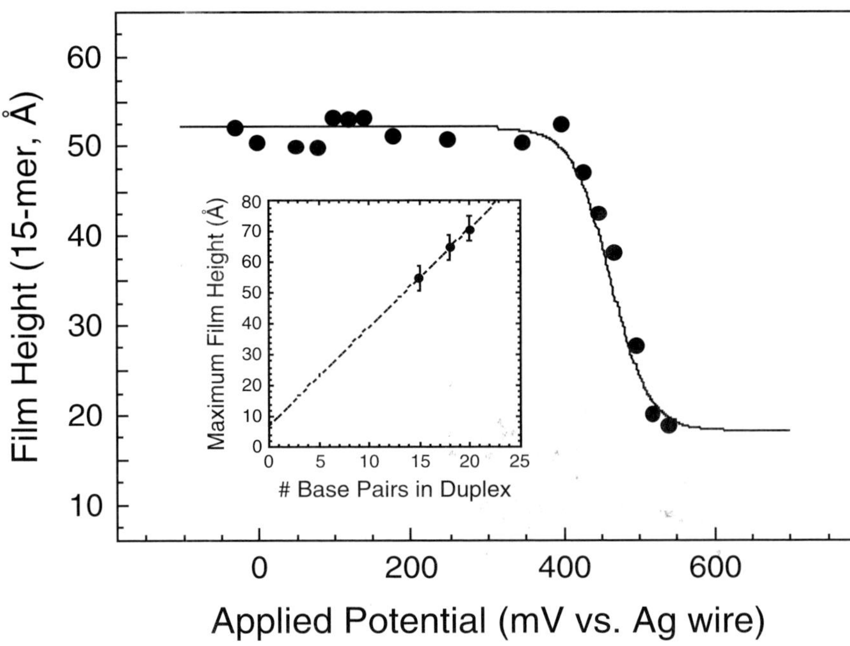

Figure 5-8. Potential dependence of the DNA film height from Fig. 5-6, as measured by AFM under electrochemical control. (For comparison, the open-circuit height is 45 Å.) The inset shows a plot of the maximum film height (measured at ~100 mV *vs.* the quasi-reference electrode) measured for duplexes possessing 15, 18, and 20 bases. These data show a slope of 3.2 Å/bp, close to the predicted value of 3.4 Å/bp. The intercept, ~7 Å, is somewhat smaller than the ~16 Å expected for a fully extended alkylthiol linker, likely due to compression of the aliphatic linker under pressure from the AFM tip.

At potentials negative of ~0.45 V (*vs.* an Ag quasi-reference electrode), the film height increases from its open-circuit value to ~55 Å (close to the value predicted for a fully extended duplex-linker conjugate oriented normal to the gold surface). As the potential is progressively scanned positive, the thickness remains relatively constant until E_{app} ~0.45 V, where the height drops dramatically to a limiting value of ~20 Å (the diameter of duplex DNA).[49] These changes are reversible, and are consistent with a morphology change that is triggered by electrostatic interactions: at voltages negative of the pzc (potential of zero charge), the negatively charged phosphate backbone is repelled from the surface causing the duplexes to stand straight up, while voltages positive of the pzc attract the phosphate groups causing the duplexes to lie down flat. Accordingly, the maximum thicknesses of films comprised of 18- and 20-base-pair duplexes, 65 and 72.5 Å, respectively, reveal a nearly ideal 3.4 Å increase in film height with each additional base pair in the duplexes.[52] Subsequent[53] AFM measurements and STM data[54,55] have indicated a similar DNA-morphology change upon application of small electric fields.

Another consequence of the anionic nature of the DNA films is the restricted transfer of ions across the monolayer-solution interface. The resulting interfacial Donnan potentials are expected to shift the apparent formal potentials of redox molecules bound within the film, by analogy to polyelectrolyte film coatings.[56] In the general case of an *n*-electron, *m*-proton redox couple (Equation 5-3), the apparent formal potential shifts as a function of both pH and cation concentration according to 5-4.

$$Ox^{+} + (n-m)C_s^{+} + ne^{-} + mH^{+} \rightleftharpoons \mathrm{Red}^{(m-n)+} + (n-m)C_f^{+} \tag{5-3}$$

$$E_{\mathrm{app}}^{\mathrm{f}} = E^{\mathrm{f}} + \frac{RT}{zF}\frac{(n-m)}{n}\ln\left(\frac{C_s}{C_f}\right) + \frac{RT}{F}\frac{m}{n}\ln\left[H^{+}\right] \tag{5-4}$$

(Here $E^{\mathrm{f}}_{\mathrm{app}}$ is the apparent formal potential, E^{f} is the formal potential in solution, z is the charge of the cation in the supporting electrolyte, C_s is the concentration of the cation in solution, and C_f is the concentration of the cation within the film, which is effectively constant at low loadings of cationic reporters.) As illustrated in Fig. 5-9, a plot of $E^{\mathrm{f}}_{\mathrm{app}}$ *vs.* log[KCl] for the reduction of $Ru(NH_3)_6^{3+}$ is linear with slope of 60 mV/log unit, in excellent agreement with the 59 mV slope predicted by eqn. 5-4 ($T = 25$ °C).

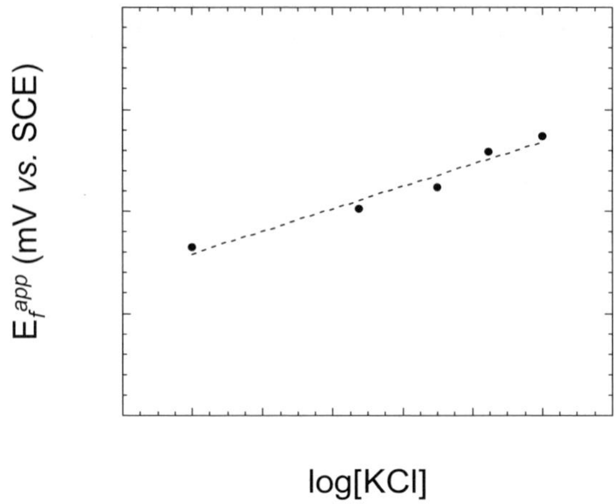

Figure 5-9. Plot of the apparent formal potential of $Ru(NH_3)_6^{3+}$ at a DNA-modified electrode. The concentration of $Ru(NH_3)_6^{3+}$ was adjusted such that Γ_{Ru} remained ~10–20 pmol cm^{-2}, or approximately 5–10% of the value at saturation.

3.2 Electrochemistry of Intercalators at DNA-Modified Electrodes

The first electrochemical study of an intercalator bound to these films involved the phenathiazine dye methylene blue (MB).[43] MB is redox active, undergoing a reversible 2e$^-$, proton-coupled reduction to leucomethylene blue (LB) in aqueous solution.[57] At pH values between 6 and 8, the reduction is coupled to a single proton transfer, and the apparent formal potential shifts by 30 mV/pH unit. At each of these pH values, the formal potential additionally shifts 30 mV/log[Na^+], consistent with Eqn. 5-4.[34] The cyclic voltammogram of 1 μM MB at a DNA-modified electrode (pH 6.5) is shown in Fig. 5-10. The pronounced response at such a low concentration and the linear increase in peak currents with scan rate indicate a surface-bound analyte. Interestingly, the peak widths (~80–100 mV, pH 7) are somewhat larger than ideal for a surface-bound 2e$^-$ Nernstian response, but decrease with higher [H^+] to limiting values of less than 60 mV at pH 5.5. This suggests that breakdown in the cooperative nature of the 2e$^-$ response within the film is primarily responsible for the broader signals; at pH values greater than ~8.5, the CV response begins to resolve into two separate waves.

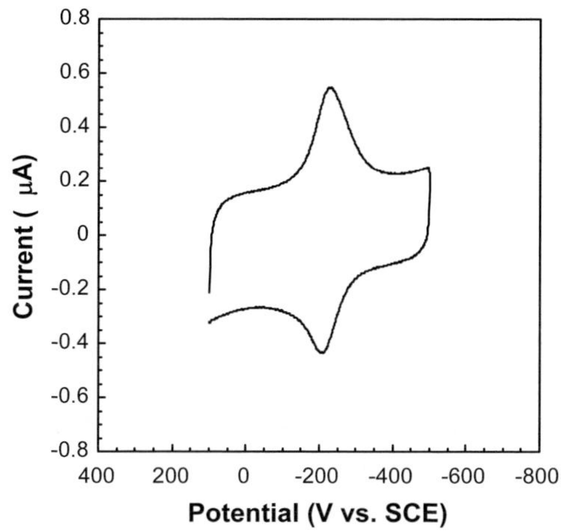

Figure 5-10. Cyclic voltammogram of MB recorded at a duplex DNA-modified electrode (sequence: 5'-SH-AGT ACA GTC ATC GCG-3') at a sweep rate of 50 mV s^{-1} in 50 mM sodium phosophate buffer, pH 6.5.

Coulometric titrations of MB at these surfaces reveal that MB binds reversibly to the films according to a simple Langmuir model, with a binding constant of $\sim 3\times10^6$ M^{-1} and a maximum surface coverage of ~55 pmol cm^{-2} (approximately one MB per duplex in the film).[43] While this binding constant agrees well with literature values obtained for MB intercalated into DNA in solution, the 1:1 stoichiometry does not: when DNA is free in solution, MB occupies up to five "neighbor excluded" sites in a 15-mer duplex.[58] The relatively low MB loadings for these electrodes has been attributed to dense duplex packing within the film, which presumably inhibits diffusion down into the monolayer. Consistent with this model, the MB:DNA binding stoichiometry increases as a function of decreasing duplex surface density, reaching a limiting value of roughly 3.5 per duplex at DNA surface densities below ~15 mol cm^{-2}.

The kinetics of MB binding to the double-stranded DNA films has been investigated, using chronocoulometry at a rotating-disk electrode to monitor the time-dependent charge due to adsorbed MB following immersion of a DNA-modified electrode into buffer solutions containing various concentrations of MB in solution.[59] As long as: (i) the number of binding sites in the film (S) is fixed, (ii) the concentration of MB in solution ($[MB]_o$)

remains constant (a condition satisfied when there is a large excess of MB relative to the number of binding sites on the surface), and (iii) the rate of convection/diffusion of MB to the electrode surface is rapid,[60] the change in Γ_{MB} as a function of time can be analyzed according to simple Langmuir kinetics to yield rate constants for MB adsorption (k_1) and desorption (k_{-1}) according to eqns. 5-5, 5-6 and 5-7:[61,62]

$$MB + S \underset{k_{-1}}{\overset{k_1}{\rightleftharpoons}} MB_I \quad (5\text{-}5)$$

$$\theta(t) = \theta_e\left(1 - \exp^{-k_{obs}t}\right) \quad (5\text{-}6)$$

$$k_{obs} = k_1[MB]_o + k_{-1} \quad (5\text{-}7)$$

In this expression, θ is the fractional coverage of MB in the film, Γ_{MB}/Γ_{max} at any given time (t) and θ_e is the fractional coverage at equilibrium, determined by $[MB]_o$ and K.

Dividing the calculated value of k_1 ($7.5(6) \times 10^5$ M^{-1} s^{-1}) by k_{-1} (0.043(8) s^{-1}) yields an equilibrium binding constant of $\sim 2\times10^6$ M^{-1}, a value close to that calculated from the binding isotherm. Not surprisingly, the actual binding constant depends both on the identity and ionic strength of the supporting electrolyte. In particular, addition of Mg^{2+} inhibits MB intercalation into DNA in solution, and titrating Mg^{2+} into the aqueous phase similarly displaces MB from the DNA-modified surfaces.[63]

3.3 Electrochemistry of Covalently Bound Intercalators

To control more precisely the location of intercalating probe molecules within the films, the chemotheraputic drug daunomycin (DM) also has been investigated at these surfaces.[40] DM possesses an amino sugar that can be covalently bound to the N2 group of guanine *via* cross-coupling reactions with formaldehyde.[69] Thus, incorporating just two adjacent GC base pairs in an otherwise A-T or inosine-C duplex[70] allows the site-specific labeling of DNA with DM. Labeled duplexes readily self assemble onto gold forming films with surface coverages and morphologies indistinguishable from those of unlabeled films. Cyclic voltammograms of all of the DM-labeled surfaces show a chemically reversible reduction of DM at –0.65 V *vs.* SCE. Doping the films with increasing percentages of DM-free duplexes results in a linear decrease in the observed electrochemical signals.[40]

Given a structural model in which the individual helices within the DNA film are oriented normal to the surface (especially at negative applied potentials),[49] the DM-electrode separations span more than 45 Å in the series of conjugates prepared. Yet based on variable scan-rate cyclic voltammetry measurements, the heterogeneous ET rates for all of the conjugates are the same.[40] Because these rates (on the order of 100 s^{-1}) are reasonable values for electron tunneling through the σ-only linker, the results suggest that ET is gated by tunneling through the thiol tether.

To test this hypothesis, a series of homologous DM-DNA-thiol conjugates were prepared where the number of intervening methylene units (n) in the thiol tether varied from 4 to 9.[64] As illustrated in Figure 5-11, the electron-transfer rate clearly decreases inversely with n. Using Laviron's model[65] to calculate the standard electron-transfer rate constants, k_s, it is apparent that the rate drops off exponentially with distance, consistent with predictions from super-exchange coupling theory. Notably, the decay constant, β, is equal to 1.0 per methylene unit, essentially identical to those previously measured for tunneling through an alkylthiol bridge to bound ferrocene (β = 1.1 per CH_2),[66] as well as to cytochrome *c* (β = 1.0 per CH_2)[67] and azurin (β = 1.0 per CH_2).[68] Assigning a β value of 1.0 per bond for the additional σ-bonds in the aliphatic linker leads to an extrapolated, zero-linker-length ET rate of ~10^8–10^9 s^{-1} – a value comparable to those found using small-molecule redox probes bound directly to metal surfaces.[66]

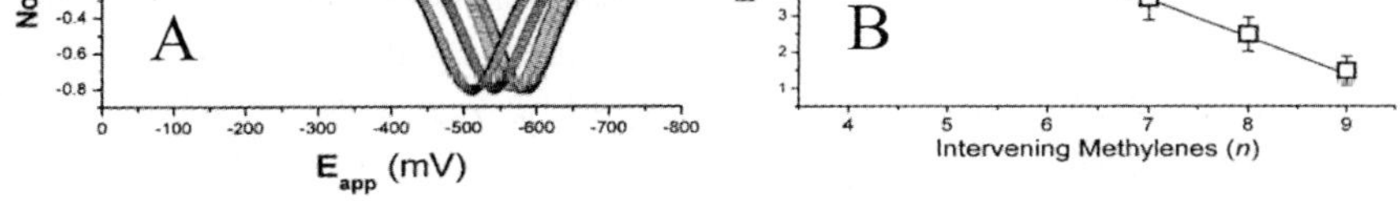

Figure 5-11. (A) Background-subtracted voltammograms of DM-DNA conjugates at ν = 1 V s^{-1} for aliphatic tethers with n = 4, 6, 7, 8, and 9. Inset: CVs (ν = 1 V s^{-1}) of DM-DNA films with identical tethers (n = 6) but different DM intercalation sites: 5'-ATC CTC AAT CAT **GG**A C-3' (top) and 5'-ATC CT**G G**AT CAT CAA C-3' (botton). (B) Plot of ln(k_s) *vs.* number of methylene units (n) in aliphatic tether. (Adapted from ref. 64).

Notably, a similar finding was reported by Heller and associates.[41] Their system involved a redox-active ortho-quinone probe molecule, PQQ, covalently attached to the top of 12 base-pair duplexes similarly self-assembled onto gold electrodes *via* alkylthiol linkers. Like the DM-labeled films, electrochemical measurements revealed a high duplex-packing density (implying an upright orientation of the individual oligonucleotides within the monolayers), and gave ET rates through the duplexes that are of the same order of magnitude as those for PQQ tethered directly to gold *via* the linker. Significantly, ET through single-stranded DNA films to which PQQ was covalently bound was too slow to give a measurable response by CV.

Perhaps the most interesting property of both DM- and PQQ-labeled films is the dramatic effect of mispaired bases within the duplexes. Incorporation of a single GA mismatch into the DNA sequence of the PQQ films drops the ET rate fully three orders of magnitude; the effect is even more striking for the DM-labeled films, where an intervening CA mismatch switches off the electrochemical response entirely. (Importantly, if the mismatch lies above the intercalated DM, the signal is unaffected.) In a key control experiment, a series of DM-labeled films featuring fully base-paired duplexes was doped with increasing fractions of CA-mismatched helices. The electrochemical signals decreased linearly with increasing percentages of mutated duplexes, indicating that the electroinactive intercalators (presumably those molecules bound to the mismatched duplexes) were not reduced by the electroactive ones. This implies that lateral charge diffusion between adjacent helices is slow on the CV timescale, ruling out gating mechanisms that involve ET to DM at "hotspots" or pinholes in the film, followed by two-dimensional charge dispersion within the monolayer.

To account for these observations, Heller has offered the intriguing proposal that the apparent conductivity of DNA monolayers results from solid-state effects within the films.[41,71] As counterions to the phosphate backbone can move freely along the helical axis of the DNA, a film in which the individual helices line up in an ordered array should exhibit a highly anisotropic dielectric response, with the dielectric constant parallel to the helices unusually large. Solid-state theory[72] predicts that the mean-free-path of an electron in such an "ionic solid" should be very large, resulting in one-dimensional conductivity along the longitudinal axis of the film. Owing to their relatively moderate ionization potentials, the DNA bases guanine and adenine have been suggested to act as the charge dopants within the films; the dramatic effects of mismatches have been attributed to electronic defects within the DNA films that limit the conductivity. Recent STM measurements of dsDNA SAMS on gold are consistent with conductivity along the helical axis, showing dramatic decreases in tunneling currents when bias voltages are adjusted to orient the helices perpendicular to the surface normal.[54,55] In line with the electrochemical measurements of intercalated reporter molecules, incorporation of a CA mismatch into the sequence appears to quench the tunneling currents significantly.

While many fundamental questions remain regarding the mechanism of electron transfer through the films, it is empirically clear that the characteristic electrochemical responses of intercalating probe molecules can signal even subtle perturbations within the double helix.[12,13,40,41,73] The possibility that similar DNA-mediated electrochemical reactions might play important roles within the cell – for example, in the cathodic protection of the genome,[74] the repair and proof-reading of DNA,[75] carcinogenesis,[75b] or even in the activation of certain DNA-binding drugs – continues to pique the interest of the chemical biology community.

3.4 Use of Intercalated Probe Molecules for Biosensing Applications

From a practical standpoint, the response of intercalated probe molecules to mismatches within the films raises the question of whether such films could be used to report mutations within a genomic sample. The essential steps of a mutational assay based on this approach would involve: (1) chemically synthesizing a single-stranded capture probe sequence that corresponds to the region of genomic DNA of interest; (2) immobilizing that capture probe on an electrode surface; (3) hybridizing a genomic sequence to the capture probe; and (4) interrogating the resulting film for the presence of a mismatch by analysis of the electrochemical response of an intercalated probe molecule. An attenuated response would signal a mismatch in the film, which, in turn, would indicate a mutation (or polymorphism) in the genomic sample. (We note that a nascent mismatch within a probe/target duplex may significantly retard the hybridization process. Thus if the intercalating probe molecule has a binding preference for double-stranded *vs.* single-stranded DNA, the electrochemical response would still signal a single-base substitution in the target even without hybridization, as a smaller signal would occur at the single-stranded surface.) Using DM (μM concentrations in the electrolyte solution) as a non-covalent intercalative reporter, single-base mismatches were indeed identified in a variety of synthetic DNA films, establishing the potential of this approach.[12]

Just as coulometric assays based on phosphate counting are limited by the surface concentrations of the electrostatically bound redox probe, so too is the electrochemical response of intercalated probes. Examination of the cyclic voltammograms of DM or MB, *e.g.,* Fig. 5-10, at duplex-modified electrodes shows that double-layer charging accounts for a significant percentage of the overall charge, further highlighting the need for selective faradaic signal amplification. To accomplish this, the reaction has been coupled to an electrochemical cycle in which MB mediates the reduction of ferricyanide in solution,[12] completely analogous to the process described using $Ru(NH_3)_6^{3+}$. Like the $Ru(NH_3)_6^{3+}/Fe(CN)_6^{3-}$ system, MB-mediated reduction of ferricyanide results in dramatically improved sensitivity over the direct-detection assay, Fig. 5-12.

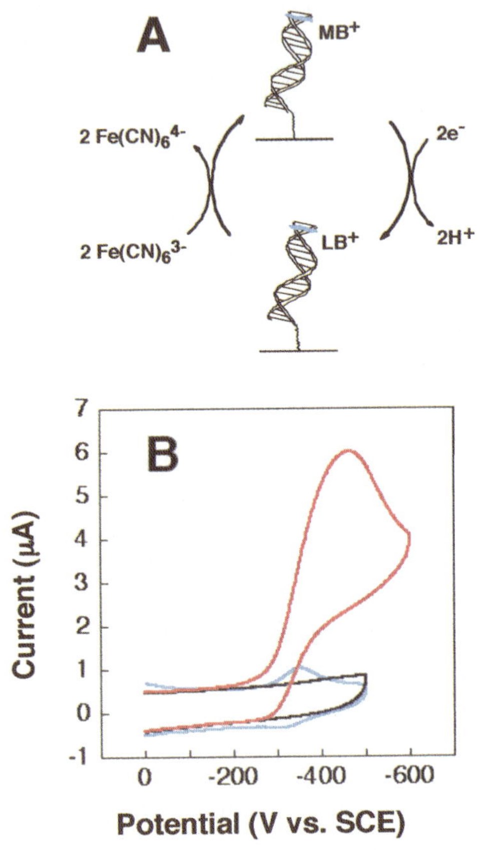

Figure 5-12. (A) Electrocatalytic scheme involving MB and ferricyanide; (B) Cyclic voltammetry (100 mV s^{-1} scan rate) at a gold electrode modified with double-stranded DNA (SH-5'-AGT ACA GTC ATC GCG-3') of 2.0 mM $Fe(CN)_6^{3-}$ (no faradaic signal), 2.0 μM MB, and 2.0 mM $Fe(CN)_6^{3-}$ and 2.0 μM MB. (Adapted from ref. 12).

Because the ultimate electrochemical substrate (ferricyanide) is in the solution phase, the current is no longer limited by the amount of redox-probe molecule that can be adsorbed onto the electrode surface. Importantly, this catalytic assay enhances the differential currents for fully matched *vs.*

mismatched films, allowing detection of all of the possible base mismatches—including the notoriously stable and difficult to detect GA mismatch— using synthetic DNAs, Fig. 5-13.[13]

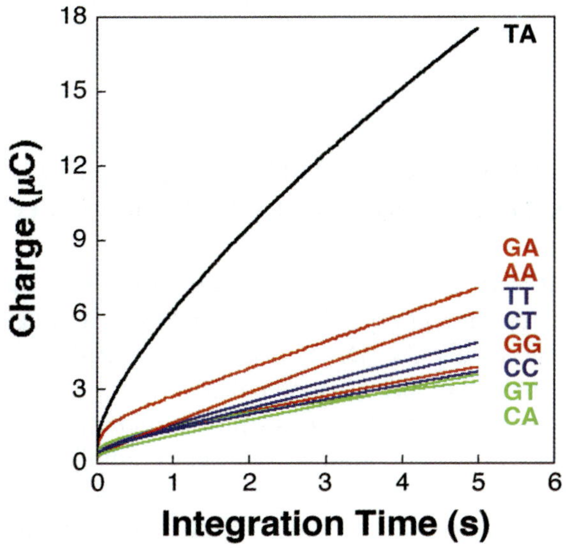

Figure 5-13. Plot showing the accumulated charge as a function of time following a potential step to –350 mV at a series of DNA-modified electrodes immersed in a solution containing 2 mM $Fe(CN)_6^{3-}$ and 0.5 μM MB. The thiol-terminated sequence (SH-5'-AGT ACA GTC ATC GCG-3') was hybridized before electrode deposition to its fully base-paired complement (TA), as well as complements that resulted in all of the possible single-base mismatches. Purine-purine, purine-pyrimidine, and pyrimidine-pyrimidine mismatches were all readily detected. (Adapted from ref. 13.)

Interestingly, DM was found to be a very poor mediator for ferricyanide reduction at the DNA-modified surfaces. As the reduction potential of DM is more than 300 mV lower than that of MB, reduced DM has an even greater driving force for turning over $Fe(CN)_6^{3-}$. To investigate this curious behavior, the reaction was analyzed at a rotating-disk electrode.[59] The results of that study suggested that the overall reaction is governed ultimately by the exchange dynamics of the mediator at the solution/film interface. As those kinetics are much faster for MB than for DM, the measured currents are correspondingly higher.

While this approach presents a potentially powerful means to identify and type human genes, the complexity and size of the genome demands that the assay be implemented in a high-throughput and versatile format. Many potential targets for genomic mapping that are medically relevant involve more than one possible SNP on a single gene, or even numerous sites on constellations of different genes. Therefore, the ability to carry out the analysis in a highly parallel fashion using microarrays is critical. Recent work adapting the assay to an addressable microelectrode array is promising; reproducible signals are obtained on electrodes as small as 30 μm, raising the possibility of carrying out the analysis on high-density arrays.[13]

4. DNA SENSING APPROACHES USING NON-DNA-BINDING ELECTROACTIVE LABELS

Several recent reports describe nucleic acid biosensing methods that rely on electroactive reporter groups that are not DNA-binding molecules.[76–78] The use of these labels for DNA detection typically requires covalent attachment because the probe molecules lack intrinsic DNA-binding capabilities. However, access to probes whose structural and electrochemical properties can be chosen for optimal efficiency amply justifies the added sample preparation time. In many of these cases, the electrochemical processes used for readout do not involve the DNA/electrode interface *per se*. Rather, the molecular-recognition properties of DNA are exploited to recruit the DNA-bound redox probes to the surface for analysis by more traditional electrochemical techniques, such as enzymatic catalysis or stripping voltammetry.

An elegant example of this approach involves tagging DNA sequences with horseradish peroxidase (HRP) for subsequent immobilization within a redox polymer film.[76,77] This amperometric “sandwich” assay relies on the formation of a chimeric probe-target-detector duplex, where an overhang created by probe-target hybridization provides a binding site for the HRP-labeled strand. The target DNA serves to recruit the HRP into the redox film, where it catalyzes the electrochemical turnover of peroxide. The assay is remarkably sensitive, able to detect as few as 3000 copies of a target sequence.[77] The HRP-based method also displays some sequence specificity, with electrical responses that are decreased by a factor of 3 in the presence of a single-base mismatch.[77]

Another recent study highlighting a redox-active label employs metal colloids to signal target-probe hybridization.[78] Magnetic beads derivatized with single-stranded DNA probe sequences were used to capture target and reporter sequences that had been labeled with nanoparticles composed of ZnS, PbS, or CdS. The beads were then physically removed from the analyte solution, and the particles were dissolved in acid. After adsorption of these ions onto solid electrodes, stripping voltammetry was used to quantify the amount of each metal present. As zinc, lead, and cadmium all undergo cathodic stripping at different potentials, this assay provides a strategy for the simultaneous analysis of several different target sequences. Excellent sensitivity was achieved, with detection limits in the femtomolar range, but the selectivity of the assay has not yet been tested. Several biosensing assays based on similar approaches have been developed and recently reviewed.[26]

5. DNA ANALYSIS USING DIRECT ELECTROCHEMICAL READOUT

Electrochemical processes involving the DNA bases themselves have also been investigated and used in DNA biosensing schemes.[14–20] Guanine (G) is the most easily oxidized base (~ +1.3 V *vs.* NHE) followed closely by adenine (~1.5 V); thymine and cytidine are significantly more difficult to oxidize, with redox potentials in excess of ~ +2.0 V (*vs.* NHE).[79] As these electrochemical processes are irreversible, this assay degrades the target DNA samples during the analysis.

Some of the first studies of DNA adsorbed onto electrode surfaces focused on the redox chemistry of adsorbed nucleotides complexed to mercury electrodes.[1,20] Single- and double-stranded DNA can be distinguished using this electrochemical handle, based on the greater affinity of single- *vs.* double-stranded sequences for mercury. A recent study showed that Factor V Leiden mutations can be detected at carbon-paste electrodes by monitoring changes in G oxidation signals in a multiplexed format.[80]

An extremely powerful approach developed by Thorp and coworkers over the last decade uses a transition-metal mediator to facilitate the catalytic oxidation of DNA.[14–19] In a typical assay, $Ru(bpy)_3^{2+}$ is electrochemically oxidized to $Ru(bpy)_3^{3+}$, which rapidly oxidizes guanine residues in DNA to regenerate the $Ru(bpy)_3^{2+}$ starting complex in a catalytic cycle, Fig. 5-1d. Multiple electrochemical turnover of DNA by $Ru(bpy)_3^{2+}$ occurs as long as a supply of G is available. By replacing all of the Gs in the probe strand with inosine (a base that pairs with C, but is more difficult to oxidize than G) the catalytic signal reports only the presence of target DNA captured at the electrode surface. Notably, by changing the metal and/or ligands of the redox mediator, its redox potential can be tuned to target specific DNA base modifications purposely incorporated into the target sequences (*e.g.*, 7-deazaguanine or 7-deazadenine) to further enhance the selectivity of the assay.[17]

The electrocatalytic reaction between $Ru(bpy)_3^{2+}$ and G has been applied to a variety of DNA detection problems.[14–19] The assay is extremely sensitive, with attomole quantities of a PCR-amplified oncogene detected at an indium tin oxide electrode.[16] Mismatched or unpaired bases within the target/probe complex can also be detected using the assay, as the kinetics of the oxidation is sensitive to base pairing. This feature permits the detection of DNA sequences that would form a bulged structure and was successfully applied to sensing a TTT deletion associated with cystic fibrosis.[15]

The $Ru(bpy)_3^{3+}$/guanine reaction can be carried out in solution, as well as many different types of electrode surfaces. It was originally developed using DNA physically adsorbed to ITO, but has also been applied to DNA attached to carbon nanotubes and polymer-modified electrodes.[18,19] As long as the DNA-electrode linkage can tolerate potentials up to ~1.3 V (the oxidation potential of $Ru(bpy)_3^{2+}$) the assay should be feasible at virtually any interface.

5.1 Non-Faradaic Electrochemical DNA Biosensing

In addition to the electrochemical approaches described above, non-faradaic processes at the DNA-electrode interface provide an alternative, and potentially straightforward route to nucleic acids analysis. Just as electric-field gradients influence DNA film morphology and hybridization dynamics, they have also been used to orient and move DNA helices in predictable ways. Asbury and van den Engh, for example, have exploited oscillating electric fields to induce large dipoles within DNA helices, which, in turn, allow the trapping of DNA molecules within regions of high field strength.[81] Coupling the oscillating fields with a static field gradient further causes a concerted movement of helices along a precise path. This type of "dipole trap" is potentially useful in the construction of miniature laboratory devices, where multiple steps (*e.g.,* buffer exchange, contaminant washing, *etc.*) would need to be carried out on the same surface.

Another method that exploits nucleic acid electrophoresis involves threading single-stranded sequences through biological nanopores using electric-field gradients.[82] Using the ion channel α-hemolysin imbedded into a lipid bilayer, an assay for DNA and RNA has been developed in which a single-stranded nucleic acid is translocated through the pore opening using small (~100 mV) bias voltages. As the sequence passes across the membrane, it temporarily blocks the movement of electrolyte ions through the channel thereby reducing the characteristic ionic current and signaling the presence of nucleic acids. The findings that the duration of the blocking event is directly proportional to the length of the sequence and that the magnitude of the blocking depends on the identity of the nucleic acid bases have raised the possibility of using this technique for single-molecule sequencing. Recent theoretical work suggest that the use of rotating electric fields may allow single-nucleotide resolution with this strategy.[83]

Nanoscale approaches based on direct conductivity measurements present another potentially powerful means to use current-based measurements for nucleic acid biosensing. In a recent application of this approach,[84] Mirkin and coworkers constructed an array of microelectrodes, all featuring 20 μm gaps in the electrical leads. Single-stranded capture strands were then deposited in the gaps, where they recruited target DNAs by hybridization. Overhang sequences on the target DNAs were subsequently bound to complementary sequences labeled with gold nanoparticles, effectively filling the gaps between the electrode leads. The nanoparticles served as nucleation sites for the deposition of silver, which, in turn, caused characteristic changes in conductivity that signaled the presence of target DNA. The use of carefully controlled hybridization conditions allowed excellent discrimination of single-base mutations within the target sequences, demonstrating the potential of this technique for mutational analysis.

Perhaps most intriguing have been nanoscale efforts to probe the direct electrical properties of DNA itself. Performing direct conductivity measurements on DNA molecules is extremely challenging, as fabrication of nanoelectrodes with sufficiently small spacings for monitoring discrete DNA sequences requires cutting-edge lithographic techniques. Efforts in this area have led to conflicting conclusions concerning the electrical properties of DNA,[85–88] and the results of these studies highlight the need for a well-defined DNA/nanolead configuration if a new method for DNA detection is to be developed.

To date, many of the direct DNA conductivity measurements have focused on the 48,502 bp double-stranded DNA from λ bacteriophage, mainly because of its commercial availability and the presence of short (12 nucleotide) single-stranded overhangs that permit attachment to surfaces modified with complementary sequences.[85–87] One of the first studies of the conductivity of λ-DNA used a low-energy electron point-source microscope (LEEPS) to monitor DNA deposited onto a 2 μm gap in a carbon foil.[85] Using this approach, high levels of conductivity were observed. However a later study, again using λ-DNA covalently attached to electrodes separated by 4 μm, produced no appreciable current, even at applied potentials up to 20 V.[86] It has been suggested that the large currents observed in the initial study originated from the use of an electron beam that produced free electrons that acted as the charge carriers.[87]

Clearly, λ-DNA is not an ideal substrate for the analysis of DNA conductivity. Because of its length, the structure is irregular and disruptions in base stacking are likely present. Any sequence longer than the persistence length of DNA (~100 base pairs) will likely lack sufficient rigidity to allow structurally well-characterized conformations. Moreover, spectroscopic studies of electronic coupling in smaller DNA oligonucleotides (12–20 base pairs) indicate that the dynamics of the stacked heterocycles inside the double helix is an important factor governing long-range charge transport.[89] How these dynamic effects are manifest in the direct conductivity experiments has not been addressed.

The use of short DNA oligonucleotides with defined electrode connectivities would provide a better-defined system and a means to establish whether the conductivity of double- *vs.* single-stranded DNA is sufficiently different to provide a handle for target sequence detection. While short oligonucleotides have been tested, the analysis was only performed in the absence of a covalent DNA-electrode linkage.[88] Current-voltage profiles consistent with those expected for a large-bandgap semiconductor were observed, but the poorly-defined DNA/electrode interface may affect the conductivity observed in an unpredictable way. In the same laboratory, longer DNA substrates (900–4500 base pairs) generated by PCR and attached via thiol linkers to electrodes exhibited insulating behavior; whether DNA of this length possesses the structural integrity to facilitate efficient charge transport remains an open question.

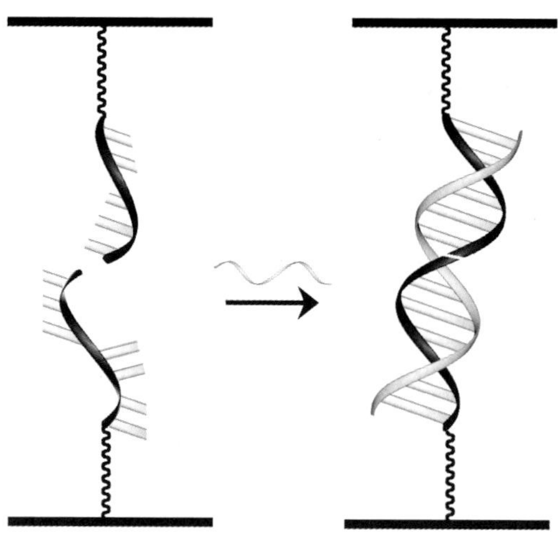

Figure 5-14. Schematic illustration of DNA detection based on the direct measurement of conductivity using nanoscale electrical leads. The binding of a target sequence to a pair of probe oligonucleotides immobilized on opposing nanoscale electrodes would provide a means to detect DNA sequences using direct conductivity measurements.

The use of conductivity-based measurements for DNA biosensing will require the basic characterization of the electrical properties of nucleic acids using improved experimental configurations (Fig. 5-14). The development of a reproducible means of measuring the conductivity of short DNA molecules attached through a well-defined linker to nanoscale electrodes, remains a challenging, but highly worthwhile objective.

6. CONCLUSIONS

Prototype DNA biosensors have been developed that rely on innovative electrochemical readout technologies. The redox chemistry of small molecules bound to nucleic acids, charge migration through the double helix, and direct conductivity measurements have all been coupled to biorecognition events to affect signal transduction in DNA-based sensors. The ever increasing need for robust and accurate assays for pathogen detection and genotyping, genomic mutation analysis, and gene expression profiling driven by advances in genomics during the past decade presents a unique opportunity for the application of basic research in nucleic acids chemistry, interfacial electrochemistry, and microengineering to the fabrication of practical sensing devices.

ACKNOWLEDGEMENTS

We wish to thank our many collaborators and students for their imagination, hard work, and insight. Work at Occidental College was supported by the David and Lucille Packard Foundation's Initiative for Interdisciplinary Research and the NIH (GM 33309). Work at Boston College was supported by GeneOhm Sciences (San Diego, CA) and the NIH (CA097945).

REFERENCES AND NOTES

1. Palecek, E., *J. Mol. Biol.* **1966**, *20*, 263.
2. Millan, K. M.; Mikkelsen, S. R., *Anal. Chem.* **1993**, *65*, 2317.
3. Millan, K. M.; Mikkelsen, S. R., *Anal. Chem.* **1994**, *66*, 2943.
4. Xu, X.-H.; Bard, A. J., *J. Am. Chem. Soc.* **1995**, *117*, 2627.
5. Wang, J.; Palecek, E.; Nielsen, P. E.; Rivas, G.; Cai, X.; Shiraishi, H.; Dontha, N.; Luo, D.; Farias, P.A.M., *J. Am. Chem. Soc.* **1996**, *118*, 7667.
6. Hashimoto, K.; Ito, K.; Ishimori, Y., *Anal. Chim. Acta* **1994**, *286*, 219.
7. Ju, H.-X.; Ye, Y.K.; Zhao, J.-H.; Zhu, Y.-L., *Anal. Biochem.* **2003**, *313*, 255.
8. Maruyama, K.; Mishima, Y.; Minagawa, K.; Motonaka, J., *Anal. Chem.* **2002**, *74*, 3698.
9. Steel, A.B.; Herne, T.M.; Tarlov, M.J., *Anal. Chem.* **1998**, *70*, 4670.
10. Steel, A.B.; Levicky, R.L.; Herne, T.M.; Tarlov, M.J., *Biophys. J.* **2000**, *79*, 975.

11. Lapierre, M.A.; O'Keefe, M.M.; Taft, B.J.; Kelley, S.O., *Anal. Chem.* **2003**, *75*, 6327.
12. Kelley, S.O.; Boon, E.M.; Barton, J.K.; Jackson, N.M.; Hill, M.G.; *Nucl. Acids Res.* **1999**, *27*, 4830–4837.
13. Boon, E.M.; Ceres, D.M.; Drummond, T.G.; Hill, M.G.; Barton, J.K., *Nature Biotech.* **2000**, *18*, 1096–1100.
14. Johnston, D.H.; Glasgow, K.C.; Thorp, H.H., *J. Am. Chem. Soc.* **1995**, *117*, 8933.
15. Ropp, P.A.; Thorp, H.H., *Chem. Biol.* **1999**, *6*, 599.
16. Armistead, P.M.; Thorp, H.H., *Anal. Chem.* **2000**, *72*, 3764.
17. Armistead, P.M.; Thorp, H.H., *Bioconjugate Chem.* **2002**, *13*, 172.
18. Holmberg, R.C.; Thorp, H.H., *Anal. Chem.* **2003**, *75*, 1851.
19. Li, J.; Ng, H.T.; Cassell, A.; Fan, W.; Chen, H.; Ye, Q.; Koehne, J.; Han, J.; Meyyappan, M., *Nano Lett.* **2003**, *3*, 597.
20. Wang J.; Cai X.; Rivas G.; Shiraishi, H.; Farias, P.A.; Dontha, N., *Anal. Chem.* **1996**, *68*, 2629.
21. Tarlov, M.J.; Steele, A.B. In *Biomolecular Films: Design, Function, and Applications*, Rusling, J.F., Ed., Marcel Dekker, Inc.: New York, 2003, Chap. 12, pp. 545–608.
22. Thorp, H. H., *Trends Biotech.* **1998**, *16*, 117.
23. Palecek, E.; Jelen, F., *Crit. Rev. Anal. Chem.* **2002**, *32*, 261.
24. Pividori, M.I.; Merkoçi, A.; Alegret, S., *Biosens. Bioelect.* **2000**, *15*, 291.
25. Wang, J., *Anal. Chim. Acta* **2002**, *469*, 63.
26. Drummond, T.G.; Hill, M.G.; Barton, J.K., *Nat. Biotech.* **2003**, *21*, 1192.
27. Murphy, L. *Curr. Op. Chem. Biol.* **2006**, *10*, 177.
28. Katz, E.; Willner, I., *ChemPhysChem* **2004**, *5*, 1085.
29. Yu, H.-Z.; Luo, C.-Y.; Sankar, C.G.; Sen, D., *Anal. Chem.* **2003**, *75*, 3902.
30. Anson, F.C.; Osteryoung, R.A., *J. Chem. Ed.* **1983**, *60*, 293.
31. Bard, A.J.; Faulkner, L.R., *Electrochemical Methods: Fundamentals and Applications*, John Wiley & Sons: New York, NY, 1980.
32. Connors, K.A. *Binding Constants: The Measurement of Molecular Complex Stability*, Wiley-Interscience: New York, NY, 1987.
33. Su, L.; Sen, D.; Yu, H.-Z., *Analyst* **2006**, *131*, 317.
34. Ceres, D.; Udit, A.K.; Hill, H.D.; Hill, M.G.; Barton, J.K., *J. Phys. Chem. B* **2007**, *111*, 663–668.
35. Heaton, R.J.; Peterson, A.W.; Georgiadis, R.M., *Proc. Natl. Acad. Sci., U.S.A.* **2001**, *98*, 3701.
36. Peterson, A.W.; Heaton, R.J.; Georgiadis, R.M., *Nucleic Acids Res.* **2001**, *29*, 5163.
37. Peterson, A.W.; Wolf, L.K.; Georgiadis, R.M., *J. Am. Chem. Soc.* **2002**, *124*, 14601.
38. Blaser, M.J., *J. Infect. Dis.* **1990**, *161*, 626.
39. Versalovic, J.; Shortridge, D.; Kibler, K.; Griffy, M.V.; Beyer, J.; Flamm, R.K.; Tanaka, S.K.; Graham, D.Y.; Go, M. F., *Antimicrob. Agents Chemother.* **1996**, *40*, 477.
40. Kelley, S.O.; Jackson, N.M.; Hill, M.G.; Barton, J.K., *Angew. Chem. Int. Ed.* **1999**, *38*, 941.
41. Hartwich, G.; Caruana, D.J.; de Lumley-Woodyear, T.; Wu, Y.; Campbell, C.N.; Heller, A., *J. Am. Chem. Soc.* **1999**, *121*,10803.
42. Jackson, N.M.; Hill, M.G., *Curr. Op. Chem. Biol.* **2001**, *5*, 209.
43. Kelley, S.O.; Barton, J.K.; Jackson, N.M.; Hill, M.G., *Bioconj. Chem.* **1997**, *8*, 31.
44. Kimura-Suda, H.; Petrovykh, D.Y.; Tarlov, M.J.; Whitman, L.J., *J. Am. Chem. Soc.* **2003**, *125*, 9014.
45. Poirier, G.E.; Pylant, E.D., *Science* **1996**, *272*, 1145.
46. Xu, S.; Liu, G.-Y., *Langmuir* **1997**, *13*, 127.

47. Xu, S.; Cruchon-Dupeyrat, S.J.N.; Garno, J.C.; Liu, G.-Y.; Jennings, G.K.; Yong, T.-H.; Laibinis, P.E., *J. Chem. Phys.* **1998**, *108*, 5002.
48. Boon, E.M.; Salas, J.E.; Barton, J.K., *Nature Biotech.* **2001**, *20*, 282.
49. Kelley, S.O.; Barton, J.K.; Jackson, N.M.; McPherson, L.D.; Potter, A.B.; Spain, E.M.; Allen, M.J.; Hill, M.G., *Langmuir* **1998**, *14*, 6781.
50. Boon, E.M.; Barton, J.K.; Sam, M.; Hill, M.G.; Spain, E.M., *Langmuir* **2001**, *17*, 5727.
51. SPM is considered to be a robust technique for accurate height measurement of features on a surface. See, for example: Lillehei, P.T.; Bottomley, L.A., *Anal. Chem.* **2000**, *72*, 189R.
52. Jackson, N.M.; McPherson, L.D.; Hill, M.G.; Spain, E.M., Unpublished results.
53. Rant, U.; Arinaga, K.; Fujita, S.; Yokoyama, N.; Abstreiter, G.; Tornow, M., *Nano Lett.* **2004**, *4*, 2441.
54. Ceres, D.; Barton, J.K., *J. Am. Chem. Soc.* **2003**, *125*, 14964.
55. Wierzbinski, E.; Arndt, J.; Hammond, W.; Slowinski, K., *Langmuir* **2006**, *22*, 2426.
56. Naegeli, R.; Redepenning, J.; Anson, F.C., *J. Phys. Chem.* **1986**, *90*, 6227.
57. Akkermans, R.P.; Roberts, S.L.; Marken, F.; Coles, B.A.; Wilkins, S.J.; Cooper, J.A.; Woodhouse, K.E.; Compton, R.G., *J. Phys. Chem.* **1999**, *103*, 9987.
58. Crothers, D.M., *Biopolymers* **1968**, *6*, 575.
59. Boon, E.M.; Barton, J.K.; Baghat, V.; Nersissian, M.; Wang, W.-D.; Hill, M.G., *Langmuir*, **2003**, *19*, 9255.
60. As described in ref. 62, this condition is satisfied when the adsorption dynamics are independent of the rate of electrode rotation.
61. Bhugun, I.; Anson, F.C., *J. Electroanal. Chem.* **1997**, *439*, 1.
62. Bourdillon, C.; Demaille, C.; Moiroux, J.; Saveant, J.-M., *J. Am. Chem. Soc.* **1999**, *121*, 2401.
63. Boon, E.M.; Wightman, M.D.; Kelley, S.O.; Hill, M.G.; Barton, J.K., *J. Phys. Chem. B*, **2003**, *107*, 11805.
64. Drummond, T.G.; Hill, M.G.; Barton, J.K., *J. Am. Chem. Soc.* **2004**, *126*, 15010.
65. Laviron, E., *J. Electroanal. Chem.* **1979**, *101*, 19.
66. Smalley, J.F.; Finklea, H.O.; Chidsey, C.E.D.; Linford, M.R.; Creager, S.E.; Ferrais, J.P.; Chalfant, K.; Zawodzinsk, T.; Feldberg, S.W.; Newton, M.D., *J. Am. Chem. Soc.* **2003**, *125*, 2004.
67. Niki, K.; Hardy, W.R.; Hill, M.G.; Li, H.; Sprinkel, J.R.; Margoliash, E.; Fujita, K.; Tanimura, R.; Nakamura, N.; Ohno, H.; Richars, J.H.; Gray, H.B., *J. Phys. Chem. B* **2003**, *107*, 9947.
68. Chi, Q.J.; Zhang, J.; Andersen, J.E.T.; Ulstrup, J., *J. Phys. Chem. B* **2001**, *105*, 4669.
69. Wang, A.H.-J.; Gao, Y.-G.; Liaw, Y.-C.; Li, Y.-K., *Biochemistry* **1991**, *30*, 3812.
70. Inosine lacks the N2 amino group necessary for the formaldehyde cross-coupling reaction.
71. Heller, A., *Faraday Discuss.* **2000**, *116*, 1.
72. Kittel, C., *Introduction to Solid State Physics,* 5th ed.; Wiley: New York, NY, 1976.
73. Boon, E.M.; Barton, J.K.; Pradeepkumar, P.I.; Isaksson, J.; Petit, C.; Chattopadhyay, J., *Angew. Chem. Int. Ed.* **2002**, *41*, 3402.
74. Friedman, K.A.; Heller, A., *J. Phys. Chem. B* **2001**, *105*, 11859.
75. Rajski, S.R.; Jackson, B.A.; Barton, J.K., *Mutat. Res.* **2000**, *447*, 49. (b) Merino, E.J., Barton, J.K., *Biochemistry* **2007**, *46*, 2805.
76. De Lumley-Woodyear, T,; Campbell, C.N.; Heller, A., *J. Am. Chem. Soc.* **1996**, *118*, 5504.
77. Campbell, C.N.; Gal, D.; Cristler, N.; Banditrat, C.; Heller, A., *Anal. Chem.* **2002**, *74*, 158.
78. Wang, J.; Liu, G.; Merkoci, A., *J. Am. Chem. Soc.* **2003**, *125*, 3214.

79. Jovanovic, S.V.; Simic, M.G., *J. Phys. Chem.* **1986**, *90*, 974.
80. Ozkan, D.; Erdem, A.; Kara, P.; Kerman, K.; Meric, B.; Hassmann, J.; Ozsoz, M., *Anal. Chem.* **2002**, *74*, 5931.
81. Asbury, C.L.; van den Engh, G., *Biophys. J.* **1998**, *74*, 1024.
82. Deamer, D.W.; Branton, D., *Acc. Chem. Res.* **2002**, *35*, 817.
83. Chen, C.-H.; Peng, E.-H., *Appl. Phys. Lett.* **2003**, *82*, 1308.
84. Park, S.-J.; Taton, T.A.; Mirkin, C.A., *Science* **2002**, *295*, 1503.
85. Fink, H.W.; Schonenberger, C., *Nature* **1999**, *398*, 407.
86. Porath, D.; Bezryadin, A.; de Vries, S.; Dekker, C., *Nature* **2000**, *403*, 635.
87. De Pablo, P.J.; Moreno-Herrero, F.; Colchero, J.; Herrero, J.G.; Herrero, P.; Baro, A.M.; Ordejon, P.; Soler, J.M.; Artacho, E., *Phys. Rev. Lett.* **2000**, *85*, 4992.
88. Storm, A.J.; van Noort, J.; de Vries, S.; Dekker, C., *Appl. Phys. Lett.* **2001**, *79*, 3881.
89. Kelley, S.O.; Barton, J.K., *Science*, **1999**, *283*, 375.

Chapter 6

CHARGE TRANSPORT OF SOLUTE OLIGONUCLEOTIDES IN METALLIC NANOGAPS – OBSERVATIONS AND SOME PUZZLES

Alexander M. Kuznetsov[1] and Jens Ulstrup[2]
[1]The A.N. Frumkin Institute of Physical Chemistry and Electrochemistry of the Russian Academy of Sciences, Leninskij Prospect 31, 119071 Moscow, Russia; [2]Department of Chemistry and NanoDTU, Technical University of Denmark, DK-2800 Lyngby, Denmark.

1. INTRODUCTION

Electron transfer (ET) through protein structures in solution, membrane environments, and in two-dimensional films at pure and modified metallic electrode surfaces have been discussed comprehensively in chapters of this volume. DNA-based molecules also transfer electronic charge (electrons or holes) through the molecular structures[1,2]. DNA-based ET in homogeneous solution is broadly restricted to the double-strand structure with dynamically induced high local stacking order as a likely pre-requisite[3,4]. DNA-conductivity has also been addressed broadly in interfacial electrochemical ET, cf. Chapter 5, and recently in scanning tunneling microscopy (STM) and nanogap electrode configurations. Interfacial electrochemical targets has involved ET across ordered self-assembled monolayers (SAMs) of the molecules[5,6] or through isolated individual (or a few) molecules[7–9], in either case, notably in aqueous buffer, as the natural environment of biological DNA-function. Observed conductivities suggest multifarious patterns. ***Both*** single- and double-strand forms conduct electrons (or holes) at the single-molecule scale. However, electrochemically controlled DNA-conductivity

O. Hammerich and J. Ulstrup (eds.), Bioinorganic Electrochemistry, 161–205.

has raised other issues no less challenging than for ET of DNA-based molecules in homogeneous solution.

The unique double-spiral structure of DNA inspired early views of DNA as an electronic conductor with a band structure resembling that of a one-dimensional semiconductor[10]. Introduction of semiconductor concepts in biological contexts with emphasis on DNA is frequently ascribed to Eley and Spivey[11] but related views were forwarded earlier[12,13]. Attempts to quantify such views theoretically by electronic band structure computations were forwarded by Ladik and others in the 1970's[14]. As for protein ET, present-day views on DNA-based conductivity rest instead on concepts and formalism from molecular ET science[15,16]. Chemical approaches to DNA-conductivity include concepts such as superexchange, "vibrationally assisted tunneling" or "hopping" between individual DNA bases or base pairs[1,2,17,18], and the concepts of small[19] and large[20] polarons. Other present-time approaches based on solid state and surface physics concepts have focused on analogies with semiconductor conductivity but are fraught with ultra-high vacuum conditions, high electric fields, and other non-biological environments detrimental to natural biological DNA function[21–23]. In comparison with proteins, DNA-based molecular electronic conductivity also displays differences which are only partly understood in spite of an impressive variability of sophisticated experimental data. Some notable phenomenological and physical differences between molecular ET patterns of proteins and DNA-based molecules are:

- DNA-based charge transport in homogeneous solution involves intercalation or in other ways tethered donor and acceptor molecular units[24] (cf. below). ET – or more commonly electronic ***hole*** transport but with increasing attention to ***electron*** transport[25–27] – is always initiated by donor or acceptor electronic excitation followed by excess positive (hole) or negative (electron) charge transmission through the ***double-strand***. The molecular excited state opens ET channels, denoted as "hopping" not immediately accessible in the ground state. ET in protein systems are also frequently, but far from always initiated by photo-excitation, cf. Chapter 1. Transmission of radical states through the peptide structure is, however, much less important.
- DNA-based hopping is remarkable by facilitating charge transfer over distances which are sometimes very long (100–200 Å) compared to all common ET notions.[28,29] Such channels are opened by the diffusive character of the hopping process, and resemble charge or excitation diffusion in other systems such as redox polymers[30] or light-harvesting antenna systems in the photosynthetic membrane systems.[31]
- High-energy radical states are also likely to be involved in one-dimensional single-molecule (or a few molecules) conductivity of DNA-based molecules in solid state configurations where the molecules are immobilized

between a pair of nanogap electrodes.[21–23] Conductivity recording is mostly in ultrahigh vacuum or other highly non-biological environment, with large, also non-biological bias voltages (≥ 1 V).

• In comparison, DNA-based molecules functionalized by suitable redox groups display efficient long-range interfacial ***electrochemical*** electron (or hole) transmission between the electrode and oligonucleotide-tethered redox groups under "mild" conditions, *i.e.* close to reversible electrochemical conditions in aqueous buffers, cf. Chapter 5 and below. This remains as one of the "puzzles" of DNA-based electronic transmission behaviour.

• As noted, DNA-based electronic conductivity rests on notions from theoretical molecular ET science such as tunneling, superexchange, hopping, polaron conductivity etc.[15,16] These notions offer a powerful rationale for a wealth of data particularly for ET of DNA-based molecules in homogeneous solution. However, as advocated,[3,4,32] this rationale cannot hold the whole story. DNA-based conductivity in solution seems to be hypersensitive to stacking disorder in the double-strands. Electronic charge (electron or hole) delocalization over a *sequence* of base pairs has, therefore, been advocated. Single-base charge localization and single-charge delocalization over a spatially extended range of the DNA-stack are interesting one-dimensional analogues of the small[19] and large[20] polaron known from the early days of solid state conductivity of polar crystalline materials.

• Charge delocalization and hypersensitivity of the conductivity to packing and lateral order may hold other clues to the observed remarkable insensitivity of interfacial ET of DNA-based molecules. Other notions such as columnar aggregation[33] and cationic condensation[34] can be other crucial factors in these intriguing observations.

• Interfacial ET and electronic conductivity are, finally, found to be extremely sensitive to the single- or double-strand state of the DNA-based molecules, with double-strand conductivity by far exceeding single-strand conductivity.[1–3,32] This is also the basis for exploiting DNA-based molecules in electrochemically based screening of biological liquids, cf. Section 4. The electronic states of immobilized single-strand molecules in ordered, almost two-dimensional crystalline domains are, however, not the same as in isolated, conformationally mobile single-strand molecules and resemble those of the ordered state in the double-strand.[5,6] This would induce higher single-strand conductivity than in the isolated, structurally disordered state of the single-strand molecules.

In addition to the fundamental importance of electronic conductivity of molecules as intriguing as DNA and proteins, electronic conductivity of DNA-based molecules holds other perspectives. Charge transfer along the DNA double-strand has been forwarded as an *in vivo* mechanism for radiation and chemical damage of the DNA-based genetic system,[35] and for possible repair mechanisms.[36] A second perspective has focused on the electronic conductivity to exploit DNA-based molecules in novel, next-generation ultra-small or even single-molecule device-like components. Such technology expectations are of two kinds.

One expectation has focused on the unique structure of DNA and hypersensitive conductivity dependence on the base composition as fundamental wires in ultra-small scale electronic circuits,[37–40]. Combination of the DNA-structure with attached redox or other functional groups would add other electronic features resembling amplification, rectification, and "switch" properties such as observed recently.[41–43] Such expectations are, however, fraught with prodigious challenges relating to the physical discrepancy between the aqueous biological media for natural DNA-function and the completely different environment required for DNA-function in putative electronic devices.

Electronic function of DNA-based molecules offers a second, still exciting but more pragmatic perspective in DNA-based biological sensing.[44–46] The DNA-based molecular targets and probes are now accommodated in real biological media. Recent studies have pointed to the feasibility of immobilizing DNA-based molecules in monolayers of controlled order on well-defined, *i.e.* atomically planar electrochemical surfaces directly in contact with the biological environment.[5,6,47] As noted, other studies have pointed to the extreme sensitivity of electrochemical signals to the single-, double-, and base pair mismatched states of the immobilized DNA-based molecular monolayers. Such observations hold an attractive combination of fundamental novel insight and novel biotechnology at ultra-small scales, extending to supra-molecular DNA-based molecular design and electronics.[37–40,48,49]

The double-strand structure of an oligonucleotide is shown schematically in Fig. 6-1. Anticipating discussion in later Sections, the molecule is shown in a upright orientation attached to an atomically planar metallic electrode surface (Au(111), cf. below) by chemisorption via a hexamethylenethiol group. Fig. 6-1 shows the four nucleobases presently in focus. We discuss first concepts and formalism of electron and hole transport of DNA-based molecules in homogeneous solution and at electrochemical interfaces. We then focus on DNA-based molecules in electrochemical nanogaps and STM in electrochemical environments (*in situ* STM). Some case examples illustrate accordance and limitations of current theoretical views of DNA-conductivity. This adds to the comprehensive overview of interfacial electrochemical ET of DNA-based molecules by O'Kelly and Hill in Chapter 5.

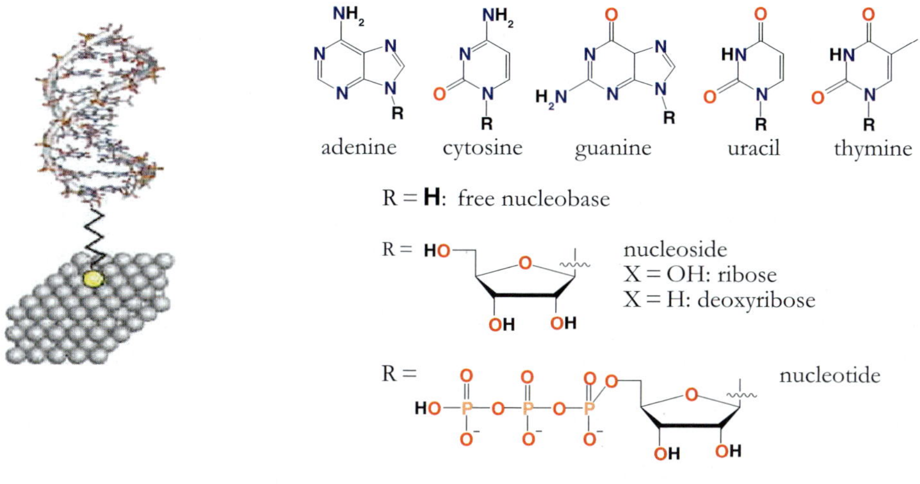

A B

Figure 6-1. Left: Schematic view of a double-strand oligonucleotide linked to an atomically planar metal surface via an oligo-methylenethiolate linker. The yellow sphere attaching the linker and the oligonucleotide to the surface is a sulfur atom. Right: The four nucleobases and their single-base nucleosides and nucleotides.

2. ELECTRON AND HOLE TRANSFER OF DNA-BASED MOLECULES IN SOLUTION

2.1 Patterns for Electron ("Hole") Hopping in DNA-Based Molecules

Broad attention to electronic charge transfer of oligonucleotides in homogeneous solution was triggered by pioneering studies of J. Barton and associates.[28,29] The initial observation that charge (in these cases "holes") could be transferred rapidly over up to several hundred Å appeared at first counter-intuitive to all expectations from molecular charge transfer. An impressive variety of other experimental studies based on creative molecular design, chemical analysis, and kinetic studies into the pico-second time ranges by several groups,[1–4,25–27,32,50–57] has followed and offered multifarious intramolecular charge transfer mapping of DNA-based double-strand molecules.

Concepts and formalism of interfacial ET of DNA-based molecules at electrochemical interfaces and in nanogap systems rest on data and theory of long-range charge transfer in homogeneous solution.[15,16] Prior to addressing interfacial charge transfer of DNA-based molecules as the focus of this Chapter, we provide a few observations on DNA charge transport in solution. These are meant to introduce some conceptual support for Sections 3 and 4. Comprehensive recent reviews of this broad field are available elsewhere.[1,2]

Crucial to most studies of intramolecular electron or hole transfer of DNA-based molecules in solution is that the ET process is triggered by photochemical excitation of a sensitizer molecule, attached covalently or by intercalation to the DNA-based molecule. Photo-excitation leads to ***electron transfer*** from nearby bases, leaving a "hole", *i.e.* a localized oxidized nucleobase or nucleobase cluster, while the excited electron on the sensitizer molecule can be annihilated by chemical reaction with solute or solvent molecules in the environment. This high-energy charge carrier injection triggers charge transport within the DNA-based molecule where electronic "holes", dominated by guanine (G) cation radicals, are transmitted along the strand by diffusive random motion. At some stage, *i.e.* some distance from the primary injection site, the migrating "hole" is trapped and further migration stopped. This mechanism has been mapped in great detail by the comprehensive work of Barton *et al.*,[3,4,28,29,32] Lewis *et al.*,[50–52] Giese *et al.*,[53,54] Michel-Beyerle *et al.*,[55,56] Kawai *et al.*,[57] and others, recently reviewed e.g. in refs. 1 and 2.

Some further notions are:

- Hole transfer is effected in sequential hole transfer steps, mostly between the guanine bases which have the lowest oxidation potentials, the base order following ($G < A < C \approx T$; G = guanine, A = adenine, C = cytosine, T = thymine).[58,59]
- In addition to externally inserted trapping sites, hole trapping occurs at sites of successive guanines, *i.e.* at GG and GGG sequences. Local oxidation potentials depend on the local environment including the bases that enclose the G-based hole carriers.[60,61] GG and GGG thus have lower oxidation potentials than G isolated between A-, C- or T-bases.
- A, C, and T represent barriers for hole transfer in the sense that hole transfer (oxidation) from G^+ to any of the other bases is endothermic, by up to half an eV. Successive hole transfer between the guanines is therefore mostly by electronic (hole) tunneling through a barrier represented by the highest occupied molecular orbital (HOMO) of the intermittent bases. The mechanism based on electronic coupling between successive guanines in the pathway, via the intermediate bases is what is denoted as ***superexchange***. In some cases oxidized states of the other bases, however, still appear as populated high-energy electronic states in the hole conduction process by

hopping,[54], as long as they are followed by subsequent exothermic steps towards a G-based hole trapping site.

• Most attention has been given to ***hole transfer. Electron transfer***, *i.e.* injection of excess electrons rather than holes is in increasing focus and has been shown to display a number of similarities to patterns of hole transfer.[25–27]

2.2 A Primer of Electron and Hole Transfer Formalism for DNA-Based Molecules

2.2.1 Superexchange and Localized Electron or HoleTransport

A suitable fundamental rate constant (k s^{-1}) form for a molecular charge (*i.e.* electron or hole) transfer step in homogeneous solution is,[15,16] cf. Chapters 1 and 8

$$k = \kappa_{el}\frac{\omega_{eff}}{2\pi}\exp\left[-\frac{\left(E_r + \Delta G^0\right)^2}{4E_r k_B T}\right] \qquad (6\text{-}1)$$

where $\kappa_{el} = \left(T_{DA}\right)^2\sqrt{\dfrac{4\pi^3}{E_r\hbar^2\omega_{eff}^2 k_B T}}$ if $\kappa_{el} << 1$; $\kappa_{el} = 1$ otherwise

ΔG^0 is the reaction free energy (driving force) of the individual charge transfer step, E_r the total local and environmental reorganization free energy, and ω_{eff} the effective vibrational frequency of the nuclear modes reorganized. κ_{el} is the electronic transmission coefficient the most important component of which is the electron exchange factor, T_{DA} that couples the electron or hole donor and acceptor state by a tunneling process via the intermediate high-energy charge conveying "matter". The "matter" is nucleobases the highest occupied (HOMO) or lowest unoccupied molecular orbitals (LUMO) of which are too far off resonance with the donor and acceptor states to become physically populated in the hopping process.

Two other observations will be useful. First, subject to the condition κ_{el} << 1, an operational form of T_{DA} is, cf. Chapters 1 and 8

$$\left(T_{DA}\right)^2 = \left(T_{DA}^0\right)^2 \exp(-\beta R) \tag{6-2}$$

where T_{DA}^0 is a constant energy quantity and R the structural distance between the donor and acceptor. The decay factor β can be related to the intermediate nucleobase properties by the following simple form, cf. Chapter 8.

$$\beta \approx \frac{2}{a} \ln\left(\frac{\Delta\varepsilon\left(\Delta G^0\right)}{\gamma}\right); \quad \left(T_{DA}^0\right) \approx \frac{\gamma_D \gamma_A}{\gamma} \tag{6-3}$$

This form illustrates the superexchange coupling via off-resonance intermediate electronic states. $\Delta\varepsilon(\Delta G^0)$ is the energy gap between the electronic donor and the intermediate HOMO or LUMO states. The quantity $\boldsymbol{a}$ is the average structural extension of an intermediate base group. In the simple form of eq.(6-3) the same average energy is taken for all the intermediate (nucleobase) states. $\Delta\varepsilon(\Delta G^0)$ varies with the driving force. In eq.(6-3) γ is, further the (average) electronic coupling between nearest-neighbour bridge group states, while γ_D and γ_A are the electronic couplings between the donor and the first bridge group, and between the acceptor and the last bridge group in the chain, respectively. The couplings and energetics have been extensively addressed by quantum chemical computations that have *i.a.* substantiated that $\beta < \Delta\varepsilon(\Delta G^0)$ (Chapters in ref. 2b) which is an obvious validity condition for eqs.(6-2) and (6-3).

The second observation is that photochemically induced electron or hole transfer involves high energies and driving forces. Charge transfer patterns that extend into the "inverted" free energy region, $\left|\Delta G^0\right| > E_r$ have indeed been observed.[51,52] These are accommodated by the following extension of eq.(6-1)[15,16]

$$k \approx \kappa_{el} \sum_{w=0}^{\infty} \left|\left(X_w \middle| X_0\right)\right|^2 \frac{\omega_{eff}}{2\pi} \exp\left[-\frac{\left(E_r + \Delta G^0 + \varepsilon_w - \varepsilon_0\right)^2}{4E_r k_B T}\right] \tag{6-4}$$

Equation (6-4) implies that excited vibrational states in local high-frequency modes characterized by the vibrational quantum numbers v and w in the initial and final electronic states, respectively become important when $\left|\Delta G^0\right| > E_r$. The vibrational energies are ε_v and ε_w, and the vibrational wave functions X_v and X_w. The initial state vibrational quantum number can be taken as $v = 0$ representative for most cases encountered. When the local nuclear modes are represented by displaced harmonic oscillators, the following equations apply

$$\kappa_{el}^{w0} = \left(T_{DA}\right)^2 \frac{1}{w!} \exp\left(-\frac{E_r^{qu}}{\hbar\omega^{qu}}\right); \; \varepsilon_p = \hbar\omega^{qu}\left(p+\tfrac{1}{2}\right);$$

$$p = v, w \tag{6-5}$$

where E_r^{qu} and ω^{qu} is the reorganization energy and frequency of the local high-frequency mode, respectively, and $\hbar$ is Planck's constant divided by 2π.

The rate constant drops with increasing $\left|\Delta G^0\right|$ in the inverted free energy region, in contrast to the rate constant increase in the "normal" region, $\left|\Delta G^0\right| < E_r$. The most important effect of the local high-frequency nuclear modes is that the rate constant falls off more slowly with increasing $\left|\Delta G^0\right|$ as excited vibrational states of the high-frequency mode(s) open new decay channels of increasing importance as $\left|\Delta G^0\right|$ increases.

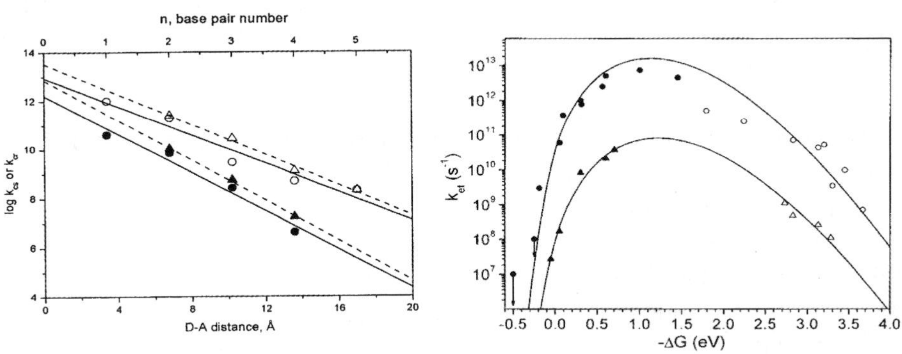

Figure 6-2. Left: Correlations between photo-induced hole transport rate constants and structural distances in synthetic capped hair-pin double-strand oligonucleotides for charge separation (open symbols) and charge recombination (filled symbols). Right: Free energy plots for charge separation (filled symbols) and recombination (open symbols) Circles and triangles represent different sensitizer molecular units. Reprinted from ref. 52 with permission.

Studies of the driving force and distance dependence between donor and acceptor centres in DNA-based systems have been based on transient spectroscopy and elucidation of trapping site yields, and offered extensive mechanistic mapping based on eqs. (1)–(5). β-values in the range 0.6–1.6 Å^{-1} testify to the suitability of a superexchange view and the use of eqs. (6-1)–(6-5) as a conceptual and formal frame. The studies of F.D. Lewis and associates based on a variety of well-defined double-strand nucleotides capped by hole-injecting sensitizer molecular groups illuminate the electronic coupling features and driving force relations in superexchange-like patterns, Fig.6-2.[50–52]

2.2.2 The Notion of Hole or Electron Hopping

Different high-energy HOMO or LUMO compositions between the G-based hole transmitters provide different rate constant patterns reflected in formally different β-values.[1,2,28,29,50–52,57,62,63] The variability of expectable superexchange-based values and particularly the commonly observed very small β-values (< 0.2 Å^{-1}) associated with (very) long-range charge transfer prompted the introduction of other mechanistic views, pioneered by Bixon, Jortner, Michel-Beyerle and their associates.[17,18] In hindsight these views follow naturally from the properties of the DNA-based molecules and their charge-injecting derivatives, as well as from the nature of the ***photochemical***, *i.e.* high-energy charge-inducing processes. These notions are that charge, particularly hole "hopping" along the double-strand DNA-stack as a photochemically induced conduction mechanism carries the charge transport. Oxidation potentials of DNA-bases are high and reduction potentials low but charge injection from excited sensitizer electronic states overcome the thermodynamic barriers, leading to facile long-range hole or electron transfer along the stacks.

Hopping along the DNA-stacks has attracted other comprehensive experimental and theoretical efforts with hopping of electronic "holes" in focus with guanine mediators as the most easily accessible cation radical states, and GG and GGG as primary "hole trappers". One-dimensional diffusion-like, random hole walk along the DNA-structure with longer time trapping at GG doublets or GGG triplets is the primary notion. Diffusion is interrupted at these sites with interfering chemical decay channels where the guanine doublets and triplets are irreversibly destroyed by chemical reaction with water.

The scheme in Fig. 6-3 shows the charge transfer and chemical schemes involved. Charge transfer along the chain is interrogated by irreversible chemical processes at the GG or GGG sites, Fig. 6-3. Interestingly, conceptually and formally this resembles the stochastic view of electron transfer in a multi-mode system.[16,64] Diffusive motion along a "slow" solvent or biological conformational nuclear mode is here interrupted by faster, undamped reactive motion along a local higher-frequency nuclear mode of the potential surface, Fig. 6-4.

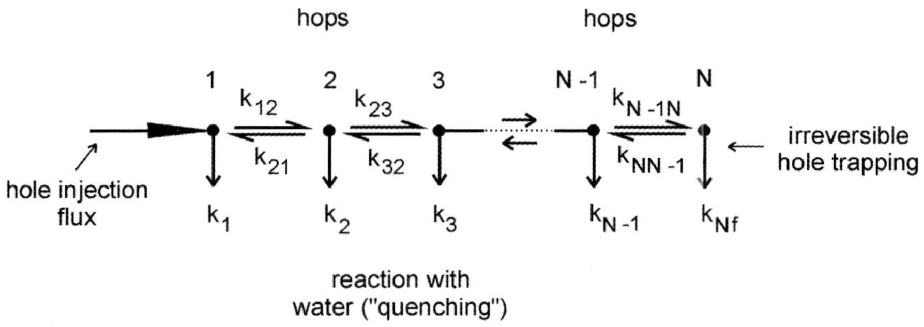

Figure 6-3. Schematic view of photo-induced electron or hole transport through the DNA-based molecular chain. Reversible random walk of the charge along the chain is interrupted by irreversible chemical degradation ("quenching") at given sites. Charge (*i.e.* electron or hole) injection at the terminals corresponds to interfacial charge transfer in the *in situ* STM or nanogap electrode configurations to be discussed in Section 4.

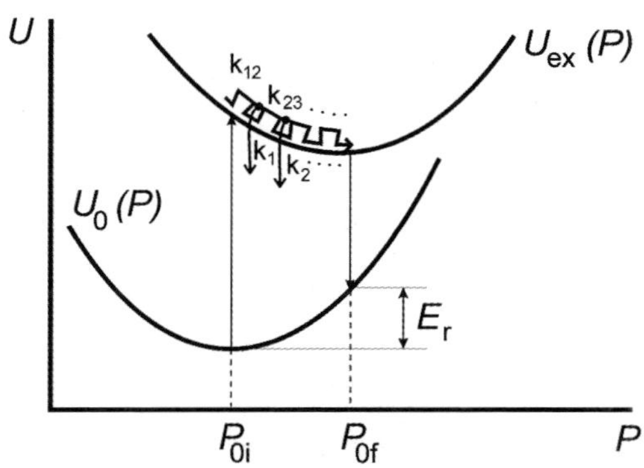

Figure 6-4. Molecular charge (electron or hole) transfer motion along a "slow" diffusion-like solvent or conformational nuclear mode with electronic reactive "spill-over" along a fast, friction-less local nuclear mode. Analogy to the diffusive electron or hole transfer shown in Fig. 6-3.

Excess charge hopping has been formalized by Bixon and Jortner,[17,18,65–67] and later by Ratner and others.[68–71] We provide elements of such a formalism in Section 5, explicitly in the novel context where DNA-based molecules are immobilized in well-defined configurations between enclosing electrochemical *in situ* STM substrate and tip electrodes or similarly configured electrochemical nanogap electrodes. As a prologue we note a simple rate constant form that incorporates the hopping mode essentials offered by Bixon, Jortner, and their associates[17,18] based on a formalism for diffusive energy transfer by Klafter and associates.[72,73] This form represents a simplified version of the scheme in Fig. 6-3 where the diffusion hops and depletion rates are each characterized by a single rate constant, k_{hop} and k_{dep}, respectively. The charge separation process is characterized by two observables, *viz.* the time for charge separation over the spatial distance corresponding to N diffusive hops, τ_{cs}^{N}, and the yield, Y^{N}, for charge transport from the initial injection site to the N'th site in competition with the depletion processes. These quantities are related, giving

$$\tau_{cs}^{N} \approx \left(k_{hop}\right)^{-1} N^{\eta}; \quad Y^{N} = \left[1 + \left(k_{hop} / k_{dep}\right)^{-1} N^{\eta}\right]^{-1} \tag{6-6}$$

The exponent $\eta \approx 2$ for weakly thermoneutral charge injection while $1 \leq \eta \leq 2$ for exothermic charge separation with $\eta \to 1$ for more strongly exothermic processes.

Equation (6-6) shows a weak, inverse distance dependence of the hopping charge constant and yield. However, eq. (6-6) also raises issues regarding the nature of the hopping sites, say whether these are confined to individual nucleobases or base pairs, or delocalized over a sequence of base pairs, to which we proceed next.

2.2.3 Polarons and Collective Base Stacking as Charge Transmitting Intermediate States

The photo-induced charge conduction mechanism in double-strand guanine-rich DNA-based molecules, and AT pairs attenuating the conductivity by off-resonance superexchange steps is broadly supported. Mechanistic mapping has been based on both direct charge transfer kinetics in the nano-, pico- and femtosecond ranges[1,2,50–52,74,75] and by comprehensive yield analysis of the trapping sites.[1,2,53–57] Kinetic analysis framed by molecular charge transfer theory[17,18,65–76] and computational efforts towards energetics and electronic coupling among individual[76] base pairs including their immediate base (pair) environment have added to this view. At the same time other observed patterns for electronic charge transfer leave "puzzles" when the charge carrier sites are referred to hopping or superexchange solely along sequences of individual bases or base pairs.

Three lines of observation have indicated that other features may have to be added:

• In hole transport studies by Barton and associates involving sophisticated designed inserted trapping sites, charge transfer and trapping was found not to follow simply the individual base (pair) energetics.[4,32] Orbital mixing delocalized over a sequence of bases was instead forwarded as charge transmitting sites. The nature of the delocalized charge transfer sites is "dynamic" with the delocalized domains forming and breaking up as charge transfer progresses. The transient nature of the delocalized charge transfer states prompts conceptual and physical relations to two other elements of molecular charge transfer theory, *viz.* the notions of gated charge (hole or electron) transfer and of the large polaron.[15,16]

• Based on computed electronic structures of groups of stacked bases at the *ab initio* level,[77] Schuster and associates[78,79] offered the suggestion that sequences, *i.e.* groups of three to five base pairs rather than individual base pairs constitute the sites for temporary excess charge trapping. Charge delocalization leads both to lower and strongly leveled energies of the trapped charge at the successive sites.

As for individual bases, trapping over a base (pair) sequence would be accompanied by configurational distortion in the nuclear environment. Solvent and high-frequency local modes would dominate in electron and hole transfer trapping at individual bases. The excess charge confinement to a single "lattice site" is the basis for the "***small polaron***" notion[19] which is, however, less important in delocalized multiple-base charge trapping where collective base stacking modes such as inter-base distances would dominate.[78–81]

The mechanism of extended charge localization is reminiscent of electronic conduction mechanisms based on the ***large polaron*** introduced in solid state physics by Pekar in the late 1940's.[20] The large polaron is, physically an excess electronic charge localized in a polar solid (or liquid) but extending over several crystal lattice sites and with a shallow energy well. The localization mechanism is based on environmental nuclear (inertial) polarization induced by the excess charge which is in turn stabilized by the electric polarization field induced by the charge itself. Large polaron-based conductivity is close to activation-less and the electronic charge drift follows adiabatically the nuclear configurational fluctuations under the influence of the external electric field. In terms of molecular charge transfer theory, large polaron conductivity belongs to the regime of strongly adiabatic charge transfer, Chapter 8 in the sense that the quantity

$$\kappa_{el} = \left(T_{DA}\right)^2 \sqrt{\frac{4\pi^3}{E_r \hbar^2 \omega_{eff}^2 k_B T}} \qquad (6\text{-}7)$$

in eq.(6-1) significantly exceeds unity (nuclear tunneling is here unimportant). The mechanism still involves localized charge carrier states, but spatially extending beyond the single base pair. The electronic-vibrational coupling, *i.e.* the nuclear reorganization free energy, E_r thus exceeds sufficiently the electron exchange coupling, T_{DA} that the excess charge is trapped in the potential wells of the individual but extended sites. A completely delocalized conduction band mechanism would emerge if E_r and T_{DA} assume comparable values.

In recent times the large polaron concept has framed for example electronic conductivity of conducting polymers.[82–84] Large polaron conductivity along the base stack with inter-base distance fluctuations as the dominating nuclear trapping modes has received support from other recent electronic structure computations on long stacks by Conwell and associates.[80,81] Notable observations are that strong inter-base distance deformations, up to half an Ångström and extending to half a dozen bases were indeed found to accompany the hole conductivity process.

Other observations are:

• The difference between the pre-stacked delocalized charge carrier states and the large polaron view is that environmental nuclear reorganization is prior to the hopping step in the former, but part of the charge relocation in the latter. In molecular charge transfer notions emphasis in the former is on "gating" and on environmental reorganization inside the base (pair) stacks in the latter. There is no need for this distinction. Both are accommodated within the comprehensive frames of molecular charge transfer theory,[15,16,64] (cf. Chapter 8).

• Configurational fluctuations towards highly non-equilibrium configurational states are crucial pre-requisites in all the mechanisms. The importance is illuminated by a recent combined Molecular Dynamics and electronic structural study.[85] Inclusion of the solvent and counter ion dynamics in suitably simplified representations leads to drastic energetic fluctuations in the charge carrying individual bases, up to almost an eV. Such conspicuous effects carry over to gating and reorganization in both hopping and superexchange mechanisms of DNA-based conductivity. This applies whether transmission is by electronic states localized on individual bases or base pairs, or by delocalized large polaron-like states.

• The need to consider gated delocalized electronic states as charge transmitting entities, with fewer hops, better electronic overlap and efficient energy gap leveling is no less important in DNA-conductivity addressed by electrochemical techniques. These have come to include the novel techniques of *in situ* STM and nanogap electrode pairs, in either case under full electrochemical potential control. These approaches are targeted towards the conductivity of DNA-based molecules directly in aqueous biological media under "mild", little invasive conditions such as low bias voltages and electrochemical overpotentials. These approaches are in contrast to DNA-conductivity studies based on vacuum conditions, with very high bias voltages (up to several volts) and no electrochemical potential control.[21–23]

3. INTERFACIAL ELECTROCHEMICAL ELECTRON TRANSFER THROUGH DNA-BASED MOLECULES

Charge transfer, viewed as electron or hole transfer along double-strand DNA-based molecules in homogeneous solution is broadly understood rooted in comprehensive data variability. Open questions remain, particularly as to the precise nature of the otherwise attractive concepts of stacking or large polaron charge transmitting states. Double-strand DNA-based molecules also transmit electrons efficiently in interfacial electrochemistry and electrochemical *in situ* STM over distances that by established molecular charge transfer concepts would be expected to be prohibitive. These "puzzles" constitute here no smaller challenges than DNA-based charge transport in homogeneous solution. In particular, electronic ***ground state*** properties now control the charge transport, in contrast to most reported studies of DNA-based electron transport in solution where ***excited electronic states*** trigger the process. Interfacial electrochemistry of nucleobases and polynucleotides was pioneered by Palaček and associates already from the 1960's (for an overview, see ref. 44) and pursued until present times. To this adds impressive efforts towards the exploitation of DNA-based electrochemical activity in the broad area of *in situ* DNA-based electrochemical sensors. These have displayed ever increasing degree of sophistication over the last decade or two,[44] with an ultimate objective to detect and control to the highest possible resolution one of the most central of life processes, the DNA-duplex formation. With the prospects of insertion of DNA-based molecules into electrochemical *in situ* STM and nanogap electrode configurations, this level of resolution can perhaps even be extended to the single molecule.

Interfacial electrochemistry and electronic conductivity of double-strand oligonucleotides is discussed in detail in Chapter 5. The overview below summarizes some additional observations of interfacial electrochemical adsorption, organization, and charge transfer, of importance to approaches towards single-molecule electrochemical conductivity and *in situ* STM of DNA-based molecules to follow in Sections 4 and 5.

3.1 Surface Immobilization and Surface Adsorbate Organization

Single- and double-strand oligonucleotides can be immobilized on gold electrodes via variable-length oligomethylene-thiol-based linker groups bound to the 5'- or 3'-ends of the oligonucleotide. Tarlov and associates,[86–89] Barton, Hill and associates,[90–93] Tornow and associates,[94,95] and Gooding and associates[46,96–98] have been prime movers in the development of surface electrochemical coverage characterization of the resulting monolayers. Methods used rest *i.a.* on spectroscopic (XPS) and voltammetric methods, cf. below. A number of issues arise in the rational approach to oligonucleotide immobilization on electrochemical, say gold surfaces of which the following stand forward, cf. Section 4:

• Aromatic nucleobases in single-strand variable-length and – composition thiol-immobilized oligonucleotides adsorb strongly on gold and other electronically "soft" electrode materials. This was studied and quantified in detail by Tarlov and associates by radio-labeling, neutron reflectivity, Fourier Transform infrared spectroscopy and XPS.[86–88] A key observation in the present context is that the layers are dense with both specific adsorption via the thiol linker and with "unspecific" adsorption via the nucleobase contacts with the electrode surface. Large parts of the oligonucleotide molecules can be replaced by the linker group analogue, 6-mercapto-1-hexanol (MCH), cf. Section 4. Partial replacement lifts the non-specifically adsorbed nucleotide moieties off the surface from random unspecific adsorption and disperses the molecules in the MCH matrix. In this way separated single-strand uncoiled and upright oligonucleotide molecules directly in solution become available for chemical reactivity, the most important of which would be duplex formation with complementary sequences in the solution contacting the surface.

• Nearly all approaches to thiolated oligonucleotide adsorption on gold electrodes have rested on ***polycrystalline*** electrode surfaces. Use of ***single-crystal***, atomically planar surfaces open other perspectives. One is that the scanning probe microscopies, STM and atomic force microscopy directly in aqueous buffer (*in situ* STM and AFM) can then be introduced to characterize and control adsorbate organization and dynamics at new levels of resolution. We shall address this in Section 4. Another perspective is that single-crystal electrodes open for voltammetric exploitation of the reductive desorption process of the Au-S linking unit[99]

$$\text{Au-SR} + e^- \rightarrow \text{Au} + {}^-\text{S-R}; \quad (\text{R} = \text{oligonucleotide residue}) \qquad (6\text{-}8)$$

This process gives sharp voltammetric peaks that offer e.g. precise surface coverage determination at well-defined surfaces. Use of single-crystal, say Au(111)-surfaces has substantiated that densely packed thiolated oligonucleotide monolayers, almost resembling a molecular "brush", cf. section 4 are formed.[1,2,100] A similar conclusion was reported by Palaček and associates who exploited the intrinsically atomically planar nature of a liquid mercury electrode surface.[101] The notion of two-dimensional condensation has been invoked for such phenomena.[34]

• The data just noted refer largely to single-strand oligonucleotide molecules. Shorter, say ≤ 10 base pair double-strand oligonucleotides otherwise stable in solution tend to dehybridize in the electric field at the electrochemical interface[102] but longer, say ≥ 15 base pair double-strand oligonucleotides adsorb in stable monolayers at (most commonly polycrystalline) electrochemical gold surfaces.[90–98] The double-strand oligonucleotide molecules are in a upright orientation in the sense that they are free of direct physical contact with the electrode surface. Several studies have illuminated this issue which holds a particular perspective in relation to the mechanism of electron exchange between the electrode and redox markers at well-defined positions along the double-strand oligonucleotide structure.

Barton, Hill, and their associates studied the electrochemical potential dependence of the apparent height (thickness) of a 15-base pair thiol-linked double-strand oligonucleotide of mixed composition (5 nm long) by electrochemical AFM.[103] The average tilt angle determined by comparison of the apparent and structural heights discloses an angle of 45° at the open circuit potential. The apparent height could be switched from 2.0 to 5.5 nm on changing the electrochemical potential from values positive to values negative of the potential of zero charge. This corresponds to a switch from molecules lying almost flat on the surface to an almost upright orientation.

Tornow, Rant and associates addressed the electrochemically controlled double-strand oligonucleotide orientation in a different way, based on fluorophor-marked surface immobilized double-strand target molecules.[94,95] The fluorescence is quenched by a nearby metal surface, giving a lower fluorescence yield the closer the metallic surface is to the fluorophor. This could be exploited to map aspects of the dynamics of adsorbed double-strand oligonucleotide molecules by the rise or lowering of the fluorescence yield as the oligonucleotide orientation is taken from an upright to a recumbent orientation by the electrostatic field variation at the electrochemical interface.

Investigations such as these are important in interfacial DNA-based electronic conductivity, because they have mapped the local environments for the electronic conductivity of the DNA-based molecules directly in the natural aqueous environment of biological DNA function. Redox markers are here crucial conductivity monitors.

3.2 DNA-Bound Redox Groups as Probes for Electrochemical DNA-Based Conductivity

Electrostatic binding or intercalation of electrochemically active redox groups to single- or double-strand oligonucleotide monolayers were introduced to quantify the oligonucleotide surface coverage.[44] The basis for the surface coverage determination is the electrostatically based stoichiometric ratio between strongly positively charged cationic redox probes such as $[Ru(NH_3)_6]^{2+/3+}$ and the negatively charged phosphate groups in the DNA-backbone. Cyclic voltammetry and chronocoulometry of the oligonucleotide-bound redox groups record the redox counter ion and therefore the phosphate coverage. These studies also touch directly on the mechanism(s) of DNA-based molecular conductivity in the electrochemical environment. Stoichiometric binding and voltammetric equivalence must necessarily involve either efficient conductivity ***through*** the oligonucleotide target molecules, be it single- or double-strand, or efficient hopping through a chain of electrostatically bound redox markers, cf. Section 4. We note presently some selected observations on single-redox group marked thiol-derivatized oligonucleotide monolayers that testify to unique electronic conductivity properties of the ordered double-strand form and the hyper-sensitivity to base mismatches or other structural disorder. Studies by Barton, Hill and their associates[90–93] and by Gooding and associates[46,96–98] illuminate these points. Other studies have been reported e.g. by Kraatz.[104] Some points with a bearing on the electrochemically controlled electronic conductivity of DNA-based molecules are:

• The electrochemical current form, equivalent to eqs.(6-1)–(6-3) is discussed in Chapter 8. The following is a useful frame

$$j(\eta) = 8e\Gamma_{ox}^{(1-\alpha)}\Gamma_{red}^{\alpha}\sqrt{\frac{\pi^3 k_B T}{E_r \hbar^2 \omega_{eff}^2}}\frac{\omega_{eff}}{2\pi}\left(\frac{\beta_{NA}}{\beta}\right)^2 \Delta_1 \times$$

$$\exp\left[-\frac{2}{a}\ln\left(\frac{\Delta\varepsilon(\eta)}{\beta}\right)R\right]\exp\left[-\frac{(E_r + e\eta)^2}{4E_r k_B T}\right]. \tag{6-9}$$

The electronic coupling between the linking group and the electrode surface is represented by the energy broadening of the linker group levels, Δ_l, Γ_{ox}

and Γ_{red} are the coverages of the oxidized and reduced form of the redox probe, respectively, e is the electronic charge and the driving force ΔG^0 is replaced by the overpotential η, *i.e.* $\Delta G^0 \rightarrow e\eta$. α is the electrochemical transfer coefficient, *i.e.* $\alpha = -k_B T d\ln j(\eta)/d(e\eta)$.

• Studies of double-strand oligonucleotides of variable length and composition, site-specifically marked by redox probes and immobilized on electrochemical (polycrystalline) Au-electrodes have been reported.[90–98,105,106] (cf. Chapter 5). Ferrocene and organic redox probes such as methylene blue and condensed aromatic quinones such as daunomycin have been redox probes. The immediately most striking observation is the virtually complete independence of the electrochemical rate constants on both the oligonucleotide length and even on the base pair composition. Reversible voltammetry was observed over a range of distances long enough to take expectably the electrochemical process well into the strongly diabatic range of weak electrode/redox probe interaction. This is completely at variance with expectations from molecular charge transfer notions such as represented, e.g. by eq. (6-9) in which both energetics and electronic couplings are strongly reflected. At the same time, ordering in the base stacking appears crucial as just a single base pair mismatch between the electrode surface and redox probe turns off the electronic communication completely. It was shown in other reports that the linker group, most conspicuously the length of the group represents a tunneling barrier with strong distance decay.[105] In still other reports interfacial electrochemical electron transfer was shown to be conducted through the base pairs and not the sugar and phosphate parts of the DNA-based backbone.[106]

• Gooding and associates addressed the issue of the electronic conductivity of DNA-based molecules with focus on the hybridization process and electrochemical "recognition surfaces".[96–98] Single-strand oligonucleotide molecules dispersed in a two-dimensional MCH diluent matrix, cf. above and Section 4 constituted a central strategy. As noted this approach has a direct bearing on electrochemical biosensing of the duplex formation. Efficient hybridization of 20-base pair hybridization could be detected by $[Ru(NH_3)_6]^{2+/3+}$-based electrostatic binding and by intercalation-based labeling using several organic and transition metal complex redox probes. These two redox probe marking principles are different. Electrostatic labeling induces voltammetric signals in ***both*** single- and double-strand forms with different intensities according with the different numbers of electrostatic binding sites (phosphate groups) in the two forms. Intercalation monitors specifically the double-strand forms unless ordered helical single-strand organization can be prepared, cf. Section 4.1.

The approach by Gooding and associates has proven sensitive to hybridization detection at the macroscopically averaged monolayer level, and single base pair mismatches could be detected. An interesting observation was that the crystallographically and in other ways known partial unwinding and kinked structure in the double-strand DNA-structure caused by *cis*-platin (*cis*-diammine-dichloro-platinum(II)) binding[107] could be directly detected by a significantly attenuated voltammetric signal of the anthraquinone-based intercalator.[108] Such data support the special conductivity properties of an ordered DNA-based stack. They also hold perspectives for DNA-based biosensing specifically based on the double-strand molecular electronic conductivity. Other important mechanistic conductivity issues such as the precise role of the redox probe, multiple redox probe binding, and electrostatic or intercalator site distributions, however, remain.

Studies such as these have addressed fundamental issues regarding DNA-based interfacial electrochemical charge transfer. The conclusions carry over to the evolving area of single-molecule DNA-based conductivity as addressed by electrochemical *in situ* STM[5,6,47,109] and DNA-target molecules inserted in nanogap electrode configurations.[110–114] Neither superexchange between the electrode and the quite distant redox probe entities nor hopping, such as represented by eqs. (6-1)–(6-3) and (6-9), however, accord easily with the observations. Charge transfer in the superexchange mode in the strongly diabatic limit of weak electrode/redox probe interaction and "normal" tunneling decay factors, say in the range 0.5-1.0 $Å^{-1}$ would display a strong distance dependence of the electrochemical current. The oxidation and reduction potentials of the nucleobases appear, further prohibitive for efficient long-range interfacial charge transfer in the hopping mode involving individual base pairs followed by intermittent full environmental vibrational relaxation. Highly base pair specific conductivity patterns would moreover be expected in both mechanisms.

Ways of circumventing these conceptual obstacles are not immediately obvious. Appealing options with potential for theoretical support include:

- The notion of collective stacking and/or large polaron charge transmitting states.[3,32,78,79] Such steps would reduce drastically the "effective" charge transfer distance and invoke sufficient energy leveling among the base pairs contributing to the spatially extended stack unit or large polaron. A shortcoming is the elusive precise nature of such states although theoretical approaches have been forwarded.[80,81]
- The strong effects of environmental configurational fluctuations on the base pair energetics can be important. Such fluctuational effects would be part of the stacking and large polaron dynamics and include "gated" charge transfer but a comprehensive theory to incorporate all these elements is some way from completion.

As noted, merits and limitations of such views carry over to approaches to single-molecule DNA-conductivity in natural biological environments, to which we proceed next.

4. DNA-BASED CONDUCTIVITY IN BIOLOGICAL MEDIA TOWARDS THE SINGLE-MOLECULE LEVEL

Interfacial electrochemical ET between metallic electrodes and redox molecules through variable-length and variable-composition DNA-based molecules has disclosed important information about the molecular conduction mechanisms, based on monolayers of molecular thickness but averaged over two-dimensional macroscopic assemblies. Important conclusions are that the molecular contact can be a controlling factor and that the conductivity is hypersensitive to base pair order and stacking. The conductivity is effectively turned off when base pair mismatches or kinks invoked by external molecular structure-modifier binding (say *cis*-platinum[108]). This view carries over in part to DNA-based conductivity at the single-molecule level but here some other modification is needed.

We discuss two approaches to single-molecule DNA-conductivity, both referring to aqueous biological media. One approach has focused on ***isolated*** individual molecules, the other one on individual molecules but now inside an ordered two-dimensional 30–60 nm laterally extended assembly, with up to at least a hundred molecules inside a domain. The systems display partly concurring conductivity properties, particularly weak distance and base composition dependence, but the origin of these patterns is different in the two cases.

4.1 *In situ* Conductivity of Isolated Single- and Double-Strand Oligonucleotides

Reports by Haiss and associates[8,9] and by Tao and associates[7,110] on molecular electronic conductivity of individual, isolated oligonucleotide molecules, in aqueous biological buffers have appeared recently. Target issues were the oligonucleotide length and base composition dependence of the conductivity. Nanoscale environments were *in situ* STM and electrochemically controlled nanogap electrode configurations. The data illuminate the multifarious DNA-based conductivity patterns. The experimental approach by Haiss and associates is based on a sophisticated *in situ* STM approach in which the STM tip is used to “catch” single molecules on the substrate surface, cf. Chapter 7. The electrochemically controlled ***single-molecule*** conductivity of ***single-strand*** oligonucleotides was found to display, strikingly very weak dependence of both the sequence length and base composition. The data were formally equally well represented by eq. (6-2) or eq. (6-6) but in the former case with unphysically small β-values, *i.e.* the range of 0.1 $Å^{-1}$. The largest conductivity was found in the guanine-based sequences with about twice the value of the other repetitive base sequences. The double-strand conductivities were found to exceed the single-strand conductivities by up to two orders of magnitude but the difference between double-strand GC- and AT-oligonucleotides was again only a factor of two or so.

Observation of single-molecule electronic conductivity of short single-strand oligonucleotides is not surprising but the patterns are at variance with expectations from either superexchange or hopping concepts. The charge transmitting LUMO or HOMO in the former case are strongly off-resonance with the Fermi levels of the enclosing electrodes and further widely different, giving expectably significant and quite different β-values, eq. (6-2). The oxidation and reduction potential of the nucleotide bases in the hopping mechanism are similarly quite different and would also give much more diverse conductivities.

A clue to these "puzzles" could be configurational fluctuations in the solvent and counter ion environments as well as in the flexible single-strand structure itself. Such effects are of two kinds. Molecular modeling and Molecular Dynamics computations of flexible string-like molecules have disclosed complex structural manifolds where increasing sequence length is not followed by corresponding average distance increases between donor and acceptor units.[111] The conformational dynamics is accompanied by corresponding fluctuations of both the electron exchange factors between successive groups and, particularly the donor and acceptor ***energetics***. An emerging view is a charge transfer process where configurational "gating" is crucial.[112–114] Energetically relatively inexpensive fluctuations along a suitable (set of) gating modes thus leads to a non-equilibrium transient configuration where electron exchange and energetics of single charge transferring units are widely different (up to a large fraction of an eV) from their values at equilibrium. In return for the gating energy expenditure, charge transfer in the transient configuration is much more facile and less distance and base composition dependent than at conformational equilibrium.

Gated charge transfer is an established notion in molecular charge transfer theory[15,16] and introduced in novel contexts of *in situ* STM based single-molecule conductivity.[8,112–115] A particular outcome is that weak distance dependence can be expected in return for additional activation (free) energy contributions from the gating mode(s). The gating mode activation energy has been observed directly for variable-length alkanedithiols in the STM gap and found to follow transparent patterns.[115] The temperature variation of the flexible single-strand oligonucleotides has not been reported but the virtually complete absence of temperature dependence of the conformationally much more rigid double-strand forms[9] is in keeping with this view. Other issues, or "puzzles", however, remain, cf. Section 4.2.

The studies by Tao and associates[7,110] both concur with those by Haiss and associates but also show some differences. Tao and associates also used electrochemically controlled interfacial environments with the target molecules confined between two nanoscale (electrochemical) electrodes. They observed that the single-molecule conductivity of $(GC)_n$-sequences showed very weak, inversely proportional length dependence, eq. (6-6) whereas $(AT)_n$-sequences followed eq. (6-2) with a decay constant $\beta = 0.43$ Å^{-1}. Single-nucleotide polymorphism in conspicuous conductivity differences between mismatched and natural sequences could also be addressed. The different observations by Haiss and associates and by Tao and associates are not necessarily in conflict. The local single-molecule environments of the target molecules in the two sets of observation are different, *viz.* a planar substrate/tip gap and a pair of nanoscale electrodes, respectively. This could affect differently the local solvent and counter ion dynamics. Ultimately this would reflect the importance of "gating", and perhaps "transient stacking" and "polaron" formation in environmentally specific electronic conductivity patterns of DNA-based molecules at single-molecule and mesoscopic spatial scales. "Bubble" formation, "transient unwinding" and other strongly non-equilibrium events[115] are other features that would contribute to the conductivity control.

4.2 Conductivity of Pure and Redox-Modified DNA-Based Molecules in Dense Monolayers

Covalent bonding via oligo-methylenethiol linkers is broadly used to attach oligonucleotide molecules to electrochemical surfaces particularly of gold, cf. Section 3. Single-strand oligonucleotides have been linked to atomically planar Au(111)-electrode surfaces in dense monolayers in this way and addressed by *in situ* STM, supported by voltammetry, interfacial capacitance data and X-ray photoelectron spectroscopy.[5,6,47] *In situ* STM image resolution is at the level of the single molecule, but as for *in situ* STM of isolated molecules, the conductivity patterns leave open issues, although for reasons likely to differ from what applies to immobilized isolated single molecules.

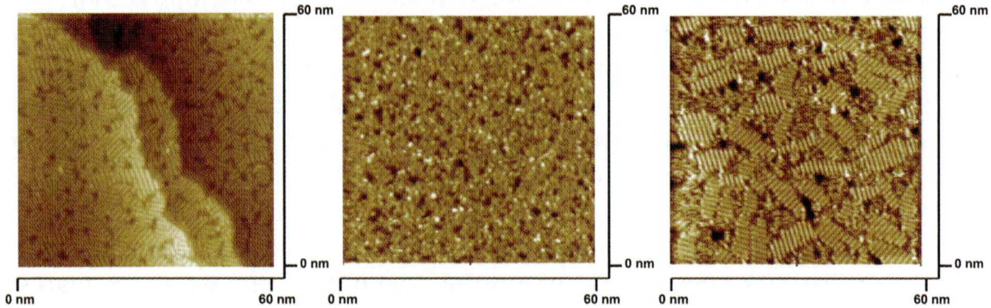

Figure 6-5. Electrochemical potential controlled reversible domain formation of a single-strand repetitive 10-adenine sequence linked to a single-crystal Au(111)-electrode surface in aqueous phosphate buffer, pH 7.6. Left: Domain formation after keeping the potential at –0.61 V (vs. saturated calomel electrode, SCE) for 40 min. Recording at –0.21 V (SCE). Middle: Disorder after potential stepping to –0.05 V (SCE). Right: Recovery of ordered domains from the disordered surface structure shown in the middle figure, after stepping the potential back to –0.61 V (SCE) and recording at –0.21 V (SCE). Reprinted from ref.. 6 with permission.

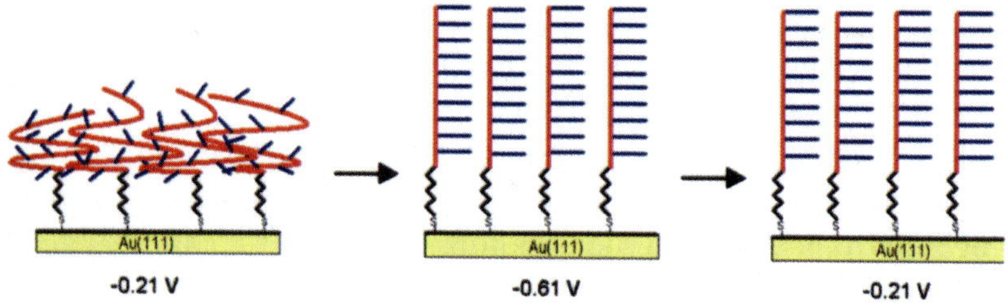

Figure 6-6. Schematic view of the electrochemically controlled order-disorder process disclosed by the *in situ* STM images in Fig. 6-5. Reprinted from ref. 5 with permission.

Figure 6-5 illustrates the data with the following observations:

- A highly dense, disordered monolayer is formed on adsorption at open circuit potentials. Shifting the potential to values well negative of the potential of zero charge triggers the formation of a highly ordered monolayer. The order-disorder process is, interestingly reversible, such as seen in Fig. 6-5. Figure 6-6 illustrates schematically the process.
- The surface pattern in the ordered phase is characterized by rows of bright contrasts each of which is a single, single-stranded oligonucleotide molecule in upright position. Such a high density at first appears unexpected.

It is as if the excess negative surface charge forces the negatively charged oligonucleotide backbone into the solution, leading to stable ordering that must necessarily involve also dense counter ion packing. The counter ions (Na^{+}, Mg^{2+}, or H^{+}) do not, however, appear in the images.

• The high density refers to single-strand oligoncucleotides. These are quite different from the double-strand forms and much longer, thinner, as well as much less rigid. Condensation of charged polyelectrolytes has been supported by Monte Carlo computations combined with models of the polyelectrolyte (oligonucleotide) as charged rods in a dielectric medium, with charge compensating counter ions[117] Fig. 6-7. Randomly scattered individual rods appear appear at low concentrations. Above a critical concentration condensation into a "brush"-like surface organization, much like the one observed appears. Condensation phenomena have been addressed comprehensively by other theoretical (statistical mechanical) and experimental approaches.[34] Correlations to columnar DNA-aggregation[33] and ultimately perhaps even chromosome packing are exciting other perspectives.

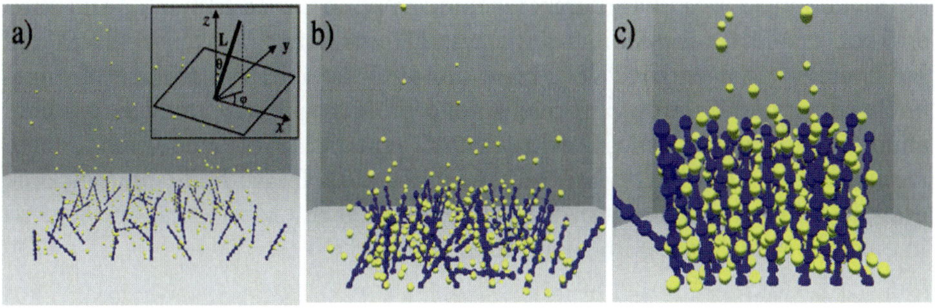

Figure 6-7. Concentration and counter ion induced surface condensation of strongly negatively charged rod-like polyelectrolyte molecular structures in aqueous solution. a) Low concentration. b) Intermediate concentration. c) High concentration. Reprinted from ref. 117 with permission.

• Condensation holds other implications. The lateral electrostatically based condensation forces compete with hydrogen bonding and stacking forces that control duplex formation. As a result, duplex adsorption is fraught with dehybridization of duplex strands otherwise stable in bulk solution.[102,118] In an applied perspective, direct adsorption of shorter (say, ≤ 15 bases) single-strand oligonucleotides is therefore not suitable as a strategy towards DNA-based biosensing as the density is far too high and prohibitive for local duplex formation.

Images and views such as in Figs. 6-5 through 6-7 illustrate the adsorption of single- and ultimately double-strand oligonucleotides. At the same time they leave "issues" encountered throughout DNA-conductivity science. For example, the *in situ* STM images, which represent directly the conductivity of a repetitive 10-adenine sequence, a single-adenine nucleotide and a 13-base symmetric mixed sequence, all linked to a Au(111)-electrode surface are indistinguishable.[5,6,47] Gating in the sense of configurational fluctuational control of the electron exchange or energetics is not an obvious rationale for the ordered dense monolayers. Fluctuational stacking and/or large polaron formation that effectively transmit electrons or holes are more attractive but also more elusive concepts. It is safe to notice that observations such as these that also relate to semi-infinite electrochemical configurations constitute a challenge where large-scale computational input would be highly desirable.

In a comparative study of calf thymus DNA and 12-base pair single- and double-strand oligonucleotide monolayers on electrochemical (polycrystalline) gold surfaces, Heller and associates introduced disordered (as opposed to crystalline) semiconductor conductivity notions as a rationale for facile long-range charge transfer through DNA-based films.[119] With such a view, a solid state medium with a small difference between optical and static dielectric constants, cf. Chapter 8, DNA-based films offer a physical environment for both significantly lowered and spatially extended excess charge localization sites. The inhomogeneous and anisotropic local environments for interfacial electrochemical ET in both macroscopic monolayer and domain-bound bound single-molecule configurations bare an interesting accordance with such views.

4.3 *In Situ* STM Conductivity Mechanisms of Pure and Redox-Marked Oligonucleotides

Interfacial ET between metal electrodes and molecular redox groups, via double-strand oligonucleotides attached to the (gold) electrode surfaces via thiol-based linker groups left fundamental issues open as to the charge transfer mechanism. *In situ* STM image patterns such as those shown in Fig. 6-5 leave somewhat the same puzzles. The almost complete absence of length and base composition dependence of the tunneling current follows expectations neither from superexchange nor from hopping via single-nucleobase or -base pairs. Conspicuous distance and base composition dependence of the tunneling energy gaps are expected in the former case, while the widely different oxidation and reduction potentials for the different bases would give widely different contributions in the latter case.

Transient extended base stacking and polaron formation have been forwarded as attractive rationales, with long-range DNA-based charge (hole or electron) transfer based on these units rather than on the base pairs. This would reduce drastically the number of hops needed for charge transfer in the shorter oligonucleotide systems addressed in electrochemistry and *in situ* STM. The spatially extended intermediate sites would be represented by lower-lying electronic states, or tunneling barrier "indentations" in the superexchange mechanism, and as temporarily or dynamically populated intermediate electronic states[15,16] in the hopping mechanism. The "invisibility" of these states remains, however, as a problem.

The role of "low-lying", *i.e.* much lower than single-base (pair) oxidized and reduced states, in interfacial charge transfer is different in the semi-infinite electrochemical environment and in the confined two-electrode *in situ* STM environment. There is, for example no indication of an electrochemical response to "stacking" or "polaron" formation in the ***absence*** of a redox marker group. The reduction potentials of these delocalized charge transmitter entities could be hidden by voltammetric signals from reductive desorption of the Au-S unit and liberation of the thiol-modified DNA-based molecules into the solution.[5,6,121] This would put focus on the importance of the Au-S linker unit as a notable contributor to the charge mediating states. It could also be that the polaron/stacked units show sufficient inhomogeneous broadening that the voltammetric response is overshadowed by capacitive background signals. Such patterns are known from protein voltammetry (e.g. ref. 120) and would be much more pronounced for polycrystalline electrodes than for single-crystal electrochemical surfaces. Single-crystal, atomically planar electrode surfaces and more sensitive voltammetric techniques such as differential pulse rather than cyclic voltammetry might open ways to detect the, so far elusive spatially extended low-lying polaron- and stacking-based charge transmitting states.

The *in situ* STM mode offers a slightly different view on the role of the intermediate polaron/stacking states. Intermediate, not too far off-resonance electronic states would still be efficient charge transmitters. Even though the overall charge transfer between substrate and tip would be initiated by an energetically unfavourable substrate to intermediate state transition, such a step would be followed by an exothermic charge transfer step from the temporarily populated intermediate state to the tip (or vice-versa). Whether by superexchange or hopping, such a mechanism would give a high tunneling current and possibly with little length or base composition dependence. Structural inhomogeneous broadening would further attenuate specific charge transfer patterns.

The difference between molecular charge transfer in the *in situ* STM mode via conceptually well-defined but physically elusive stacking/polaron states and well-defined low-lying and accessible redox states is illuminated by observations shown in Figs. 6-8 and 6-9.[47,121]

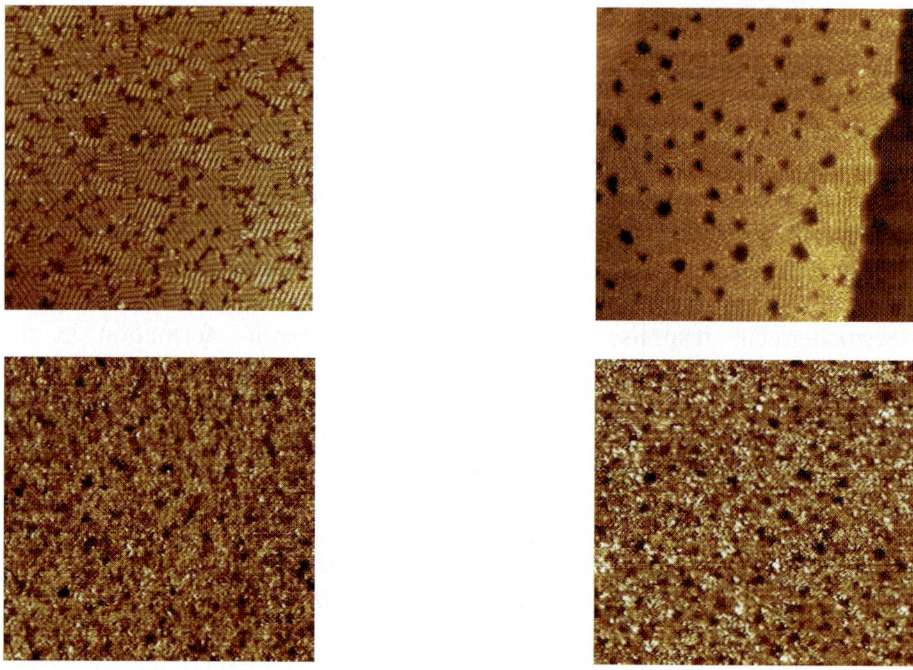

Figure 6-8. 60×60 nm^2 *in situ* STM images of pure and mixed symmetric single-strand 13-base oligonucleotide/MCH monolayers (HS-$(CH_2)_6$-5'-CGC ATT ATT ACGC) on single-crystal Au(111)-electrode surface. 10 mM TRIS/50 mM NaCl, pH 7.6. Electrochemical potential control (in the range –0.21 to –0.26 V (SCE)). Tunneling currents in the range 100–400 pA. Bias voltage +0.02 V. Top left: Domain order in pure single-strand oligonucleotide layer. Top right: Domain order in pure MCH. Bottom left: Mixed oligonucleotide/MCH monolayer. Bottom right: Same mixed oligonucleotide/MCH monolayer with 0.4 mM $[Ru(NH_3)_6]^{3+}$ added to the solution. Reprinted from ref. 47 with persmission.

Single-strand oligonucleotides can be dispersed in a two-dimensional "matrix" of a molecule that resembles the linker unit, in Figs. 6-8 and 6-9, ω-mercaptohexanol (MCH). Such a step offers exposed single-strands accessible to duplex formation and other (bio)chemical reactivity.

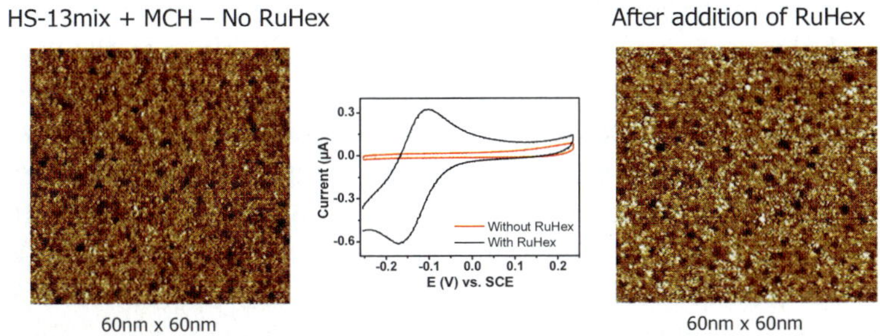

Figure 6-9. Left: *In situ* STM image of 13-mix/MCH mixed monolayer at a Au(111)-electrode surface. Same conditions as in Fig. 6-8. Right: *In situ* STM image of the same mixed monolayer after addition of 0.4 mM $[Ru(NH_3]_6]^{3+}$ to the solution. Middle: Cyclic voltammogram of the layer shown to the right. Reprinted from ref. 47 with permission.

The oligonucleotide (here hexamethylene-thiol linked 13-base 5'-CGC ATT ATT ACGC) and MCH separately form the same kind of highly ordered monolayer as shown in Fig. 6-5 but the order is completely eroded on mixing, Fig. 6-8. Addition of the long-time used redox marker, $[Ru(NH_3)_6]^{2+/3+}$ that binds electrostatically to the phosphate groups in the oligonucleotide backbone induces a new *in situ* STM pattern. Almost randomly scattered, molecular-size strong *in situ* STM contrasts now appear. These are caused by a new mechanism in which bound $[Ru(NH_3)_6]^{2+/3+}$ units transmit electrons (or holes) in a very efficient two-step interfacial electrochemical charge transfer sequence. Elements of this mechanism are summarized in Section 5.1 and discussed in Chapter 8. A key point presently is that the difference between the strong $[Ru(NH_3)_6]^{2+/3+}$-based contrasts and the equally well-defined but weaker contrasts from the unmodified oligonucleotide molecules, Figs. 6-5 and 6-8, illustrate directly the physical difference between a two-step redox mechanism via a "conventional" redox marker couple and the polaron/stacking electronic states. Figs. 6-9 and 6-10 offers a further illustration. When $[Ru(NH_3)_6]^{2+/3+}$ is added to the ordered DNA-based monolayer, adsorption ***also*** occurs but now only at the domain boundaries.[121] These are the only regions in the densely packed domains where the negatively charged phosphate groups are accessible to adsorption of large positively charged counter ions, here the $[Ru(NH_3)_6]^{2+/3+}$ couple, and therefore also to duplex formation.

Images such as those in Figs. 6-8 through 6-10 offer obvious insight into DNA-based electronic conductivity. They also offer clues to the merits and limitations of the use of redox markers for the determination of the electrochemical electrode coverage by the DNA-based molecules. We rather conclude, however, by noting still another recent observation that illustrate fundamentals of electronic conductivity of pure and redox-marked DNA-based molecules. *In situ* STM-based electronic conductivity of single redox molecules is discussed in Chapters 7 and 8. In contrast to non-redox molecules, ***redox*** molecules can be brought to display novel features of strong "amplification", "switch" and "rectifier" features in narrow ranges of the tunneling current/overpotential and current/bias voltage relations, at the level of the ***single molecule***. The $[Ru(NH_3)_6]^{2+/3+}$ couple in the DNA-based local molecular environment can be brought to display similar features.

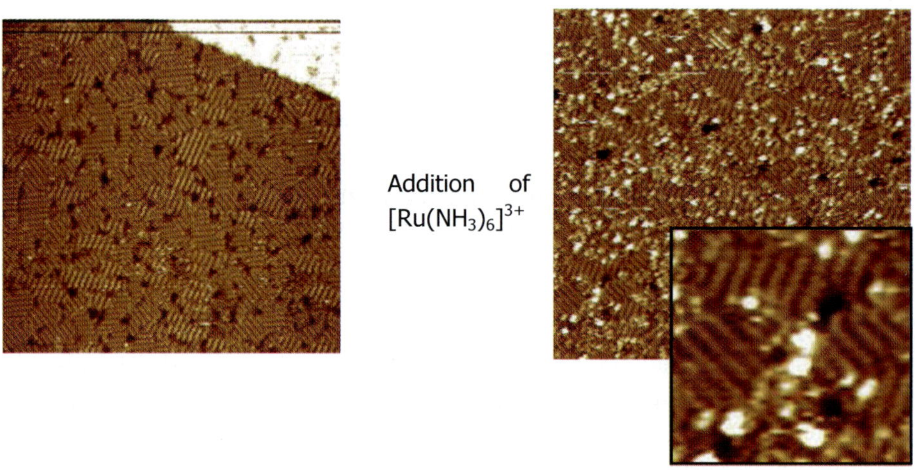

Figure 6-10. 60 × 60 nm^2 *in situ* STM image after addition of 1mM RuHex in 10mM Tris/50mM NaCl (pH 7.6) to a ordered pure single-strand 13-mix monolayer. Inset: 13×13 nm^2 zoom-in. Image recorded 30 minutes after addition of $[Ru(NH_3)_6]^{3+}$. $V_{bias} = -0.15$ V, $I_t = 0.35$nA, $E_{sample} = -0.25$V (SCE). 10 mM Tris + 50 mM NaCl, pH 7.6. Reprinted from ref. 121 with permission.

We can allocate the appellation "single-molecule DNA-transistor" to such a system. In contrast to all previously suggested DNA-based molecular electronics, this system operates in the natural aqueous environment for biological DNA-function. The perspective in the present context is, however, perhaps rather that experimentally sophisticated single-molecule electronics data add important support to the subtle electronic conductivity features of

DNA-based molecules in their natural environment. We offer briefly in Section 5 some further concepts and implications of such observations.

5. TWO- AND MULTI-STEP HOPPING IN DNA-BASED MOLECULES IN THE *IN SITU* STM GAP

A physical and formal rationale for "spectroscopic" tunneling current/overpotential and current/bias voltage patterns of single-molecule redox systems is given in Section 3.2.2 of Chapter 8 with two-step electron (hole) transfer between substrate and tip via a single redox level specifically in focus. The tunneling current/overpotential relation is particularly informative. The current maximum arises when the redox level is brought into the energy window between the Fermi levels of the enclosing electrodes by the overpotential. Continued overpotential rise, however, shifts the redox level further to a position below the two Fermi levels, causing the tunneling current to drop again. Such a view also applies to the redox-marked DNA-based molecules in the *in situ* STM configurations. The view would also apply to pure, unmarked DNA-based molecules if ***well-defined*** low-lying, spatially extended but still temporarily populated stacking or polaron states transmit the current.

Neither the chemical composition of the redox marked molecules nor the number of real or putative polaron/stacking sites in the unmodified molecules are known. The electrostatic $[Ru(NH_3)_6]^{2+/3+}$/oligonucleotide stoichiometry and the oligonucleotide lengths are suggestive of either a small number, say two-to four, of charge transmitting states, or a distribution of sites along the oligonucleotide chain. The attractive simple two-step formalism in Chapter 8 is therefore in need of extension to multi-step hopping. Conceptual and formal frames along these lines for ***three-step*** hopping via donor-acceptor molecules were offered in ref. 122 with focus on molecular electronic rectification. We summarize here views on the two limiting *in situ* STM configurations with a single intermediate hopping site that offers detailed electronic and energetic insight, and with an arbitrary number of such sites. The latter also offers important insight, although averaged and therefore of a somewhat "congested" nature.

5.1 A Two-step *In Situ* STM Electronic Transition via a Single DNA-based Redox Marker

We refer to Fig. 6-6 and Section 3.2. in Chapter 8 for a discussion of the physical nature of this mechanism. The following simple tunneling current form, cf. Chapter 8 catches presently the essence[42,43,123–125]

$$i_{tunn}^{symm} = \frac{1}{2} e n_{o/r} \frac{\omega_{eff}}{2\pi} \exp\left(-\frac{E_r - eV_{bias}}{4k_B T} \right) \left\{ \cosh\left[\frac{\left(\frac{1}{2}-\gamma\right) eV_{bias} - e\eta}{2k_B T} \right] \right\}^{-1}$$

(6-10)

V_{bias} is the bias voltage and γ ($0 < \gamma < 1$) a parametric representation of the bias voltage distribution in the tunneling gap. Equation (6-10) gives a maximum when $\eta = \eta_{\max} = \left(\frac{1}{2}-\gamma\right) V_{bias}$, *i.e.* close to the equilibrium potential in symmetric *in situ* STM contacts (γ = ½). $\boldsymbol{n_{o/r}}$ is the number of electrons (holes) transferred in a single *in situ* STM electronic event. This quantity can be large, giving a "boost" of up to ***several hundred*** electrons transferred in a narrow overpotential range around the equilibrium potential of the redox molecule when the coupling between the redox molecule and the electrodes is strong. This is a crucial difference between interfacial electrochemical ET and ***single-molecule*** *in situ* STM-based electrochemistry. Similar, but attenuated and broadened effects would be expected when suitably well-defined polaron/stacking states operate.

5.2 *In Situ* STM of Pure and Redox-marked DNA-Based Multi-Step Hopping – a Model

A formalism for multi-step hopping through DNA-based molecules or other molecular "wires" in the *in situ* STM gap or between a pair of electrochemical nanogap electrodes at the level of detail as for homogeneous solution, is not available. Formal elements discussed below illustrate a simplified view of electrochemically controlled hopping through a chain of N molecular groups. Quenching reactions are disregarded, and all groups except the terminal groups "1" and "N" are taken to be identical, cf. Fig. 6-3. These assumptions can both be relaxed. We confine the discussion to the length and bias voltage dependence of the tunneling current at fixed overpotential, here taken as the equilibrium potential. The kinetic equations for the populations, P_1, P_n (n = 3, … N-2), P_{N-1} and P_N are then

$$\frac{\partial P_1}{\partial t} = k_1^{el}\left(1-P_1\right) - k_1^{el}P_1 - k_1P_1 - k_{12}P_1 + k_{21}P_2\left(1-P_1\right)$$

$$\frac{\partial P_2}{\partial t} = k_{12}P_1 - k_{23}P_2 + k_{32}P_3 - k_{21}P_2\left(1-P_1\right)$$

$$\frac{\partial P_n}{\partial t} = k_{n-1n}P_{n-1} - k_{nn-1}P_n - k_{nn+1}P_n - k_nP_n + k_{n+1n}P_{n+1}\text{; n = 3, ...N–2}$$

(6-11)

$$\frac{\partial P_{N-1}}{\partial t} = k_{N-2N-1}P_{N-2} - k_{N-1N-2}P_{N-1} + k_{NN-1}P_N - k_{N-1N}P_{N-1}\left(1-P_N\right)$$

$$\frac{\partial P_N}{\partial t} = k_{-N}^{el}\left(1-P_N\right) - k_{NN-1}P_N - k_N^{el}P_N + k_{N-1N}P_{N-1}\left(1-P_N\right)$$

with the following observations:

• The nature of the charge transmitting groups is left unspecified. Efficient hopping through individual base pairs would need large overpotentials. Spatially extended stacking/large polaron states or a chain of redox markers such as $[Ru(NH_3)_6]^{2+/3+}$ accord better with the assumptions and would attenuate electronic and energetic differences between the units.

• We shall assume that the (electron or hole) population of all groups except the terminal groups is small. This implies that forward charge transfer dominates, *i.e.* $k_{12} << k_{21}$ and $k_{NN-1} << k_{N-1N}$. Together with the set of electrochemical notions, this assumption extends previous approaches[66] to hopping in molecular wires. The rate constants then take the form

$$k_{nn+1} = k^0_{nn+1} e^{\frac{e\alpha\gamma_{nn+1}V_{bias}}{k_BT}} \quad ; \quad k_{n+1n} = k^0_{n+1n} e^{-\frac{e(1-\alpha)\gamma_{n+1n}V_{bias}}{k_BT}} \tag{6-12}$$

$\gamma_{nn+1} = \gamma_{n+1n}$ are the fractions of the voltage drops in the $n \rightarrow n+1$ and $n+1 \rightarrow n$ transitions inside the chain. The quantity α coincides approximately with the electrochemical transfer coefficient taken as α = ½ in the following. Equation (6-12) can then be given the shorter form

$$k_{nn-1} = k^0 u(V); \quad k_{n+1n} = k^0 \frac{1}{u(V)} \; ; \; u(V) = e^{\frac{e\gamma V_{bias}}{2k_BT}} \tag{6-13}$$

where all the γ's inside the chain have been given the same value, γ. Further,

$$k_1^{el} = k_1^0 e^{\frac{e\gamma_1 V_{bias}}{2k_BT}} \; ; \; k_{-1}^{el} = k_1^0 e^{-\frac{e\gamma_1 V_{bias}}{2k_BT}} \; ; \; k_N^{el} = k_N^0 e^{\frac{e\gamma_R V_{bias}}{2k_BT}} \; ; \; k_{-N}^{el} = k_N^0 e^{-\frac{e_R V_{bias}}{2k_BT}} \tag{6-14}$$

or

$$k_1^{el} = k_1^{el0} u^{(1)}(V); \; k_N^{el} = k_N^{el0} u^{(N)}; \; k_{-1}^{el} = \frac{k_1^{el0}}{u^{(1)}}; \; k_{-N}^{el} = \frac{k_N^{el0}}{u^{(N)}} \tag{6-15}$$

The steady-state tunneling current (scaled to units of the electronic charge e) is

$$j = k_1^{el0} u^{(1)}(V)(1-P_1) - k_1^{el0} \frac{1}{u^{(1)}(V)} P_1 = k_N^{el0} u^{(N)}(V) P_N - k_N^{el0} \frac{1}{u^{(N)}(V)} (1-P_N) \approx$$

$$k^0 \left(P_n u(V) - P_{n+1} \frac{1}{u(V)} \right); \quad n = 2,3 \ldots N-2 \tag{6-16}$$

The relations for n = 1 and N–1 are different, cf. eq.(6-11). Further,

$$j = k_{12}^{0} u(V) P_1 - k^0 \frac{1}{u(V)} P_2 \left(1 - P_1\right);$$

$$j = k^0 u(V) P_{N-1} \left(1 - P_N\right) - k_{NN-1}^{0} \frac{1}{u(V)} P_N \tag{6-17}$$

All derivatives in eq.(6-11) vanish under steady state conditions. We then obtain from eq.(6-17)

$$P_N = \frac{j + k_N^{el0} \frac{1}{u^{(N)}(V)}}{k_N \left(u^{(N)}(V) + \frac{1}{u^{(N)}(V)} \right)}; \; P_{N-1} = \frac{j + k_{NN-1}^{0} \frac{1}{u(V)} P_N}{k^0 u(V) \left(1 - P_N\right)} \tag{6-18}$$

It is convenient to measure the current and all rate constants in units of the rate constant k^0. The second equation in eq.(6-18) can then be written formally as

$$P_{N-1} = u(V)^2 A\left(j\right) + \frac{1}{u(V)} jU_0 = \frac{j + k_{NN-1}^{0} \frac{1}{u(V)} P_N}{u(V)\left(1 - P_N\right)} \tag{6-19}$$

$$U_0 = 0; \quad A\left(j\right) = \frac{j + k_{NN-1}^{0} \frac{1}{u(V)} P_N}{u(V)^3 \left(1 - P_N\right)} \tag{6-20}$$

Inserting eq.(6-18) into the third eq.(6-11) for $n = N–1$ we obtain

$$P_{N-2} = A\left(j\right) + \frac{1}{u(V)} jU_1; \; U_1 = 1 \tag{6-21}$$

Continuing this procedure we obtain

$$P_{N-m} = \frac{1}{u(V)^{2m-4}} A\left(j\right) + \frac{1}{u(V)} jU_{m-1}; \quad m = 1, 2, \ldots N - 2 \tag{6-22}$$

where the recurrence relationship

$$U_m\left(u(V)\right)=1+\frac{1}{u(V)^2}U_{m-1}\left(u(V)\right) \tag{6-23}$$

applies. For m = N–2

$$P_2\left(j,u\right)=\frac{1}{u^{2N-8}}A\left(j\right)+\frac{1}{u}jU_{N-3}\left(u\right) \tag{6-24}$$

Inserting eq.(6-24) into the first eq.(6-11) we obtain

$$P_1\left(j,u(V)\right)=\frac{k_1u(V)^2+\frac{1}{u(V)}P_2\left(j,u(V)\right)}{k_1\left(u^{(1)}(V)+\frac{1}{u^{(1)}(V)}\right)+k_{12}^0u(V)+\frac{1}{u(V)}P_2\left(j,u(V)\right)} \tag{6-25}$$

Using this expression in the first eq.(6-11) we obtain, finally the equation for the dependence of the tunneling current on *u(V)* , *i.e.* on the bias voltage

$$j+k_1^0\left(u^{(1)}\left(V\right)+\frac{1}{u^{(1)}\left(V\right)}\right)P_1((j,u(V)-k_1^0P_1\left(j,u(V)\right)=0 \tag{6-26}$$

Examples of the dependences of reduced current $I=j/\left(k_1^0u^{(1)}(V)\right)$ on the potential are shown in Fig. 6-11 and discussed below.

5.3 *In Situ* STM Hopping Current Dependence on Probe Length and Bias Voltage

Equations (6-11) through (6-26) offer a scheme for the simplified *in situ* STM or electrochemical nanogap hopping mode of DNA-based molecules or other molecular "wires" conducting by this mechanism. The steady state tunneling current is given by eq. (6-25) in terms of the population of the first charge transmitting group, $P_1(V)$. The latter is found by the iterative scheme, ending by the input of the controlling rate constant and bias voltage parameters in eq. (6-25). The scheme has focused on the explicit ***bias voltage*** and chain length dependence of the tunneling current. Extension to the electrochemical ***overpotential*** dependence has not yet been explored but is in principle straightforward by inclusion of the overpotential dependence of the interfacial rate constants k_1 and k_N, *i.e.* $k_1^{el0} \rightarrow k_1^{el0} f(\eta)$ and $k_N^{el0} \rightarrow k_N^{el0} f(\eta - V_{bias})$ in the scheme where the function $f(\eta)$ or $f(\eta - V_{bias})$ follows patterns described in Chapter 8.

Figure 6-11 shows a representative example of the *in situ* STM tunneling current dependence on the bias voltage through variable-length oligonucleotides for the hopping conductivity mechanism. The current is normalized with respect to $k^0 exp(eV_{bias}/2k_BT)$, and the bias voltage is in units of k_BT (= 25 mV).

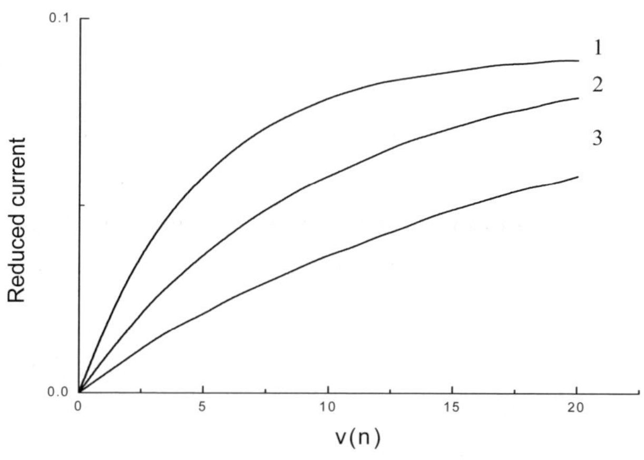

Figure 6-11. Tunneling current/bias voltage relations, normalized to $k_1^0 \exp(\gamma eV / 2k_BT)$, calculated by the iterative scheme in eqs. (6-11) through (6-26). $k_1^0 = k_N^0 = 1$; $k_{12}^0 = k_{NN-1}^0 = 0.1$. The three correlations correspond to different numbers of charge transmitting groups: Top: N = 5. Middle: N = 10. Bottom: N = 20.

The figure shows:

- The normalized current rises approximately linearly with increasing bias voltage but reaches a plateau for higher values. The plateau corresponds to exponential bias voltage dependence of the un-normalized current. The exponential bias voltage dependence reflects essentially the dependence of the rate of charge injection into the molecular chain. The plateau is reached faster, *i.e.* at lower bias voltages the shorter the oligonucleotide chain, as if the full bias voltage dependence carries though faster for the shorter chains.

• in contrast to the bias voltage dependence, the distance dependence is weak. Increase of the chain length from five to ten units only lowers the current by about 30 % with another 30 % on chain length increase from ten to twenty units. In exponential decay, eqs. (6-2) and (6-3), this gives only 0.07 per unit. A crude numerical accordance with eq. (6-6) would give η = 0.25 which is lower than the limiting value for homogeneous solution, cf. eq. (6-6).

The hopping conductivity scheme above represents formally the data for *in situ* STM imaging and up to a point single-molecule conductivity of DNA-based molecules as observed. The issues of the energetics and the nature of the charge (electron or hope) transmitting states are, however, left open.

6. SOME CONCLUDING OBSERVATIONS

The discussion above has addressed the exciting and still controversial issue of the electronic, *i.e.* electron or hole conductivity of DNA-based molecules at electrochemical interfaces in contact with the aqueous environment of biological DNA-function. DNA-based electronic conductivity in homogeneous solution has been mapped in great detail by a variety of innovative experimental work. Conditions for the prevalence of superexchange or hopping modes have been identified but the physics of the conduction seems to go beyond individual base pairs as hopping or low-lying superexchange charge transmitting units. Spatially extended base pair stacks accompanied by environmental nuclear relaxation reminiscent of large polaron states introduced as a mechanism for excess electron trapping in polar media in the late 1940's are attractive other notions. Computational approaches have disclosed features of one-dimensional analogues of the large polaron along the double-strand DNA-stacks.

DNA-based interfacial electronic conductivity at electrochemical surfaces poses the same as well as other challenges. Voltammetry of redox-marked immobilized DNA-based molecular monolayers again points to the notable independence of the distance and base composition of the charge transmitting molecular DNA-frame. At the same time the electrochemical charge transmission is hypersensitive to double-strand structural order for efficient long-range electronic charge transfer. *In situ* STM and conductivity of DNA-based molecules in electrochemical nanogap electrode configurations add other observations and puzzles. One is the control of highly ordered packing and condensation of DNA-based molecules which can be mapped with close to molecular resolution by *in situ* STM. Another one is that weak distance and base (pair) composition dependence of the tunneling current is also here a feature but possibly with different origins rooted either in the role of extended stacking or large polaron charge transmitting states, or in strong environmental configurational fluctuational effects on the base pair energetics. A third observation is that single-strand oligonucleotides can form ordered domains on atomically planar electrochemical metal (Au(111)) surfaces where the molecules show the same weak distance dependence of the electronic conductivity as otherwise broadly ascribed solely to the structurally ordered double-strand form. This observation may offer other new approaches to the understanding of the electronic conductivity of DNA-based molecules in aqueous media.

With focus on DNA-based function in natural aqueous biological media, electronic DNA-conductivity is supported as a feature of radiation damage and repair mechanisms.[35.36] Perhaps this would include the issue of aggregation and columnar stacking.[33] Interfacial electrochemical ET and DNA-based electronic conductivity is, further a central concept in the whole wide range of electrochemical biosensing based on pure and structurally modified DNA-based molecules as developed by Scheller,[126,127] Wang,[44] Palaček,[44] Gooding,[46,96–98] Barton and Hill[128,129] (See also Chapter 5), Willner[40] (See also Chapter 3) and others. These perspectives are beyond the scope of this Chapter but discussed in Chapters 3 and 5 of the present volume as well as elsewhere.[44] Conductivity along DNA-stacks is also the underlying concept of notions such as DNA-based molecular "wires", "DNA-templated electronics and networks", or "DNA-nanostructures for computing".[37–40] Such notions are exciting but the operational medium for natural DNA-function is aqueous biological media. As it appears presently, the importance of dynamic gating processes in long-range DNA charge transfer under "mild" conditions, such as extended base stacking and large polaron formation is associated with the aqueous environment and hard to reconcile with ambient gas, ultra-high vacuum or cryogenic environments in DNA-based electronics. From the observed energetic leveling caused by structural pre-organization and gating, the otherwise exquisitely base pair specific electronic properties of the molecules would also be significantly attenuated.

Once a note of caution is expressed, novel DNA-based molecular design and architectures have, however, reached levels where the emerging supramolecular structures can perhaps become robust to different electronically working environments. Redox- and fluorophor-labeled molecules offer steps towards "device-like" function where both structural organization and, as noted, additional sophisticated electronic function can be added to the DNA-based molecules.[37–39,48,49,130,131] New electronic conductivity mechanisms based on hopping via a chain of ***well-defined*** redox probes bound in the DNA-backbone were noted above.[47,121] Such perspectives extend to new sophisticated structures in which efficient hopping sites in the form of a sequence of well-defined transition metal sites have been inserted ***inside*** the DNA-structure.[132,133] Strategic chemical synthesis will also develop towards extension of these notions to two- and three-dimensional structures (e.g. ref. 38.

Combination of strategically designed DNA-based molecular structures with nanoscale solid state structures such as carbon and semiconductor nanotubes[134] or metallic and semiconductor nanoparticles[37–40] into nanoscale hybrid structures, constitute other approaches to novel biological nanoscale materials with interesting properties. DNA-conjugates with metallic or semiconductor nanoparticles is such a concept which can be combined with natural DNA-function in biological media. Adsorption of DNA-based molecules on ***electrochemically controlled*** semiconductor nanotubes, again in aqueous media of natural DNA-operation is likely to become another evolving area of nanoscale DNA-based science. Chemical reactivity such as the prime chemical DNA-process of duplex formation, approaching the level of resolution of the single molecule would become closer along such lines. The understanding of the fundamental electronic conductivity phenomenon of the DNA-based target molecules and their inorganic metallic or semiconductor complementary moieties in electronic hybrid function offers, however, prodigious challenges in forthcoming new experimental and theoretical interfacial bioelectrochemistry.

ACKNOWLEDGEMENT

This work has been supported by The Danish Research Council for Technology and Production Sciences (Contract No. 26-00-0034), the EU strp Programme CIDNA (Contract No. NMP4-CT-2003-505669) and Russian Foundation for Basic Research (Grant No.06-03-32193).

REFERENCES

1. *Charge Transfer in DNA: From Mechanism to Application*, Wagenknecht, H.-A., Ed., Wiley-VCH: Weinheim, 2004.
2. (a) *Long-Range Charge Transfer in DNA I*, Schuster, G.B., Ed., *Topics in Current Chemistry*, **2004**, *236*. (b) *Long-Range Charge Transfer in DNA II*, Schuster, G.B., Ed., *Topics in Current Chemistry*, **2004**, *237*.
3. O'Neill, M.A.; Barton, J.K., *J. Am. Chem. Soc.* **2004**, *126*, 11471-11483.

4. O'Neill, M.A.; Barton, J.K. In *Charge Transfer in DNA: From Mechanism to Application*, Wagenknecht, H.-A., Ed., Wiley-VCH: Weinheim, 2004, Chapter 2, pp. 27–75.
5. Wackerbarth, H.; Grubb, M.; Zhang, J.; Hansen, A.G.; Ulstrup, J., *Angew. Chem.* **2004**, *43*, 198–203.
6. Wackerbarth, H.; Grubb, M.; Zhang, J.; Hansen, A.G.; Ulstrup, J., *Langmuir* **2004**, *20*, 1647–1655.
7. Xu, B.; Zhang, P.; Li, X.; Tao, N., *Nano Lett.* **2004**, *4*, 1105–1108.
8. Van Zalinge, H.; Schiffrin, D.J.; Bates, A.D.; Haiss, W.; Ulstrup, J.; Nichols, R.J., *ChemPhysChem* **2006**, *7*, 94–98.
9. Van Zalinge, H.; Schiffrin, D.J.; Bates, A.D.; Starikov, E.B.; Wenzel, W.; Nichols, R.J., *Angew. Chem. Int. Ed.* **2006**, *45*, 5499–5502.
10. Gutman, F.; Lyons, L.E., *Organic Semiconductors*, Wiley: New York, 1967, Chapter 8.
11. Eley, D.D.; Spivey, D.I., *Trans. Faraday Soc.* **1962**, *58*, 411–415.
12. Jordan, P., *Naturwissenschaften* **1938**, *26*, 693–694.
13. (a) Szent-Györgyi, A., *Nature*, **1941**, *148*, 157–159; (b) Szent-Györgyi, A., *Bioelectronics. A Study in Cellular Regulations, Defense and Cancer*, Academic Press: New York and London, 1968.
14. Ladik, J., *Quantenbiochemie für Chemiker und Biologen* (Quantum Biochemistry for Chemists and Biologists), Ferdinand Enke: Stuttgart, 1972.
15. Kuznetsov, A.M., *Charge Transfer in Physics, Chemistry and Biology*, Gordon & Breach: Reading, 1995.
16. Kuznetsov, A.M.; Ulstrup, J., *Electron Transfer in Chemistry and Biology: An Introduction to the Theory*, Wiley: Chichester, 1999.
17. Jortner, J.; Bixon, M.; Langenbacher, T.; Michel-Beyerle, M.E., *Proc. Nat. Acad. Sci. USA* **1998**, *95*, 12559–12765.
18. Bixon, M.; Giese, B.; Wessely, S.; Langenbacher, T.; Michel-Beyerle, M.E.; Jortner, J., *Proc. Nat. Acad. Sci. USA* **1999**, *96*, 11713–11716.
19. Holstein, T, *Ann. Phys.* **1959**, *8*, 343–389.
20. Pekar, S.I., *Untersuchungen über die Elektronentheorie der Kristalle* (Investigationson the Electronic Theory of Crystals), Akademie Verlag: Berlin, 1954.
21. Fink, H.-W.; Schönenberger, C., *Nature* **1999**, *398*, 407–410.
22. Porath, D.; Bezryadin, A.; de Vries, S.; Dekker, C., *Nature* **2000**, *403*, 635–638.
23. Cohen, H.; Nogues, C.; Ullien, D.; Daube, S.; Naaman, R.; Porath, D., *Faraday Discussions* **2006**, *131*, 367–376.
24. Chapters in refs. 1 and 2.
25. Behrens, C.; Burgdorf, L.T.; Schwögler, A.; Carrell, T., *Angew. Chem. Int. Ed.* **2002**, *114*, 1841–1844.
26. Carrell, T.; von Meltzer, M. In *Charge Transfer in DNA: From Mechanism to Application*, Wagenknecht, H.-A., Ed., Wiley-VCH: Weinheim, 2004, Chapter 3, pp. 77–91.
27. Wagenknecht, H.-A., *Angew. Chem.* **2003**, *42*, 2454–2460.
28. Turro, N.J.; Barton, J.K., *J. Biol. Inorg. Chem.* **1998**, *3*, 201–209.
29. O'Kelley, S.; Barton, J.K., *Science* **1999**, *283*, 375–381.
30. Andres, P.R.; Schubert, U.S., *Adv. Mat.* **2004**, *16*, 1043–1068 and references cited therein.
31. Van Grondelle, R.; Novoderzhkin, V.C., *PCCP* **2006**, *8*, 793–807.
32. Shao, F.; Augustyn, K.; Barton, J.K., *J. Am. Chem. Soc.* **2005**, *127*, 17445–17452.
33. Kornyshev, A.A.; Leikin, S., *Phys. Rev. Lett.* **2001**, *86*, 3666–3669.
34. Netz, R., Andelman, D., *Phys. Repts.* **2003**, *380*, 1–95.

35. Nunez, M.E.; Rajski, S.R.; Barton, J.K., *Meth. Enzymol.* **2000**, *319*, 165–188.
36. Boon, E.M.; Livingston, A.L.; Chmiel, N.H.; David, S.S.; Barton, J.K., *Proc. Nat. Acad. Sci. USA* **2003**, *100*, 12543–12547.
37. Park, S.-J.; Lazarides, A.A.; Mirkin, C.A.; Brazis, P.W.; Cannewurf, C.R.; Letsinger, R.L., *Angew. Chem. Int. Ed.* **2000**, *39*, 3845–3848.
38. Chapters in: *Nanobiotechnology: Concepts, Applications and Perspectives*, Niemeyer, C.M.; Mirkin, C.A., Eds., Wiley-VCH: Weinheim, 2004.
39. Chung, S.W.; Ginger, D.S.; Morales, M.W.; Zhang, Z.F.; Chandrasekhar, V.; Ratner, M.A.; Mirkin, C.A., *Small* **2005**, *1*, 64–69.
40. Katz, E.; Willner, I., *Angew. Chem. Int. Ed.* **2004**, *43*, 6042–6108.
41. Xiao, X.; Nagahara, L.A.; Rawlett, A.M.; Tao, N., *J. Am. Chem. Soc.* **2005**, *127*, 9235–9240.
42. Albrecht, T.; Guckian, A.; Ulstrup, J.; Vos, J.G., *Nano Lett.* **2005**, *5*, 1451–1455.
43. Albrecht, T.; Guckian, A.; Kuznetsov, A.M.; Vos, J.G.; Ulstrup, J., *J. Am. Chem. Soc.* **2006**, *128*, 17132–17138.
44. *Electrochemistry of Nucleic Acids and proteins: Towards Electrochemical Sensors for genomics and Proteomics*, Paleček, E.; Scheller, F.; Wang, J., Eds., Elsevier: Amsterdam, 2005.
45. Drummond, T.G.; Hill, M.G.; Barton, J.K., *Nature Biotechnol.* **2003**, *21*, 1192–1199.
46. Gooding, J.J., *Euroanalysis* **2002**, *14*, 1149–1156.
47. Grubb, M.; Wackerbarth, H.; and Ulstrup, J., *J. Am. Chem. Soc.* **2006**, *128*, 7734–7735.
48. Gothelf, K.V.; LaBean, T.H., *Org. Biomol. Chem.* **2005**, *3*, 4023.
49. Seeman, N.C., *Quart. Rev. Biophys.* **2005**, *38*, 373–381.
50. Lewis, F.D.; Letsinger, R.L.; Wasielewski, M.R., *Acc. Chem. Res.* **2001**, *34*, 159–170.
51. Lewis, F.D.; Kalgutkar, R.S.; Wu, Y.; Liu, X.; Liu, J.; Hayes, R.T.; Miller, S.E.; Wasielewski, M.R., *J. Am. Chem. Soc.* **2000**, *122*, 12346–12351.
52. Lewis, F.D.; Wasielewski, M.R., In *Charge Transfer in DNA: From Mechanism to Application*, Wagenknecht, H.-A., Ed., Wiley-VCH: Weinheim, 2004, Chapter 4.
53. Giese, B.; Amaudrut, J.; Köhler, A.-K.; Spormann, M.; Wessely, S., *Nature* **2001**, *412*, 318–320.
54. Giese, B., In *Long-Range Charge Transfer in DNA I*, Schuster, G.B., Ed., *Topics in Current Chemistry*, **2004**, *236*, 27–44.
55. Meggers, E.; Michel-Beyerle, M.E.; Giese, B., *J. Am. Chem. Chem. Soc.* **1998**, *120*, 12950–12955.
56. Davis, W.B.; Hess, S.; Naydenova, I.; Haselsberger, R.; Ogrodnik, A.; Newton, M.D.; Michel-Beyerle, M.E., *J. Am. Chem. Soc.* **2002**, *124*, 2422–2423.
57. Kawai, K.; Majima, T., In *Long-Range Charge Transfer in DNA I*, Schuster, G.B., Ed., *Topics in Current Chemistry*, **2004**, *236*, 117–137.
58. Steenken, S.; Telo, J.P.; Novais, H.M.; Candeias, L.P., *J. Am. Chem. Soc.* **1992**, *114*, 4701–4709.
59. Seidel, C.A.M.; Schulz, A.; Sauer, M.H.M., *J. Am. Chem. Soc.* **1996**, *100*, 5541–5553.
60. Lewis, F.D.; Liu, X.; Hayes, R.T.; Wasielewski, M.R., *J. Am. Chem. Soc.* **2000**, *122*, 12037–12038.
61. Voityuk, A.A.; Jortner, J.; Bixon, M.; Rösch, N., *Chem. Phys. Lett.* **2000**, *324*, 430–434.
62. Fukui, K.; Tanaka, K., *Angew. Chem. Int. Ed.* **1998**, *37*, 158–161.
63. Harriman, A., *Angew. Chem. Int. Ed.* **1999**, *38*, 945–949.
64. Kuznetsov, A.M., *Stochastic and Dynamic Views of Chemical reaction Kinetics in Solutions*, Press Polytechniques et Universitaires Romandes: Lausanne, 1999.
65. Bixon, M.; Jortner, J., *Chem. Phys.* **2002**, *281*, 393–408.
66. Bixon, M.; Jortner, J., *Chem. Phys.* **2005**, *319*, 273–282.
67. Bixon, M.; Jortner, J., *Chem. Phys.* **2006**, *326*, 252–258.

68. Berlin, Yu.A.; Burin, A.L.; Ratner, M.A., *J. Phys. Chem. A* **2000**, *104*, 443–445.
69. Grozema, F.C.; Siebbeles, L.D.A.; Berlin, Y.A.; Ratner, M.A., *ChemPhysChem* **2002**, *3*, 536–539.
70. Berlin, Yu.A.; Kurnikov, I.V.; Beratan, D.; Ratner, M.A.; Burin, A.I., In *Long-Range Charge Transfer in DNA II*, Schuster, G.B., Ed., *Topics in Current Chemistry*, **2004**, *237*, 1–36.
71. Senthilkumar K.; Grozema F.C.; Guerra C.F.; Bickelhaupt, F.M.; Lewis, F.D.; Berlin, Y.A.; Ratner, M.A.; Siebbeles, L.D.A., *J. Am. Chem. Soc.* **2005**, *127*, 14894–14903.
72. Bar-Haim, A., Klafter, J., Kopelman, R., *J. Am. Chem. Soc.* **1997**, *119*, 6197–6198.
73. Bar-Haim, A.; Klafter, J., *J. Phys. Chem. B* **1998**, *102*, 1662–1664.
74. Wan, C.; Fiebig, T.; O'Kelley, S.; Treadway, C.R.; Barton, J.K., Zewail, A.H., *Proc. Nat. Acad. Sci. USA* **1999**, *96*, 6014–6019.
75. O'Neill, M.A., Becker, H.-C.; Wan, C.; Barton, J.K.; Zewail, A.H., *Angew. Chem. Int. Ed.* **2003**, *42*, 5896–5900.
76. Rösch, N.; Voityuk, A.A., In *Long-Range Charge Transfer in DNA I*, Schuster, G.B., Ed., *Topics in Current Chemistry*, **2004**, *236*, 37–72.
77. Saito, I.; Nakamura, T.; Nakatani, K.; Yoshioka, Y.; Yamaguchi, K.; Sugiyama, H., *J. Am. Chem. Soc.* **1998**, *120*, 12686–12687.
78. Henderson, P.T.; Jones, D.; Hampikian, G.; Kan, Y.; Schuster, G.B., *Proc. Nat. Acad. Sci. USA* **1999**, *96*, 8353–8358.
79. Schuster, G.B.; Landman, U., In *Long-Range Charge Transfer in DNA II*, Schuster, G.B., Ed., *Topics in Current Chemistry*, **2004**, *237*, 139–161.
80. Conwell, E.M., Rakhmanova, S.V., *Proc. Nat. Acad. Sci. USA* **2000**, *97*, 4556–4560.
81. Conwell, E., In *Long-Range Charge Transfer in DNA I*, Schuster, G.B., Ed., *Topics in Current Chemistry*, **2004**, *236*, 73–101.
82. Brédas, J.-L.; Street, G.B., *Acc. Chem. Res.* **1985**, *18*, 309–315.
83. Stafström, S.; Brédas, J.-L.; Logdlund, M.; Salaneck, M.R., *J. Chem. Phys.* **2001**, *99*, 7938–7945.
84. Davis, W.B.; Svec, W.A.; Ratner, M.A.; Wasilelewski, M.R., *Nature* **1998**, *396*, 60–63.
85. Voityuk, A.A.; Siriwong, K.; Rösch, N., *Angew. Chem. Int. Ed.* **2004**, *43*, 624–627.
86. Herne, T.M.; Tarlov, M.J., *J. Am. Chem. Soc.* **1997**, *119*, 8916–8920.
87. Steele, A.B.; Herne, T.M.; Tarlov, M.J., *Biocon. Chem.* **1998**, *10*, 419–423.
88. Petrovykh, D.Y.; Kimura-Suda, H.; Whitman, L.J.; Tarlov, M.J., *J. Am. Chem. Soc.* **2003**, *125*, 5219–5226.
89. Petrovykh, D.Y.; Pérez-Dieste, V.; Opdahl, A.; Kimura-Suda, H.; Sullivan, J.M.; Tarlov, M.J.; Himpsel, F.J.; Whitman, L.J., *J. Am. Chem. Soc.* **2006**, *128*, 2–3.
90. O'Kelley, S.; Jackson, N.M.; Hill, M.G.; Barton, J.K., *Angew. Chem. Int. Ed.* **1999**, *38*, 941–945.
91. Jackson, N.M.; Hill, M.G., *Curr. Opin. Chem. Biol.* **2001**, *5*, 209–215.
92. Treadway, C.R., Hill, M.G.; Barton, J.K., *Chem. Phys.* **2002**, *281*, 409–428.
93. Boal, A.K.; Barton, J.K., *Bioconjug. Chem.* **2005**, *16*, 312–321.
94. Rant, U.; Arinaga, K.; Fujita, S.; Yokoyama, N.; Abstreiter, G.; Tornow, M., *Langmuir* **2004**, *20*, 10086–10092.
95. Arinaga, K.; Rant, U.; Tornow, M.; Fujita, S.; Abstreiter, G.; Yokoyama, N., *Langmuir* **2006**, *22*, 5560–5562.
96. Wong, E.L.S.; Gooding, J.J., *Austr. J. Chem.* **2005**, *58*, 280–287.
97. Wong, E.L.S.; Chow, E.; Gooding, J.J., *Langmuir* **2005**, *21*, 6957–6965.
98. Mearns, F.J.; Wong, E.L.S.; Short, K.; Hibbert, D.B.; Gooding, J.J., *Electroanalysis* **2006**, *18*, 1971–1981.
99. Finklea, H.O., *Electroanal. Chem.* **1996**, *19*, 109–335.

100. Wackerbarth, H.; Marie, R.; Grubb, M.; Zhang, J.; Hansen, A.G.; Chorkendorff, I.; Christensen, C.B.; Boisen, A.; Ulstrup, J., *J. Solid. St. El. Chem.* **2004**, *8*, 474–481.
101. Ostatna, V.; Paleček, E., *Langmuir* **2006**, *22*, 6481–6484.
102. Heaton, R.J.; Peterson, A.W.; Georgiadis, R.M., *Proc. Nat. Acad. Sci. USA* **2001**, *98*, 3701–3704.
103. O'Kelley, S.; Barton, J.K.; Jackson, N.M.; McPherson, L.D.; Potter, A.B.; Spain, E.M.; Allen, M.J.; Hill, M.G., *Langmuir* **1998**, *14*, 6781–6784.
104. Long, Y.-T.; Li, C.-Z.; Sutherland, T.C.; Chama, M.; Lee, J.S.; Kraatz, H.-B., *J. Am. Chem. Soc.* **2003**, *125*, 8724–8725.
105. Drummond, T.G.; Hill, M.G.; Barton, J.K., *J. Am. Chem. Soc.* **2004**, *126*, 15010–15011.
106. Liu, T.; Barton, J.K., *J. Am. Chem. Soc.* **2005**, *127*, 10160–10161.
107. Jamieson, E.R.; Lippard, S.J., *Chem. Rev.* **1999**, *99*, 2467–2498.
108. Wong, E.L.S.; Gooding, J.J., *J. Am. Chem. Soc.*, submitted for publication.
109. Ceres, D.M.; Barton, J.K., *J. Am. Chem. Soc.* **2003**, *125*, 14964–14965.
110. Hihath, J.; Xu, B.; Zhang, P.; and Tao, N., *Proc. Nat. Acad. Sci. USA* **2005**, *102*, 16979–16983.
111. Haiss, W.; van Zalinge, H.; Bethel, D.; Ulstrup, J.; Schiffrin, D.J.; Nichols, R.J., *Faraday Disc.* **2006**, *131*, 253–264.
112. Kornyshev, A.A.; Kuznetsov, A.M.; Ulstrup, J., *Proc. Nat. Acad. Sci. USA* **2006**, *103*, 6799–6804.
113. Kornyshev, A.A.; Kuznetsov, A.M., *ChemPhysChem* **2006**, *7*, 1036–1040.
114. Kornyshev, A.A.; Kuznetsov, A.M., *Chem. Phys.* **2006**, *324*, 276–279.
115. Articles in *J. Biol. Phys.* **2005**, *31*, 233–644, *Special Issue Proceedings ICBP-2004*, L.Matsson, Ed.
116. Wackerbarth, H.; Zhang, J.; Grubb, M.; Hansen, A.G.; Ooi, O.B.; Christensen, H.E.M.; Ulstrup, J., In *Electrochemistry of Nucleic Acids and proteins: Towards Electrochemical Sensors for genomics and Proteomics*, Paleček, E.; Scheller, F.; Wang, J., Eds., Elsevier: Amsterdam, 2005, Chapter 15, pp. 485–516.
117. Fazli, H.; Golestanian, R.; Hansen, P.L.; Kolahchi, M.R., *Europhys. Lett.* **2006**, *73*, 429–435.
118. Wackerbarth, H.; Grubb, M.; Wengel, J.; Chorkendorff, I.; Ulstrup, J., *Surf. Sci.* **2006**, *600*, 122–127.
119. Hartwich, G.; Caruana, D.J.; de Lumley-Woodyear, T.; Wu, Y.; Campbell, C.N.; Heller, A., *J. Am. Chem. Soc.* **1999**, *121*, 10803–10812.
120. Chi, Q.; Zhang, J.; Nielsen, J.U.; Friis, E.P.; Chorkendorff, I.; Canters, G.W.; Andersen, J.E.T.; Ulstrup, J., *J. Am. Chem. Soc.*, **2000**, *122*, 4047–4055.
121. Grubb, M.; Wackerbarth, H.; Wengel, J.; Ulstrup, J, *Langmuir* **2007**, *23*, 1410–1413.
122. Kuznetsov, A.M.; Ulstrup, J., *J. Chem. Phys.* **2002**, *116*, 2149–2165.
123. Kuznetsov, A.M.; Ulstrup, J., *J. Phys. Chem. A*, **2000**, *104*, 11531–11540. Errata: *J. Phys. Chem. A* **2001**, *105*, 7494.
124. Zhang, J.; Chi, Q.; Kuznetsov, A.M.; Hansen, A.G.; Wackerbarth, H.; Christensen, H.E.M.; Andersen, J.E.T.; Ulstrup, J., *J. Phys. Chem. B* **2001**, *105*, 1131–1152.
125. Medvedev, I.G., *Russ. J. Electrochem.* **2005**, *41*, 227–240.
126. Lisdat, F.; Ge, B.; Scheller, F.W., *Electrochem. Comm.* **1999**, *1*, 65–68.
127. Warsinke, A.; Stöcklein, W.; Leupold, E.; Micheel, E.; Scheller, F.W., In In *Electrochemistry of Nucleic Acids and proteins: Towards Electrochemical Sensors for genomics and Proteomics*, Paleček, E.; Scheller, F.; Wang, J., Eds., Elsevier: Amsterdam, 2005, Chapter 14, pp. 451–483.
128. Drummond, T.G.; Hill, M.G.; Barton, J.K., *Nature Biotechnol.* **2003**, *21*, 1192–1199.
129. Boon, E.M.; Salas, J.E.; Barton, J.K., *Nature Biotechnol.* **2002**, *20*, 282–286.

130. Rothemund, P.W.K., *Proc. Int. Conf. on Computer-Aided Design ICCAD 2005, IEEE/ACM*, pp. 470–478.
131. Rothemund, P.W.K., **Nature 2006**, *440*, 297–302.
132. Tanaka, K.; Tengeiji, A.; Kato, T.; Toyama, N.; Shionoya, M, *Science* **2003**, *299*, 1212–1213.
133. Clever, G.H.; Carrell, T., *Angew. Chem. Int. Ed.* **2007**, *46*, 250–253.
134. Patolsky, F.; Zheng, G.; Lieber, C.M., *Anal. Chem.* **2006**, *78*, 4261–4269.

Chapter 7

IN SITU STM STUDIES OF IMMOBILIZED BIOMOLECULES AT THE ELECTRODE-ELECTROLYTE INTERFACE

RICHARD J. NICHOLS,[1] WOLFGANG HAISS,[1] DAVID G. FERNIG[2], HARM VAN ZALINGE,[1] DAVID J. SCHIFFRIN,[1] AND JENS ULSTRUP[3]
[1]The Chemistry Department, University of Liverpool, Liverpool L69 7ZD, UK; [2]The School of Biological Sciences, University of Liverpool, Liverpool L69 7ZB, UK; [3]Department of Chemistry and NanoDTU, Technical University of Denmark, DK-2800 Lyngby, Denmark.

1. INTRODUCTION

The immobilization of biological molecules at the solid-liquid interface is a growing interdisciplinary field, which combines biochemistry and biotechnology with physical science disciplines such as surface physics and chemistry, electrochemistry and nanoscale science. Monolayers of biological molecules such as proteins and DNA fragments at the solid-liquid interface have a wide number of important applications. These include, immobilised DNA fragments which find application in DNA-"chips", biofilms providing support for cells or tissue culture, biocompatible surfaces for medical implants and sensors based on redox proteins and enzymes. This ability to assemble biomolecules at surfaces, in which the orientation and supramolecular organization can be controlled, offers exciting possibilities for control of biological function as well as sensitive monitoring of activity.

Monolayers and films of biomolecules on surfaces have been prepared in a variety of environments, including the liquid-liquid, solid-liquid, solid-gas and solid-ultra high vacuum interfaces. The interface between two immiscible liquids can be a useful model system for biomolecular adsorption at model membranes, but this interface is generally difficult to interrogate

O. Hammerich and J. Ulstrup (eds.), Bioinorganic Electrochemistry, 207–247.

with high-resolution techniques. The solid-UHV interface can on the other hand be analysed with a wide variety of molecular specific techniques. However, the removal of water from this interface means, in many cases, that the structural integrity as well as the useful function of biomolecules can be lost. The solid-liquid environment conveys a number of advantages. Biomolecular adsorption at this interface can be achieved by direct adsorption from the liquid phase or by transfer of films formed by the Langmuir-Blodgett (L-B) technique. In the latter case the biomolecular film is first spread in a precisely controlled way at the liquid-gas interface to form Langmuir films, which are then transferred to solid-substrates using the L-B method. The film pressure and the L-B emersion parameters can be used to command a certain degree of control over the adlayer which is formed at the solid surface. Direct adsorption at the solid-liquid interface can also be controlled in a sensitive manner. Under electrochemical conditions the substrate potential and related surface charge density may be used to control the assembly and structure of adlayers. For instance, it has been shown that the structure of phospholipid monolayers can be controlled in such a way, and a description will be given in this chapter of how the structural transitions in monolayers of DNA and RNA bases can be induced by electrochemical potential control.

The last ten years have seen many important advances in the immobilization and characterisation of biomolecules at solid-liquid interfaces. The use of well defined atomically flat single crystal electrodes and probe microscopy methods combined with other state-of-the-art surface characterization techniques have given important insights into biomolecular adsorption. This article deals with selected applications of biomolecular adsorption at solid-electrolyte interfaces, focusing on the use of scanning tunneling microscopy (STM). It is clear that STM, as well as atomic force microscopy (AFM) have opened a horizon in the investigation of biomolecular adsorption, by allowing imaging down to the single molecule level. Examples are given here, for a range of biologically interesting molecules, of the use of STM in elucidating structural aspects of the adsorption as well as adlayer dynamics. We also address relevant theoretical aspects of STM imaging. A discussion is further given of new methods for measuring single molecular conductance, which are likely to find future application in fundamental biomolecular science and single biomolecule detection. In the concluding section of this chapter selected recent results in the emerging field of nano-biotechnology are presented. In this respect it is noted that fundamental knowledge gained about biomolecular adsorption at solid-liquid interfaces is expected to help in the design of such nanostructures. This opens up exciting prospects for the biosensing down to the single molecule level and the assembly of nanoscale devices for biotechnological applications.

2. DNA AND RNA BASES

The two types of nucleic acids DNA and RNA are long polymers consisting of sugar units joined together by phosphate links, with each sugar linked to a base. The sequence of bases is extremely important, as it constitutes the genetic code. The four principal bases found in DNA are the substituted pyrimidines, cytosine and thymine, and the two substituted purines, adenine and guanine. In RNA thymine is replaced by uracil. These five bases are shown in Fig. 7-1. These elementary bases are hydrogen substituted at the R position, while in DNA and RNA they are attached to the sugar ribose via the N-1 and N-9, respectively for the pyrimidine and purine bases.[4,5]

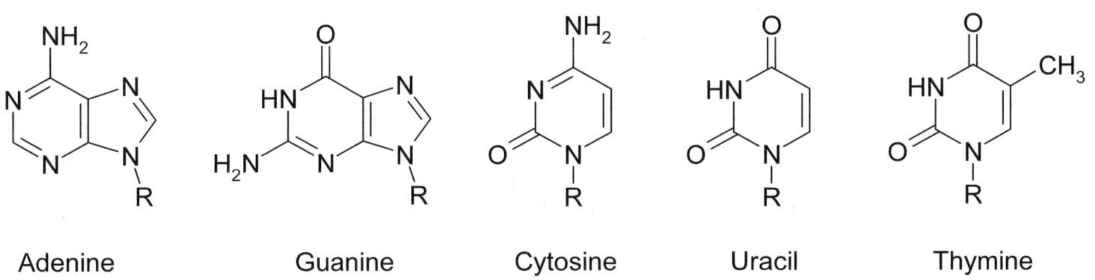

Figure 7-1. The DNA and RNA Bases

There has been considerable interest in recent years in the formation of condensed films of purine and pyrimidine bases at the solid-liquid interface. It is well recognised that non-covalent affinities between base pairs play a prevalent role in determining nucleic acid conformation and functionality. Likewise, there has been interest in the role of substrate and non-covalent intermolecular interactions in the configuration of ordered monolayers of purine and pyrimidine bases. There is also more general interest in the interaction of bases with metal surfaces and metal complexes. In the latter case it is noted that the biological role of nucleic acids and certain nucleotides are dependent on metal ions, particularly Mg^{2+}, Ca^{2+}, Zn^{2+}, Mn^{2+}, Cu^{2+} and Ni^{2+}.[4] Also certain metal complexes, notably of platinum, have the anti-tumour activity, which is linked to their ability to bind to bases on DNA.[4] On a different note, the possibility that purine-pyrimidine arrays assembled on naturally occurring mineral surfaces might act as possible templates for biomolecular assembly has been discussed by Sowerby *et al.*

These authors also invoked such monolayers in a possible model for the origin of life.[6,7]

It has been realised for a considerable time that purine and pyrimidine bases, along with a host of other organic molecules, form two-dimensional condensed monolayers at the mercury-aqueous electrolyte interface, whose structure depends on electrode potential. In an electrochemical environment the potential applied to the electrode can be used as a sensitive means of controlling the structure and formation of the condensed phase. Phase transitions in organic films at electrode surfaces have been of particular interest to electrochemists and they can be readily monitored using a variety of established electrochemical techniques.[8] For instance, the condensation of organic adlayers at electrode surfaces is generally typified by the appearance of capacitance pits, in capacitance *vs.* potential measurements, and defined voltammetric peaks in cyclic voltammograms, which may be related to two-dimensional phase transitions in the adlayers. Although a great deal has been learned about the thermodynamics and kinetics of the condensation of purine and pyrimidine bases at the mercury|electrolyte interface, it was not until the introduction of well-defined single crystalline substrates and the implementation of modern spectroscopic and imaging techniques that a detailed molecular scale description of adlayer formation was possible. In the following section a selection of the *in situ* STM studies of the purine bases, adenine and guanine, and the pyrimidine bases, cytosine, uracil and thymine are reviewed.

2.1 Adenine and Guanine

Adenine and guanine, as well as thymine, represent some of the earliest examples of ordered 2-D phases of organic monolayers imaged with STM at high resolution. In these studies adlayers were formed by a "sizzling" technique in which a droplet of the aqueous DNA base was applied to freshly cleaved molybdenum disulfide or highly ordered pyrolytic graphite sample (HOPG) and then vaporised.[9–11] From molecular resolution images of guanine on MoS_2, Heckl *et al.* concluded that that intermolecular hydrogen bonding determined the structure of the imaged two-dimensional arrays.[11] Since these early studies there has been much work in imaging, analysing and refining models for such planar arrangements of purine and pyrimidine bases, particularly on MoS_2 and HOPG surfaces. [6,7,11–19] A model for the molecular arrangement of adenine proposed by Sowerby *et al.* on the basis of molecular modelling and energy minimisation is shown in Fig. 7-2. In this model, which Sowerby *et al.* have related to STM images on MoS_2, the molecules adopt a planar arrangement on the surface, with hydrogen bonds being represented by dashed lines.[12,16] As will be discussed below, hydrogen bonding networks are also important in the formation of ordered two-dimensional adlayers of nucleic acid bases at electrochemical interfaces.

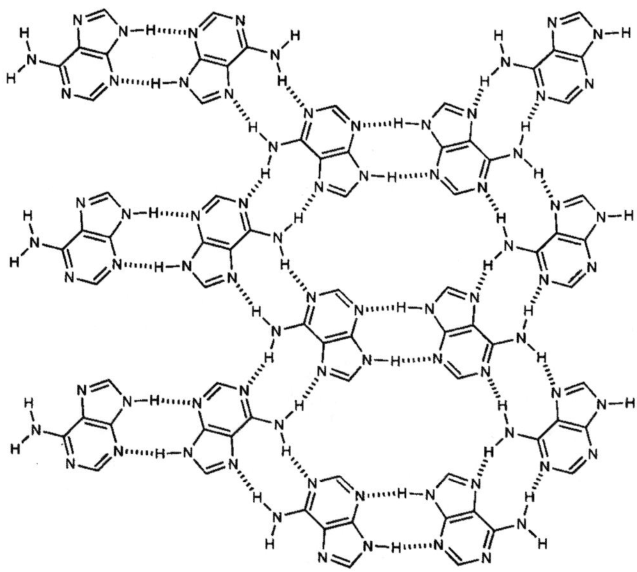

Figure 7-2. Model for the two-dimensional ordered phase of adenine on MoS_2 proposed by Sowerby *et al.*[12,16] Reprinted from ref. 16 with permission.

In a number of pioneering experiments it was also shown that guanine and adenine on HOPG could be imaged with STM under electrochemical conditions.[20–22] Other ground-breaking experiments followed in which adenine, thymine, guanine and cytosine were adsorbed on Au(111) and imaged with *in situ* STM.[23] As well as enabling adlayer structure to be studied as a function of electrode potential, the *in situ* studies have also facilitated imaging at a variety of electrolyte pH values. Tao *et al.* have varied electrolyte pH for guanine adsorbed on HOPG, to confirm the role of hydrogen bonding in the monolayer structure.[24] In this study they first formed a guanine monolayer on HOPG from 0.1 mM guanine + 0.1 M NaCl. They were able to image this monolayer *in situ* with high resolution AFM and proposed the structure shown in Fig. 7-3 composed of chains of hydrogen bonded guanine held together by van der Waals interactions.[23] While imaging, they then added 10 μL of 0.1 M NaOH to the *in situ* cell which already contained ~100 μL of 0.1 M NaCl. Dissolution of the ordered phase was observed after addition of the NaOH. This was interpreted as arising from a severe disruption of the hydrogen bonding network through loss of the N1-H·····O6 hydrogen bonds resulting from abstraction of the proton at the N1 position. This demonstrates the role of hydrogen bonding in forming the structure. Interestingly, after a few minutes of further imaging a new adlayer was formed. Since N1-H·····O6 hydrogen bonding is no longer possible at these high pH values, they proposed that the deprotonated N1 forms alternative hydrogen bonds with the amine groups in a different adlayer structure, which nevertheless also gives a two-dimensional lattice of flat lying guanine.

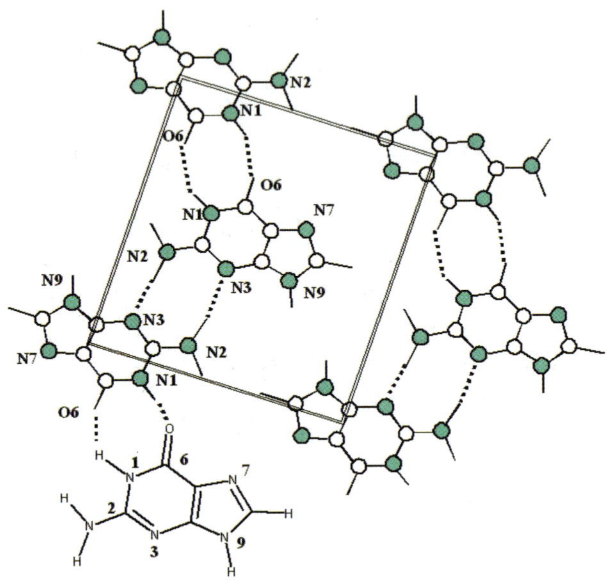

Figure 7-3. Model proposed for the 2-D hydrogen bonded network of guanine on HOPG in 0.1 M NaCl, derived from *in situ* STM images.[23] The parallelogram shows the unit cell. Adapted from ref. 23 with permission.

2.2 Cytosine, Uracil and Thymine

In situ STM has provided detailed structural models for the adsorption of cytosine, uracil and thymine at a number of electrode surfaces.[15,23,25–30] These studies have included thymine adsorbed on Au(111), Au(100) and Au(210),[1,23,27] uracil adsorption on Au(111), Au(100) and Ag(111).[1,25,26,28,31] These pyrimidine bases show a particularly interesting phase behaviour, which can be seen from cyclic voltammograms (CV). A CV for the adsorption of thymine on Au(111) is shown in Fig. 7-4.[27] Four potential regions are marked on this voltammogram. At the negative end, a disordered phase (Region I) is observed; this is followed by a 2-dimensional phase transition to a so-called physisorbed phase (Region II). A substantial current peak is observed (Region III) that is finally followed by the formation of a condensed, stable chemisorbed phase (Region IV). A similar pattern is followed for the cyclic voltammetry of uracil on Au(111).

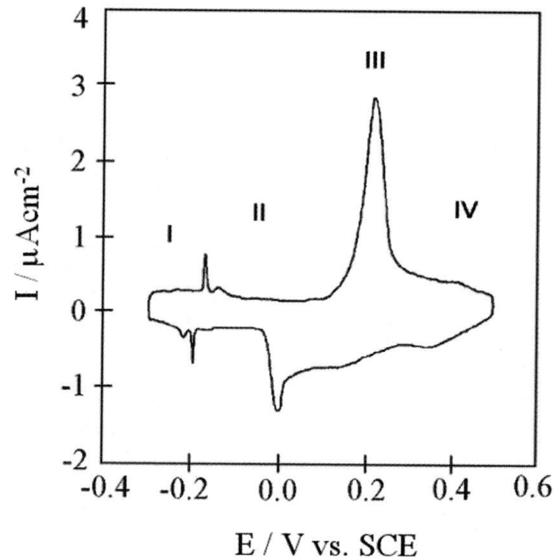

Figure 7-4. A cyclic voltammogram of Au(111) in 5 mM thymine and 100 mM $KClO_4$, recorded at 5 mV s^{-1} after potential controlled immersion of the electrode at the negative potential limit and an initial waiting period of 3 minutes at the same potential. Reprinted from ref. 27 with permission.

Analysis of cyclic voltammograms for thymine and uracil adsorption on Au(111) have revealed the formation of the ordered phase found in region II to be both pH independent and temperature sensitive. This indicates that both uracil and the thymine are in their protonated form in this potential region and are rather weakly adsorbed.[32,33] Detailed analysis of this physisorbed region has been provided through chronocoulometric[33] and *in situ* STM investigations.[1,27–29] Chronocoulometry has furnished quantitative thermodynamic parameters such as film pressure, Gibbs surface excess and Gibbs energy of adsorption, which indicate a weak chemisorption.[33] *In situ* STM images of this phase show a 2-dimensional pattern of regularly shaped "features".[1,26–29] These highly periodic features were interpreted as a resolution of individual molecules. On the basis of the assumption that each individual feature represented an adsorbed uracil or thymine molecule, unit cells have been constructed for both systems.

The interplay between substrate-adsorbate and adsorbate-adsorbate interactions in the structure of the phase formed in Region II has been analysed.[28] This has been achieved by comparing structural parameters obtained by *in situ* STM for a number of different surface crystallographic orientations, including thermally reconstructed and potential-induced Au(111)-$(p\times\sqrt{3})$, exhibiting large and small domains respectively and the hexagonal and (1x1) reconstruction of Au(100). An STM image and corresponding model for the adlayer are shown in Fig. 7-5.

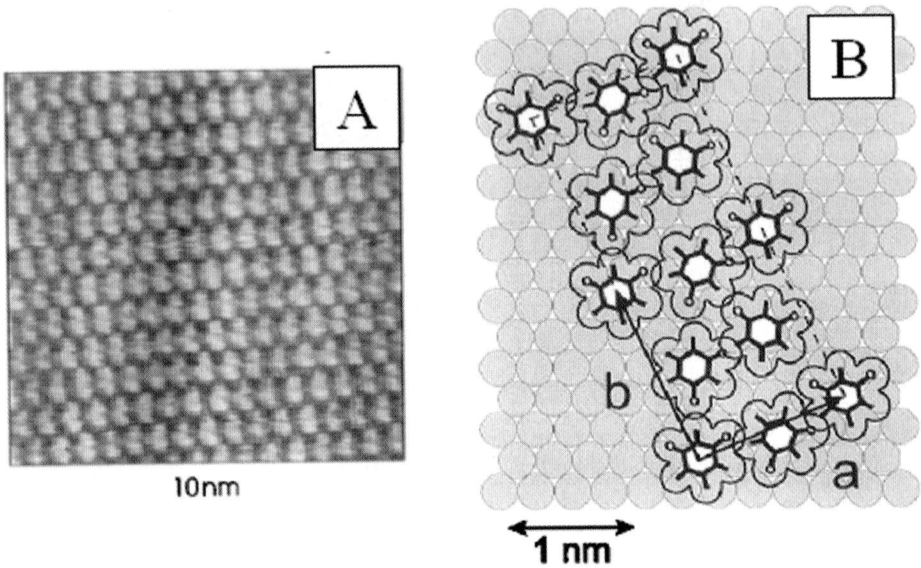

Figure 7-5. In situ STM image and model of the physisorbed phase of uracil. Reprinted from ref. 28 with permission.

The parameters for the primitive unit cell are a = (12.2 ± 0.6) Å, b = (15.1 ± 1.0) Å and α = (90 ± 2)°. In this model the uracil molecules lie flat on the gold surface, opening the possibility for hydrogen bonding interactions to stabilize the structure. Wandlowski *et al.* have noted that this structure bears similarities to the hydrogen bonded two-dimensional structure within crystalline uracil.[28] Notably, slight variations were observed between the unit cell parameters, for adlayers on Au(111)-$(p\times\sqrt{3})$ and the hexagonal and (1x1) reconstruction of Au(100). This indicates that lateral interactions play a determining role in the adlayer structure with a lesser influence of the substrate geometry. STM studies of uracil and thymine adlayers on a number of other surfaces, both at the solid-liquid, solid-air and vacuum interface, have also led to an observation of two-dimensional arrays, for instance uracil and thymine on highly ordered pyrolytic graphite (HOPG), molybdenum disulfide and on Cu(111) in UHV.[15,18,34] In each case hydrogen bonding was distinguished as a major factor in adlayer ordering.

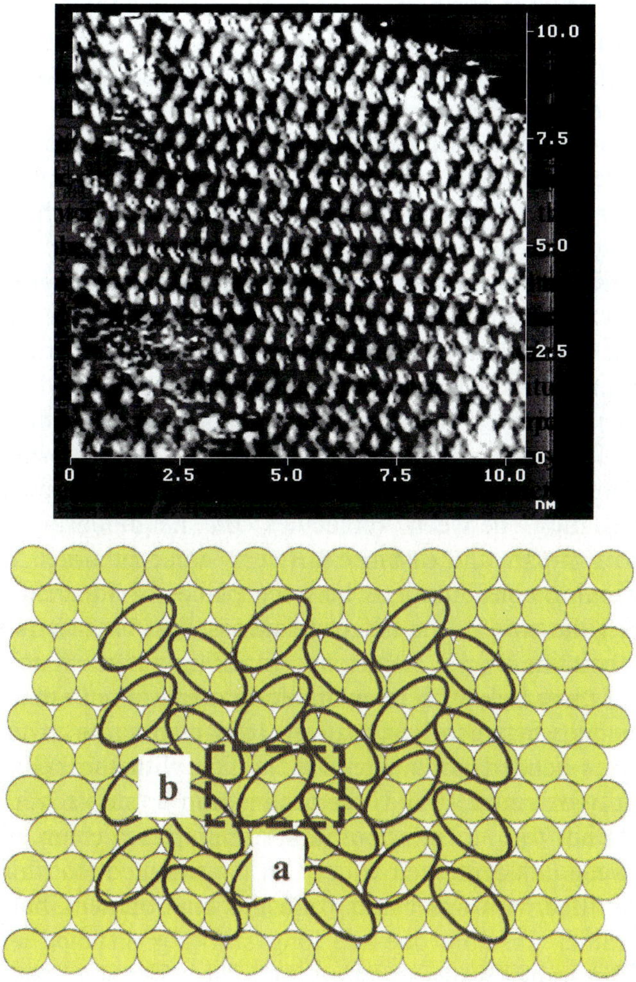

Figure 7-6. (a, top) STM image of the chemisorbed phase of uracil on Au(111) at +700 mV (SCE) in 6 mM uracil + 10 mM $HClO_4$ + 25 mM $KClO_4$.[1]; (b, bottom) Adsorption model for chemisorbed uracil.[1,28] See ref. 1 for further STM images of uracil on Au(111).

In Region III, Fig. 7-4, there is a transition between the so-called physisorbed and the chemisorbed phases, which is characterised by a large current flow associated with a partial charge transfer. In addition pH dependent studies have revealed that this process involves the abstraction of one proton.[23,30] The deprotonation of both uracil and thymine could occur at either the N(1) or N(3) positions. An *in situ* STM image of the chemisorbed phase is shown in Fig. 7-6a, recorded at 700 mV (*vs.* SCE) for Au(111) in 6 mM uracil, 10 mM $HClO_4$ and 25 mM $KClO_4$. A $c(\sqrt{3}\times 3)$ has been proposed for the structure by Dretschkow *et al.*[28] In Fig. 7-6a each spot, which has a distinctive elliptical shape, is considered to be an individual uracil molecule. These spots in the STM image appear to mirror an end-on profile of chemisorbed uracil molecules. The molecules align along rows and a rotation between adjacent rows is apparent. The observation of uniform contrast in this image would favour identical adsorption sites for molecules on adjacent molecular rows, which on the basis of the commensurate unit cell dimensions could be 3-fold hollow sites.

The adsorption model for the chemisorbed phase, described by a $c(\sqrt{3}\times 3)$ structure, which contains 2 uracil molecules per unit cell, is shown in Fig. 7-6b. In this model vertically oriented molecules stack with their molecular planes facing each other and individual rows are rotated by 45° with interdigitation between molecules on neighbouring rows. It is interesting to compare the distance between adjacent uracil molecules on rows with stacking configurations between planes of adjacent base pairs of DNA. In DNA the core of the helix is composed of base pairs that are held together by both hydrogen bonds and stacking interactions. The energy resulting from this stacking is comparable in magnitude to that produced by hydrogen bonding between base pairs.[35] It is instructive in this respect to compare fully extended polynucleotide strands with the helix structure of DNA. The fully extended strands are arranged in a “stepladder” structure in which they do not form a helix, but rather line up in a chain with a spacing of 6.8 Å between adjacent nucleotides.[35] It is presumed that this rather large gap would be filled with water in the aqueous environment, but the resulting structure is rather unstable due the long stacking distance and the rather fragmented long-range water structure.[35] On the other hand, the double helix is formed by a simple clockwise twist of the stepladder structure and results in a much closer base pair spacing of 3.4 Å.[35] It is interesting to note that the 5 Å spacing along the stacked rows in uracil on Au(111) is intermediate between the spacing in the helix structure of DNA (3.4 Å) and that in the step laddered polynucleotide strands (6.8 Å). This would imply that the stacking forces in uracil on Au(111) are considerably less that those in DNA, leading to the conclusion that there are other more significant energetic contributions to the formation of this adlayer. Clearly, these result from interactions between uracil and the gold substrate in the form of chemisorption bonds.

Although STM can give rather precise details on the adlayer structure, it does not give direct information on the nature of the chemisorption between the pyrimidine bases and the gold substrate. Spectroscopic methods, in the form of *in situ* infrared spectroscopy and *ex situ* XPS have greatly helped in this respect. In the case of thymine adsorption on Au(111), XPS has shown that one of the nitrogen atoms is chemically modified, but only in the case of the chemisorbed film.[27] This implies that the deprotonated nitrogen directly binds to Au(111). *In situ* infrared spectroscopy has provided further details. The two strongest bands of thymine in aqueous solution in the mid-infrared are at 1703 cm^{-1} and 1672 cm^{-1} and these two bands have major contributions from the stretching vibrations of C2=O and C4=O, respectively (see Fig. 7-1 for the thymine structure). Figure 7-7 shows an *in situ* IR spectrum of the chemisorbed phase of thymine on Au(111).[1,36] Contributions of solution bands have been omitted from this spectrum and two clear bands from the adsorbate are observed at 1655 and 1596 cm^{-1}. These bands are assigned to the two respective C=O stretches of chemisorbed thymine.[1,36] Both stretching frequencies are considerably red-shifted when compared with the solution bands for thymine. Similar spectral data has been obtained for uracil.[1] On the basis of *in situ* IR, XPS and electrochemical data, proposed models for uracil and thymine chemisorption are shown in Fig. 7-8.[1,36] The deprotonated N3 as well as both exocyclic oxygens face in towards and interact strongly with the metal surface.

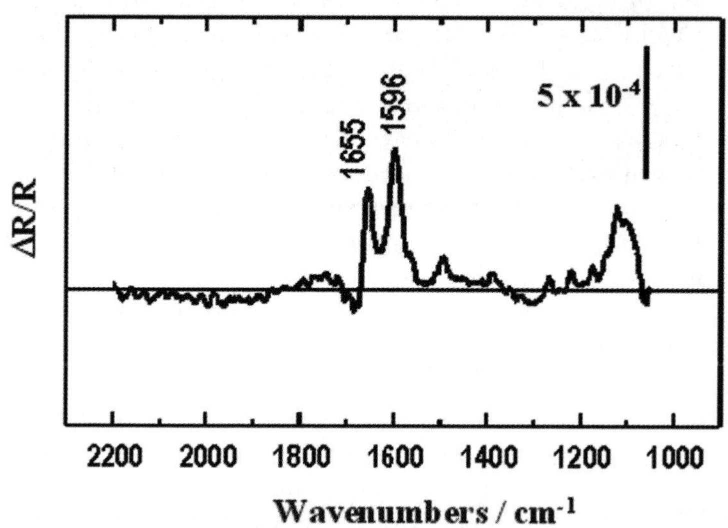

Figure 7-7. In situ IR spectrum of chemisorbed thymine at 0.8 V (*vs*. SCE) on Au(111) in 10mM $HClO_4$ + 10 mM thymine. Reprinted from ref. 36 with permission.

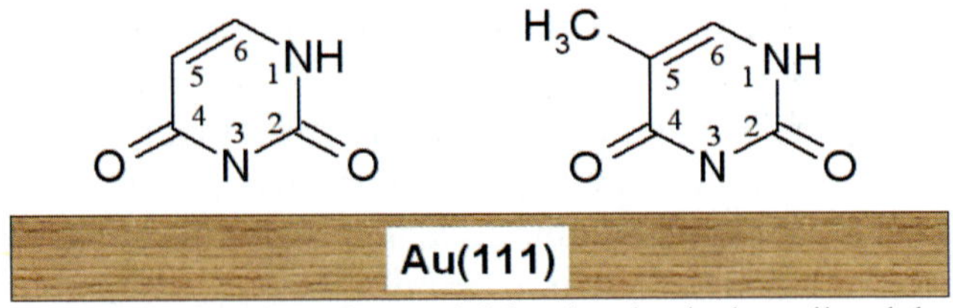

Figure 7-8. A proposed adsorption model for chemisorbed uracil and thymine.[1,36]

The dependence of adlayer structure on the molecular structures has been examined in a comparative study of the 3 pyrimidine bases, uracil, thymine and 3-methyl uracil.[1] *In situ* STM imaging of thymine shows that an adlayer geometry similar to uracil is adopted, with the exception that the adlayer is expanded in one unit cell dimension.[1] An *in situ* STM image of thymine chemisorbed on Au(111) from 10 mM thymine and 10 mM $HClO_4$ is shown in Fig. 7-9A.

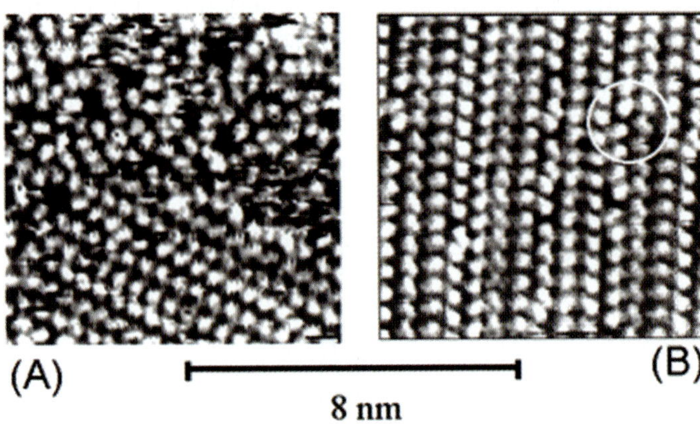

Figure 7-9. An *in situ* STM image of thymin on Au(111), recorded at +700 mV vs. SCE, after 1 (A) and 3 (B) temperature annealing cycles. Reprinted from ref. 1 with permission.

This image was recorded in region IV, Fig. 7-4, and shows both ordered phase regions and disordered regions. In small ordered domains this adlayer appears at first sight to be described by a $c(4\times\sqrt{3})$ unit cell. This would imply that when compared with uracil there is an extension along one unit cell direction to accommodate the steric requirement of the methyl group, which is substituted at the 5-position in thymine but not present in uracil. It is apparent also that there is a significantly lower ordering of thymine adlayers when compared to uracil adlayers.[1] The STM image in Fig. 7-9A was recorded after ramping the temperature *in situ* in the STM cell from 20°C to 80°C and then returning to room temperature for STM imaging.[1] After performing this "annealing" cycle two further times the order in the overlayer was substantially improved, Fig. 7-9B, and further details can be obtained about the adlayer structure. This illustrates the use of *in situ* thermal annealing to improve the ordering of molecular films at the solid-liquid interface.[1]

It is clear from Fig. 7-9B that thymine exhibits a more complex chemisorbed phase than uracil. Although, in common with uracil, it maintains a constant $\sqrt{3}$ displacement between neighbouring molecules along the rows, the separation of adjacent rows is variable. This uneven inter-row spacing has been accredited to different stacking possibilities of the thymine molecules.[1] If one designates the C-5 methylated side of the molecule as the "tail" and the non-methylated side as the "head", there are a number of possible configurations of adjacent rows, including head-to-head, head-to-tail and tail-to-tail.[1] Due to the spatial requirement of the methyl group the former configuration would be expected to give rise to the largest inter-row spacing. In spite of these different inter-row arrangements, it is apparent from STM imaging that the spacing along rows remains constant for a certain distance until a defect is reached which can result in either a decrease or increase in the separation between rows. Such a switch in spacing is marked by the white circle in Fig. 7-9B. The fact that these transitions in inter-row spacing occur only occasionally along a certain row illustrates that good alignment along rows enhances favourable π-stacking interactions. The intrinsic complexity of these thymine structures when compared to uracil and the steric hindrance to ordering provides a molecular scale rationalisation for the kinetic inhibition of forming well-ordered large domain structures of thymine.[1]

3. CYSTEINE

Cysteine is one of the 20 common amino acids found in protein molecules (or 21 if selenocysteine is included). There has been interest in studying the interaction of amino acids with solid surfaces since they are expected to play a crucial role in the chemisorption of proteins and polypeptides. The redox electrochemistry of immobilized proteins, chiefly metalloenzymes, is of interest for biosensing applications[37] and also fundamental studies of electron transfer. Self-assembled monolayers of cysteine have also themselves been used for the immobilization of proteins.[38] The molecular structure of cysteine is shown in Fig. 7-10. In common with all other amino acids the alpha carbon is covalently bonded on one side to an amino ($-NH_2$) group and the other side to a carboxylic acid group (–COOH). Importantly, cysteine has a thiol side chain. It is well recognised that alkane thiols strongly chemisorb to gold surfaces. Likewise there is interest in the chemisorption of cysteine to gold surfaces, as it may provide a model for the attachment of proteins with peripheral cysteine residues to surfaces. Proteins may also be produced with such cysteine residues by site directed mutagenesis to aid surface binding.

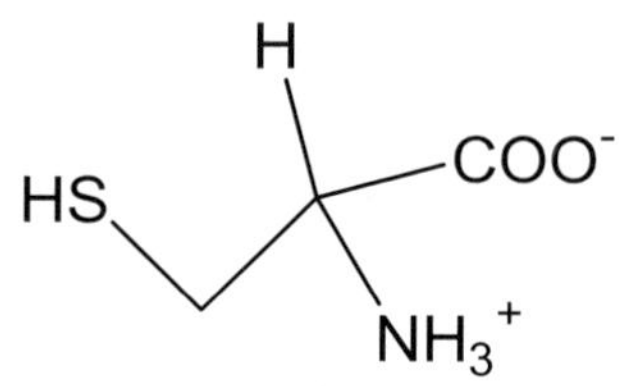

Figure 7-10. Cysteine

There have been a number of *in situ* STM studies of cysteine adsorption on Au(111).[39–42] Dakkouri *et al.*[39] have obtained a $(\sqrt{3}\times\sqrt{3})R30^o$ structure for the monolayer on Au(111) in an acidic electrolyte (0.1 M $KClO_4$ + 1 mM $HClO_4$), giving a coverage of 1/3. Zhang *et al.* [40,41] on the other hand propose a $(3\sqrt{3}\times 6)R30^o$ structure of clusters of 6 cysteine molecules, with a coverage of about 0.17. They obtained images in ammonium acetate buffer (pH 4.6). Xu *et al.*[42] have proposed yet a different structure for cysteine on Au(111) in 0.1 M $HClO_4$ with a $(4\times\sqrt{7})R19^o$ unit cell and a coverage of 0.16. These differences may arise, at least in part from different adsorption and imaging conditions. The charge properties of the adsorbate are likely to play a major role in structure formation, through electrostatic and hydrogen bonding interactions. In this respect it is noted that cysteine is an overall neutral zwitterion in ammonium acetate buffer (pH 4.6), while the amino group is protonated in lower pH solutions such as 0.1 M $HClO_4$.

On a larger scale Zhang *et al.*[40] have observed a superstructure pattern in *in situ* STM images of cysteine adsorbed on Au(111), which can be used to establish the repetition in the structure and hence the unit cell. Along the $[110]$ direction, which is parallel to Au atomic rows, a period of 17 Å was observed, while perpendicular to the atomic row direction a period of 15.2 Å was measured. This gives unit cell dimensions of 6 times the diameter of a gold atom (a) and $3\sqrt{3}a$, respectively. The unit cell is thus described as $(3\sqrt{3}\times 6)R30^o$. However, this unit cell dimension does not fully detail the adlayer, since the number of cysteine molecules per unit cell is required. This has been determined with reference to electrochemical reductive desorption measurements and analysis of the STM contrast within the unit cell. In the former case voltammograms of cysteine adsorbed on gold in 0.1 M NaOH show a sharp desorption wave around –700 mV (*vs*. SCE), which can be attributed to cleavage of the Au-S bond through a one-electron electrochemical reduction process. The coulometric charge associated with this process has been used to determine the coverage of $(4.0 \pm 0.4) \times 10^{-10}$ mol cm^{-2}. This translates to a coverage of (41 ± 4) Å^2 per cysteine molecule. From this value it has been concluded that each unit cell contains six cysteine molecules. Interestingly, they have also observed different types of hole-like defects in the structures imaged. The smallest hole at about 40 Å^2, is close to the size of a single cysteine molecule. This also supports the idea that the unit cell is composed of clusters of cysteine molecules.

In developing a model of the arrangement of cysteine molecules within the unit cell Zhang *et al.* have made certain assumptions about contrast in high-resolution STM images.[40] They have tentatively assumed that the contrast of spots in the STM image may be expected to decrease in the order sulfur>amine>hydroxyl. In making this assumption they have noted that this should be applied cautiously since it is on the basis of *ex situ* STM of various molecular adsorbates rather than imaging in aqueous electrolyte solutions. The model that they have proposed also allows for an arrangement of cysteine molecules that facilitates hydrogen bonds and electrostatic interactions between carboxylic acid and amine moieties. The model, comprising clusters of 6 cysteine molecules, and an *in situ* STM image are shown in Fig. 7-11.

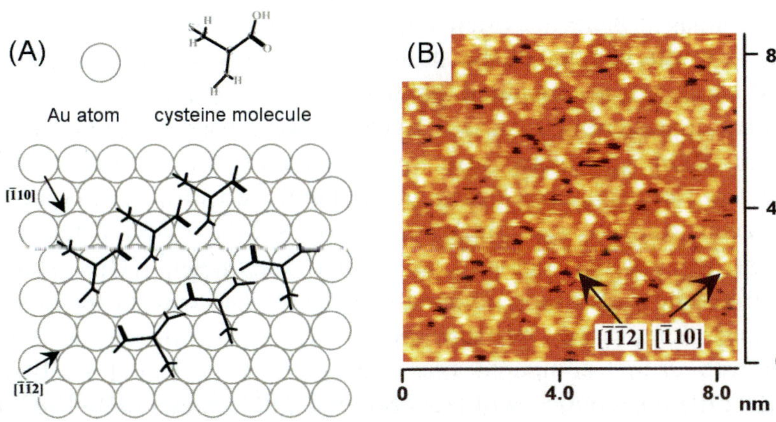

Figure 7-11. (A) A proposed superstructure of six cysteine molecules on Au(111) based on hydrogen bonding and (B) an *in situ* STM images of cysteine on Au(111) in 50 mM NH_4Ac (pH 4.6) + 1.0×10^{-6} mol cm^{-3} cysteine at 0.1 V *vs.* SCE. Reprinted from ref. 40 with permission.

In this model two rows of three cysteine molecules face each other. The amine group of a given cysteine is placed close to a carboxylic acid group on a neighbouring molecule. This is expected to support the formation of hydrogen bonds. Amine-carboxylic acid group interactions could also occur across the pair of rows, giving rise to a cluster stabilized by intermolecular hydrogen bonding.

4. DNA AND OLIGONUCLEOTIDES

DNA was one of the first biomolecules to be imaged by scanning probe microscopy. One of the original motivations for imaging DNA and DNA fragments at surfaces was the anticipation that high-resolution images may furnish the possibility of sequencing down to the individual molecule level.[43–47] Both STM and AFM are attractive tools for imaging DNA, since they can be used to image DNA in its fully hydrated condition and are able to investigate conformational states. AFM has now been widely applied to investigate DNA curvature and condensation at surfaces, the mapping and sizing of DNA molecules, DNA-protein interactions and binding forces between complementary strands.[48,49]

Electrochemical STM has also been used to image oligonucleotides. Lindsay *et al.* have imaged several synthetic oligonucleotides adsorbed on positively charged Au(111) electrodes.[50] Images of single stranded and double stranded molecules have been obtained in electrolyte. In the high-resolution images there is evidence of internal structure and image features are related to the helicity of the strands. From these images and models of the STM contrast it is concluded that the double helices have a major groove periodicity consistent with a 36° twist.[50] This imaging demonstrates the ability of STM to resolve structural characteristics of differing DNA fragments *in situ* in aqueous electrolyte and under potentiostatic control.

In a recent study, Wackerbarth *et al.* have used *in situ* STM to image the formation of ordered domains of DNA fragments.[3] This study has demonstrated how electrode potential can be used to control the adsorption mode and long-range domain order of adsorbed oligonucleotides. Wackerbarth *et al.* compared the adsorption on Au(111) of two thiol modified oligonucleotides with ten adenine nucleobases (HS-10A) and a single adenine base (HS-A) to which a hexamethylene thiol linker was covalently attached at the 5'-end.[3] XPS and reductive desorption voltammetry clearly showed that both oligonucleotides adsorb through a Au-S bond, with a coverage corresponding to an upright or tilted orientation. An *in situ* STM image recorded at –0.21 V (*vs*. SCE) after adsorbing HS-10A on Au(111) at the open circuit potential is shown in Fig. 7-12 (left). Although oligonucleotide adsorption is evident, no long-range structural order is seen. After lowering the potential to –0.61 V another image was recorded and ordered $(\sqrt{3}\times 4)R30^o$ domains were observed, Fig. 7-12 (right).

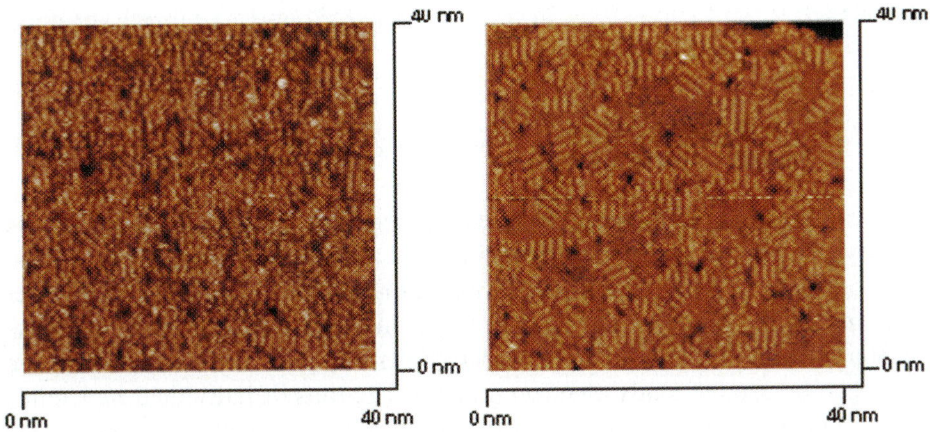

Figure 7-12. In situ STM images of HS-10A in 0.01 M phosphate buffer (pH 7.1). The oligonucleotide was adsorbed under open circuit potential and imaged at the sample potential –0.21 V *vs.* SCE (left image). By decreasing the sample potential to –0.61 V *vs.* SCE domain formation, displayed by areas with ordered structure, was observed (right image). Reprinted from ref. 3 with permission.

On the basis of this STM imaging Wackerbarth *et al.* have proposed an electrostatic model for this electrochemically induced surface ordering of the DNA fragments.[3] This is schematically depicted in Fig. 7-13.

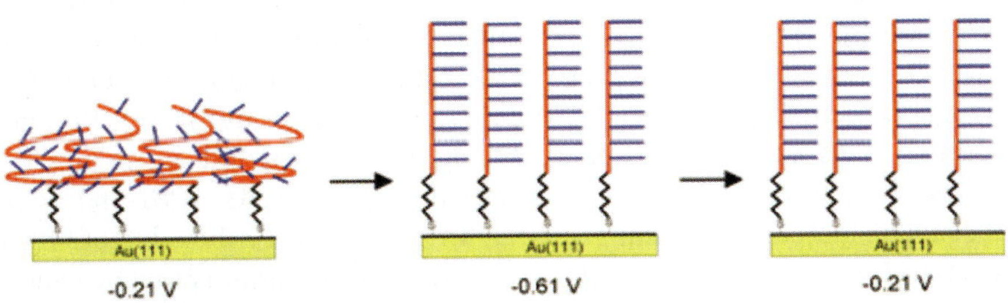

Figure 7-13. Orientation of HS-10A oligonucleotide adsorbed in ordered domains on the Au(111)-electrode. Reprinted from ref 3 with permission.

At –0.21 V the oligonucleotide chains are in a disordered state. Upon decreasing the potential to –0.61 V, the increased electric field at the electrode-electrolyte interface induces a conversion between the coiled state and an elongated state, in which the chain takes on an upright and ordered configuration. This study by Wackerbarth *et al.* hence demonstrates the use of electrochemical potential to control assembly and conformation of biomolecules and the ability of *in situ* STM to image the dynamics of such processes.[3]

5. PORPHYRINS

There is a rich electrochemistry associated with metalloporphyrins. Iron porphyrins are an essential component of the oxygen binding proteins, haemoglobin and myoglobin, and important cofactors in electron-transfer enzymes, while related cobalt corrin found in vitamin B_{12} is essential for its carbon transfer abilities. In novel experiments Tao has demonstrated how the STM can be used to study electron transfer reactions of individual molecules adsorbed on surfaces from aqueous electrolyte solution.[2] In these experiments monolayers composed of Fe(III)-protoporphyrin IX ("FePP", Fig. 7-14, left) and protoporphyrin IX ("PP", Fig. 7-14, right) were coadsorbed on a highly ordered pyrolytic graphite (HOPG) electrode. These mixed monolayers formed regular two-dimensional ordered structures, with the plane of the porphyrin parallel to the HOPG surface.[2]

FePP exhibited a reversible one-electron transfer reaction at –0.48 V (*vs.* SCE), while PP was electroinactive in the potential region studied. Figure 7-15 shows STM images (labelled A–E) and corresponding height profiles (labelled F–J) of an FePP molecule imbedded in an ordered array of PP molecules.[2] The upper image, acquired at –0.15 V, which is significantly away from the reduction of FePP shows little height difference between the FePP molecule in the centre of the image and surrounding PP molecules. However, image C, recorded close to the reduction potential for FePP shows a much larger apparent height and hence image contrast for the FePP molecule. This increase in apparent height is interpreted as arising from alignment of the Fermi level of the metal electrode with the lowest unoccupied molecular orbital (LUMO) of FePP. On further decreasing the potential, Fig. 7-15E, these energy levels are moved out of resonance and the uniform contrast is restored. As well as providing a new method for studying electron transfer at the single molecule level these observations demonstrate how structurally similar adsorbate molecules can be identified on the basis of their differing redox properties. We address these data again in Section 8. Other data and theoretical approaches are addressed in Chapter 8.

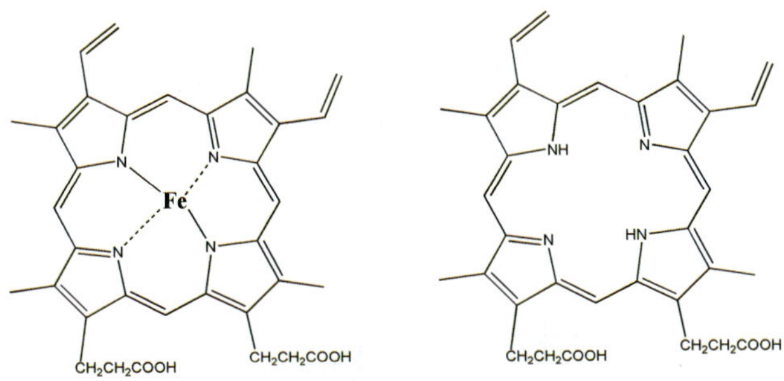

Figure 7-14. Porphyrin structures: Fe(III)-protoporphyrin IX ("FePP", left hand side) and protoporphyrin IX ("PP", right hand side).[2]

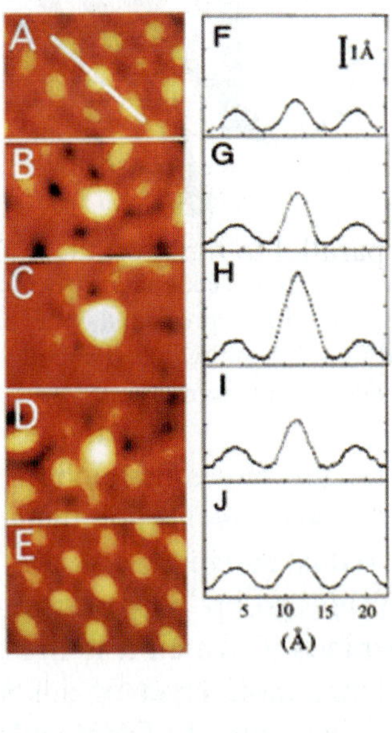

Figure 7-15. In situ STM images of a FePP molecule embedded in an ordered array of PP molecules on graphite under electrochemical potential control. The substrate potentials were 0.15 (A), 0.30 (B), 0.42 (C), 0.55 (D) and 0.65 V (E), respectively. (F)–(J) are the corresponding plots of the cross sections along the white line indicated in (A). Reprinted from ref. 2 with permission.

6. METALLOPROTEINS

A model of the structure of the blue single copper centre metalloprotein *Pseudomonas aeruginosa* azurin, is shown in Fig. 7-16.

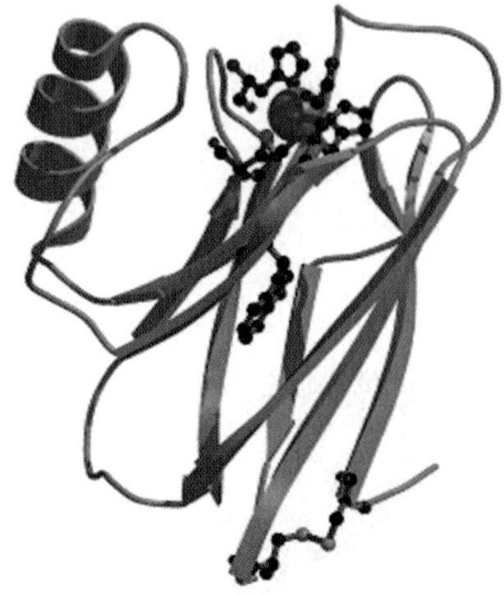

Figure 7-16. Structure of azurin. Protein secondary structure is represented by β-strands (ribbons with arrows) and a single α-helix (represented by a coil). The amino acid side chains coordinating the copper ion and those of the two cysteine residues forming a disulfide bond are also shown. Reprinted from ref. 53 with permission.

Azurin has been more thoroughly investigated by *in situ* STM than any other protein, in efforts to understand its immobilization and electron transfer mechanisms through this redox active protein.[51–60] Azurin is adsorbed on Au(111) through its disulfide group and electron transfer to the copper centre is facilitated by β-strands to which it is linked. Figure 7-17 shows an *in situ* STM image of a dense monolayer of adsorbed azurin molecules in ammonium acetate buffer. The contrast of molecularly resolved *in situ* STM images of azurin has been investigated as a means of probing the mechanism of electron transfer through the immobilized biomolecule. Figure 7-18 shows images recorded as the potential was swept from ca. –0.05 to 0.1 V, relative to the equilibrium redox potential of the $Cu^{2+/+}$ centre. Groups of azurin molecules have been highlighted in images (a)–(c) to identify features between images recorded at successively increasing electrode potential. Focusing on the larger ringed set of features in image (b), it can be seen that the number of spots, attributed to individually resolved azurin molecules, decreases from 6 to 5 in image (c). However, the contrast is markedly reduced in image (d), showing the blurring that is apparent when the $Cu^{2+/+}$ equilibrium potential is crossed.

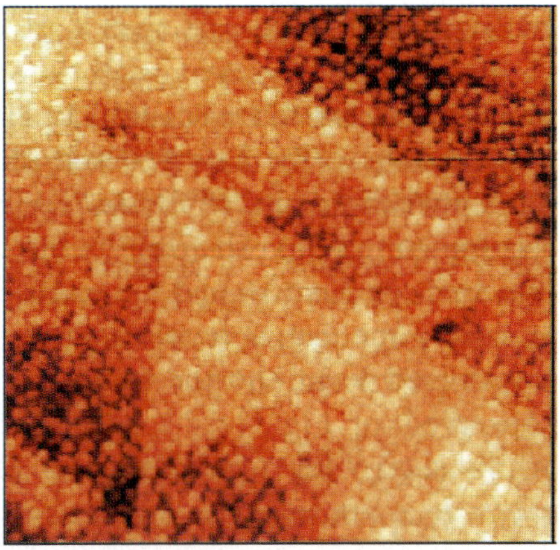

Figure 7-17. An *in situ* STM image of azurin adsorbed on Au(111). Image dimensions 200 nm x 200 nm. Reprinted from ref. 53 with permission.

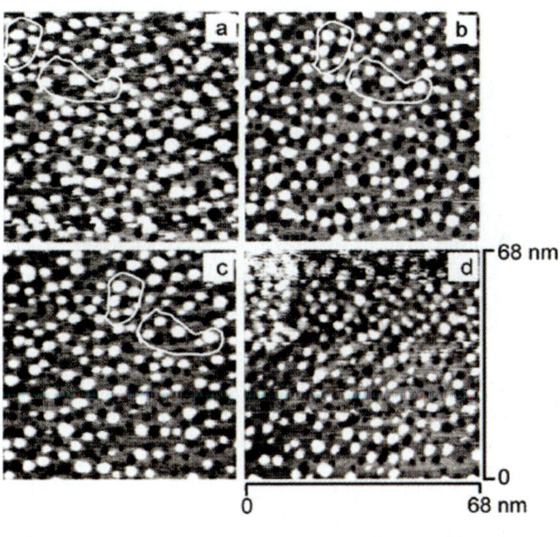

Figure 7-18. A sequence of *in situ* STM images of azurin adsorbed on Au(111) in 50 mM NH_4Ac (pH 4.6). Taken for increasing positive substrate potentials: –0.05 (a), 0.0 (b), +0.05 (c) and +0.10 V (d). Potentials quoted are relative to the equilibrium redox potential of the $Cu^{2+/+}$ centre. Note that the image contrast has decreased in images (c) and especially (d), which are above the equilibrium potential at the copper redox site (see the text). Reprinted from ref. 57 with permission.

These *in situ* STM images of azurin contain "spectroscopic" information reflecting the positions of respective energy levels and the mechanisms of electron transfer. A brief qualitative description of the model proposed by Zhang *et al.* to rationalise these contrast changes is given here, followed by further discussion in Section 8.[57] Figure 7-19 shows the energy levels involved; the Fermi levels of the substrate (ε_s, left) and the tip (ε_{tip}, right) and the energy levels corresponding to the oxidised and reduced state of azurin (ε_{ox} and ε_{red}), in the gap between tip and substrate. The bias voltage between tip and substrate is also shown (eV_{bias}). The drop in the contrast occurs as the substrate potential is made more positive and the oxidised and reduced levels are raised sufficiently above the Fermi level of the substrate. In this state azurin is in its stable oxidised form and the reduction in tunneling current and hence contrast observed in Fig. 7-18d is attributed to the need for thermal activation for this energy level configuration. By comparison, when the electrode potential is at a lower value and the occupied (reduced) and vacant (oxidised) redox levels lie between the substrate and tip Fermi levels, activationless coherent electron transfer prevails giving rise to the strong contrast observed in Figs. 7-18a and 7-18b.

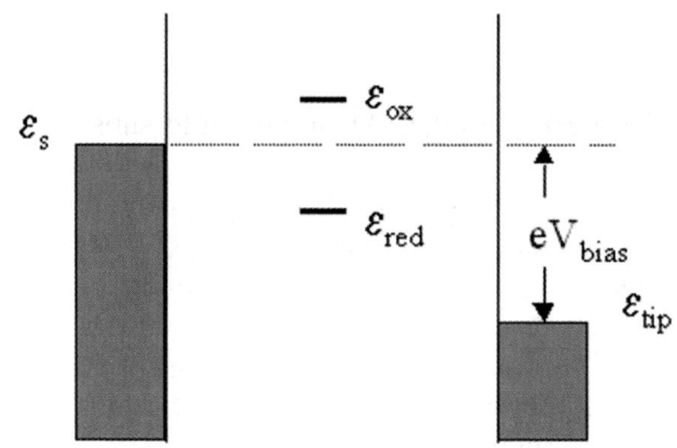

Figure 7-19. Energy level diagram for tunneling in an STM configuration involving a redox active molecule. The energy levels involved are shown; the Fermi levels of the substrate (ε_s, left) and the tip (ε_{tip}, right) and the energy levels corresponding to the oxidised and reduced state of the redox active molecule (ε_{ox} and ε_{red}), in the gap between tip and substrate.

Although no other technique is able to deliver such details of electron transfer at the single molecule level some limitations have been noted. Protein layers need to be stable, requiring strong adsorption and at most limited mobility during imaging. In this respect, adsorption through the disulfide, as for azurin, seems particularly favourable. Ideally, tip and field effects should be minimised since they could alter the electron transfer pattern. Bias voltage should be limited, since as in optical spectroscopy, the narrower the probing energy, which in this case is linked to the tip bias voltage, the better the spectroscopic resolution. However, a certain threshold bias voltage ($V_{bias} \geq 0.2$ V) was required to image azurin, potentially reducing "spectroscopic resolution".[57] Nevertheless, these results show that *in situ* STM, in which both the tip-substrate bias and the electrode potential can be varied, offers unique insights into long-range electron transfer through metalloproteins.

7. MEASURING SINGLE MOLECULE CONDUCTANCE

It has been shown that both a conducting probe AFM as well as a scanning tunneling microscope can be used to measure single molecule conductance (SMC). These new techniques have promising applications in biophysics. These probe microscopy techniques rely on forming molecular wires between the tip and substrate surface. In the technique introduced by Cui *et al.* the molecular wires are formed between the substrate surface and a gold nanoparticle which is subsequently probed by a conducting AFM tip.[61] On the other hand the technique developed by Tao *et al.* leads to direct formation of molecular wires in a break junction configuration after direct metallic contact between a gold STM tip and gold substrate is broken.[62] By contrast in the techniques developed by Haiss *et al.* the STM tip is not directed into the gold surface, instead the tip is brought into very close proximity with the surface with consequent formation of molecular bridges between tip and substrate.[63,64] Molecules of biological interest which have been studied by these techniques include carotene, studied by the method of Cui and Lindsay,[65,66] peptides studied by the method of Xu *et al.*,[67,68] and DNA studied by the methods of Xu *et al.*[69] and Haiss *et al.*[70] The following text illustrates these methods with the examples of carotene and DNA.

In the single molecule conductance determinations of Cui *et al.*, the molecule under investigation was embedded at high dilution into a matrix of alkanethiol molecules attached to a gold electrode.[61] The molecules were then connected to gold nanoparticles so that a molecular bridge was formed between the gold electrode and the nanoparticle. Since the molecule under examination was terminated by thiol groups at both ends, this served to provide good electrical contact at both ends of the molecules. The alkanethiol matrix, on the other hand, was chosen since it does not make chemical contact with the gold nanoparticle and provides a high resistance medium in which the target molecules are separated, enabling determination of single molecule electrical properties. These electrical measurements were performed by making direct electrical contact between the conducting AFM probe and the nanoparticle. A schematic representation of this measurement technique is shown in Fig. 7-20.

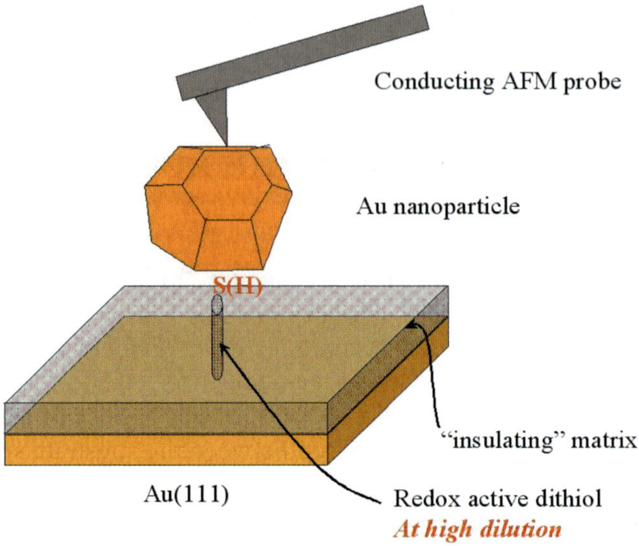

Figure 7-20. An illustration of the conducting AFM developed by Cui *et al.*[61] for measuring single molecule conductance.

This conducting AFM method has been used to measure the single molecule conductance of a number of molecules including alkanedithiols,[61] conjugated "molecular wires" and carotene.[65,66] The I-V curves recorded were observed to cluster around integer multiples of a fundamental curve. This phenomenon has been related to discrete groups of conductance, corresponding to 1, 2 and 3 target molecules being isolated between the Au nanoparticle and surface. Single molecular conductance was then calculated by fitting the fundamental I-V curve at low bias (the linear I-V region).

The conducting AFM method has revealed interesting details of the single molecule conductance of carotene molecules.[65,66] The electronic properties of carotenoids are of interest since they play a role in their biological function, especially their optical properties which are a manifestation of their delocalised electron structure. The structure of the dithiolated carotenoid used in a study by Ramachandran *et al.* is shown in Fig. 7-21.[66] A synthetic molecular wire 2,5-di(phenylethynyl-4'-thioacetyl)benzene, also appraised by conducting AFM is also shown in Fig. 7-21.[66] The measured low bias resistance of the carotenoid molecule was found to be (4.9 ± 0.2) GΩ. After taking into account the influence of contact resistance, first principle calculations based to tunnel transport showed this value to be consistent with electron tunneling dominating transport, despite the relatively long length of this molecule (3 nm). This molecule has 28 carbons separating the thiol ends, yet its measured resistance is less than that of the 12 carbon long dodecanedithiol (R = 8.3 ± 1 GΩ). Its resistance is also significantly less than the synthetic "molecule" wire shown in Fig. 7-21 (R = 8.3 ± 1 GΩ). These studies have hence given insight into electron transport through a carotenoid molecule, allowing quantification of its single molecule conductance values and assessment of its behaviour as a "molecular wire".[66]

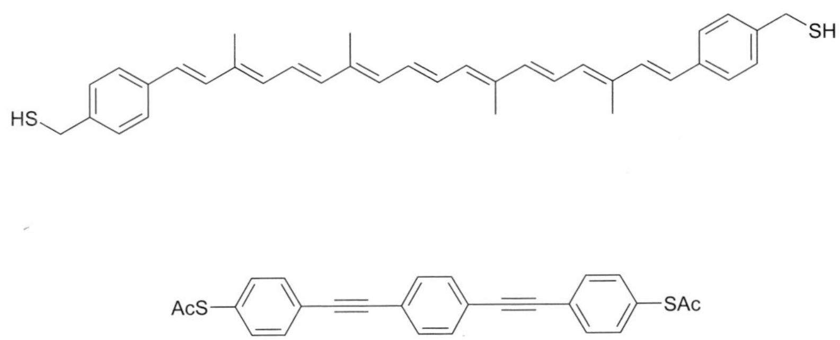

Figure 7-21. Molecular structures the dithiolated carotenoid and a synthetic molecular wire 2,5-di(phenylethynyl-4'-thioacetyl)benzene whose single molecule conductance has been determined by conducting AFM.[66]

The conductance of DNA molecular wires has been measured by both the methods of Xu *et al.* and Haiss *et al.*[69,70] Both these methods rely on the formation of molecular bridges between a gold STM tip and surface, however the former method requires the formation of mechanically formed break junction produced by repeatedly directing the STM into the substrate surface followed by cyclic retraction and rupture of the metallic junction. This method is described in reference.[62] The so-called "I(s)" method of forming molecular wires between a gold STM tip and substrate developed by Haiss *et al.* is depicted in Fig. 7-22 and described briefly below with the simple organic molecule nonanedithiol taken for the purpose of illustration.[63,64] A gold STM tip is brought into close proximity to a gold substrate, avoiding direct mechanical contact. The gold surface contains a low coverage of the analyte molecule with thiol groups at both ends. These serve the purpose of ensuring appropriate chemical, and hence, electrical contact between both ends of the molecule to the tip and the surface, respectively. After positioning the tip, the feedback loop of the STM is switched off and the tunneling current is measured as the tip is retracted from the surface. The complete current-bias voltage response can also be recorded by retracting the tip in a stepwise manner and recording current-bias voltage sweeps (I-V_{bias}) sweeps at each distance step.

Typical current-distance (I-(s)) curves recorded during this procedure are also shown in Fig. 7-22 (right), in the presence (A) and absence (B) of molecular wire formation, in this case nonanedithiol (NDT). In the absence of molecular wires (B) the current decays rapidly as the tip is retracted from the surface, while in the presence of a single NDT molecule a current plateau (of height I_w) is observed. The molecular conductance can be calculated from this. Single molecular bridges are distinguished from multiple bridges through histogram plots of I_w (see below for "I(t)" method). A complete I-V_{bias} response can be obtained by recording over a wide range of V_{bias} values or by performing I-V sweeps at distances in the current plateau region. An ohmic response is observed for simple alkanedithiol molecules over a wide bias window (-1 to +1 V), with a logarithmic dependence on methylene chain length, as theoretically expected.[63]

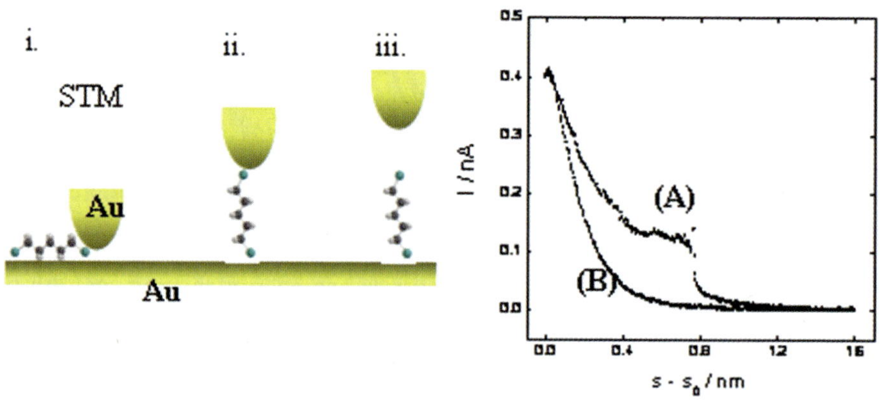

Figure 7-22. Left: The STM technique (called "I(s)" method) of forming molecular wires between a Au tip and substrate for electrical property measurement. Right: Typical I(s) scans performed in the presence (A) and the absence (B) of one molecule of nonanedithiol (NDT) between tip and substrate.

Haiss *et al.* have also developed an alternative approach that also employs an STM.[63] When the STM tip is placed in close proximity to a low coverage of the dithiol molecule, characteristic jumps are observed which are attributed to the attachment or detachment of the analyte molecule to or from the STM tip. The molecular conductance is calculated from the applied tip-substrate bias and the value of the current jumps. In order to improve accuracy and also to distinguish between single molecule and multiple attachment/detachment events the jumps are statistically rendered in the form of histogram plots. Such a histogram of current jumps is shown in Fig. 7-23 for nonanedithiol on Au(111). Current peaks I_1, I_2 and I_3 correspond to 1, 2 and 3 nonanedithiol molecules bridging the gap between tip and substrate, and from these data a molecular conductance of 0.5 nS is calculated for nonanedithiol which is good agreement with the I(s) method described above.[63]

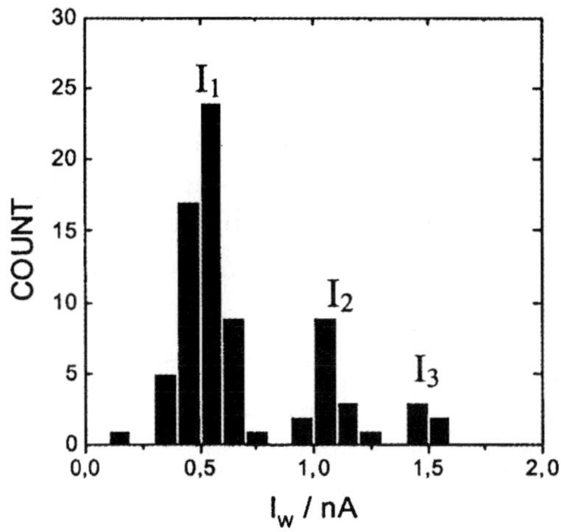

Figure 7-23. Histogram of I(t) jumps, recorded on Au(111) having a small surface coverage of NDT. Reprinted from ref. 63 with permission.

The method of Xu *et al.* and the I(t) method of Haiss *et al.* described above have been used to determine the single molecule conductance of variety of double stranded (ds) and single stranded (ss) oligonucleotide sequences. Xu *et al.* conducted studies of ds-DNA sequences of the type 5'-$(GC)_n$-3'-thiol linker with n= 4, 5, and 7 and 5'-CGCG$(AT)_m$CGCG-3'-thiol linker with m = 0, 1 and 2.[69] They observed that the conductance depends on the DNA sequence and length. In particular, the conductance exhibited a weak (algebraic) distance dependence for a poly(GC) duplexes while for $(AT)_m$ domains an exponential dependence with length of the domain was measured with a decay constant of 0.43 $Å^{-1}$.[69] This illustrates the differing electrical properties of (GC) and (AT) domains. Particularly for the ds-$(GC)_n$ domains the conductance decreases very slowly with length. This is consistent with such domains being much more "conductive" than for example alkanes or peptides.[69]

van Zalinge *et al.* have conducted a comparative study of single and double stranded oligodeoxynucleotides (ODNs) to demonstrate that charge transport via the stacked base pairs in DNA dominates the conductance rather than transport via the backbone.[70] Figure 7-24 shows the conductance values for the double stranded ODNs $(A)_{15}$-$(T)_{15}$ and $(G)_{15}$-$(C)_{15}$ alongside a number of homopolymeric single stranded ODNs, ranging in length from 3 to 7 bases.[70] Even though these duplexes are 15 base pairs long they have a conductance comparable to 4 or 5 base long ssODNs. The conductance of the double stranded structures is orders of magnitude higher than that for single stranded materials with similar numbers of bases (see extrapolated lines for ss-data in Fig. 7-24). This observation clearly demonstrates that the interaction between the base pairs and stacking plays a vital role in charge transport through ds-DNA.[70] Puzzles, however, remain as the distance decay of the conductivity, particularly for the guanine and cytosine based oligonucleotides, appears abnormally weak. Configurational fluctuations, important at the single molecule level but unimportant for macroscopic ensembles, has been forwarded as a possible reason.[70]

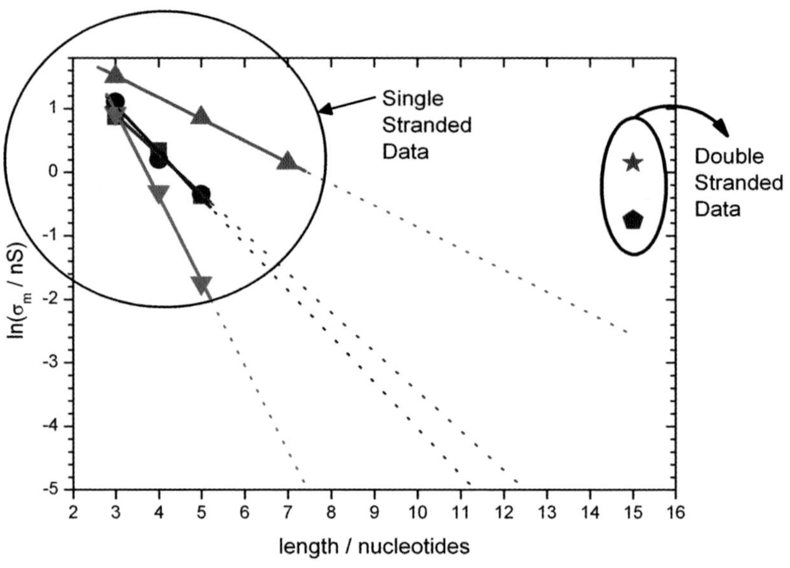

Figure 7-24. Dependence of the single molecule conductance of the oligonucleotides on the number of bases, ■ ss-oligo-A, ● ss-oligo-T, ▲ ss-oligo-G and ▼ ss-oligo-C. Also shown are the conductances of double stranded $(A)_{15}$-$(T)_{15}$ (⬟) and $(G)_{15}$-$(C)_{15}$ (★). The dotted lines are extrapolations of the length dependence in order to compare the shorter single stranded oligonucleotides to the longer double stranded compounds.[70]

Haiss *et al.* have also demonstrated how single molecule conductance can be measured in electrolyte solution as a function of electrode potential.[64] This offers the ability to perform more defined measurements of single molecule electrical properties in which the alignment of the Fermi level of the substrate and energy levels in the molecular adsorbate can be precisely controlled. Measurements of this kind have been made for the redox active molecule 6-[1'-(6-mercaptohexyl)-[4,4']bipyridinium]-hexane-1-thiol iodide (called "6V6"). The resulting dependence of molecular conductance for 6V6 on electrochemical potential is shown in Fig. 7-25.[64] The conductance increases from 0.5 nS at 0.1 V, where the molecule is in its oxidised state, to 2.8 nS at –0.65 V, where it is in its reduced state. Although this electrochemical SMC method has not as yet been applied to biomolecular adsorbates, it does offer the possibility to measure the electrical properties of dithiolated biomolecules, *in situ* in an aqueous environment and as a function of electrochemical potential. In this respect it may be helpful in resolving conduction mechanisms in DNA or amino acids. Such a configuration of biomolecules extended between two molecular contacts may also find applications in single molecule sensing, where, for instance, the attachment of a ligand to a receptor molecular wire may be detected by the change in electrical response.

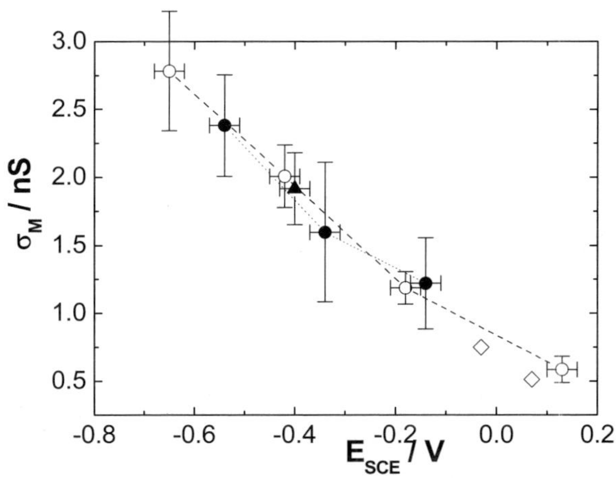

Figure 7-25. Measurements of the single molecule conductance of 6V6 molecular wires performed under electrochemical control in 0.1 M phosphate buffer solution. The electrode potential dependence of the conductance at constant tip-sample bias (U_T = 0.2 V) of a single 6V6 molecule is shown.[64]

8. STM IMAGING AND ELECTRON TRANSPORT THROUGH BIOMOLECULAR ADSORBATES

STM and *in situ* STM images of molecules and biomolecules at the single-molecule level of resolution such as shown in previous Sections, record the electrical (tunneling) current through the molecules, or the single-molecule conductivity. This is different from atomic force microscopy (AFM) which records, more directly the physical surface topography. Recording of the electronic conductivity offers novel insight into the electronic function of the molecules, posing at the same time considerable challenges and need for theoretical support. This includes novel concepts as to the molecular mechanisms of electron transport trough the (bio)molecular adsorbates, and novel theoretical and computational efforts.

Mechanistic views and theoretical formalism of molecular STM processes are addressed in some detail in Chapter 8. Views of single molecule electron transport are rooted in theories of interfacial electrochemical electron transfer[71,72] but offers new theoretical features and even *phenomena*. At the same time puzzles remain, resolution of which requires substantial computational efforts in the form of molecular dynamics and quantum chemical computations. Efforts along such lines are only just beginning.[73] We provide here a few observations of immediate relevance to the data and images shown above.

8.1 Single-Molecule Electronic Conductivity by Superexchange

Two major limiting notions constitute convenient theoretical frames for data such as those shown in Figs. 7-9, 7-11, 7-12, 7-17, 7-18, and 7-22 to 7-25. Molecules with no redox groups provide routes for tunneling through barriers constituted by the lowest-lying unoccupied molecular orbitals (LUMOs), or for hole transport, through the highest occupied molecular orbitals (HOMOs). LUMO (or HOMO) energies can be regarded as direct contributions to a tunneling barrier counted from the Fermi energies of the enclosing negatively or positively biased electrode, respectively. The tunneling current can be expanded as[74]

$$i_{tunn} = \frac{2e}{\pi\hbar}\frac{\Delta_{substr}\Delta_{tip}}{\Delta_i^2}\int_{\varepsilon_{Fsubstr}}^{\varepsilon_{Ftip}} d\varepsilon \prod_i \left(\frac{\Delta_i^2}{\left[\left(\varepsilon - \varepsilon_{Fsubstr}\right) - \varepsilon_i\right]^2 + \Gamma_i^2} \right) \qquad (7\text{-}1)$$

Δ_{substr} and Δ_{tip} are the electronic couplings between the terminal groups of the molecule and the substrate and tip, respectively, and Δ_i the electron exchange coupling between the i'th and (i+1)'th nearest neighbour groups. ε_i is the energy of the i'th group counted from the substrate Fermi level, $\varepsilon_{Fsubstr}$, the substrate here being taken as the negatively biased electrode. Γ_i is, further, the exchange couplings for all the interactions in which the i'th group is engaged. $2\pi\hbar$ is finally Planck's constant and *e* the electronic charge. The integration is with respect to ε is over the energy window between the two Fermi energies (see Fig. 7-19).

Equation 7-1 offers a reference for the tunneling current in terms of the level energies and electronic exchange couplings. Reference to simpler tunneling views can be appreciated if all Δ_i's and ε_i's take the same values, Δ and ΔE, respectively. Equation 7-1 then reduces to

$$i_{tunn} \approx \frac{2e}{\pi\hbar}\frac{\Delta_{substr}\Delta_{tip}}{\Delta^2}\left(eV_{bias}\right)\left(\frac{\Delta}{\Delta E}\right)^{2N}, \text{ or}$$

$$i_{tunn} \approx \frac{2e}{\pi\hbar}\frac{\Delta_{substr}\Delta_{tip}}{\Delta^2}\left(eV_{bias}\right)\exp\left[-\frac{2}{a}\ln\left(\frac{\Delta E}{\Delta}\right)r\right] \qquad (7\text{-}2)$$

where *N* is the number of groups, *a* the average spatial extension of each group, and *r* the tunneling gap width. Equation 7-2 gives the exponential distance decay known from tunneling theory in different contexts, cf. Chapter 8, with faster decay the larger the energy gap and the weaker the electronic coupling between the groups. Exponential distance decay has been observed in a number of reported cases of monolayer or single-molecule conductivity,[75,76] including variable-length dithiols[63] and variable-length and base composition oligonucleotides,[69,70] however, with noted puzzles remaining, cf. above.

The following simplified variant of eqn 7-1 applies for a single intermediate group,

$$i_{tunn} \approx \frac{2e}{\pi\hbar}\left(eV_{bias}\right)\frac{\Delta_{substr}\Delta_{tip}}{\left[\left(\varepsilon-\varepsilon_{Fsubstr}\right)-\varepsilon_b\right]^2+\Gamma_b^2} \approx \frac{2e}{\pi\hbar}\frac{\Delta_{substr}\Delta_{tip}}{\left[\Delta E(\eta)\right]^2} \qquad (7\text{-}3)$$

where the explicit dependence of the energy gap, $\Delta E(\eta)$ on the overpotential, η, is indicated. Δ_{substr} and Δ_{tip} now represent the coupling of the intermediate group to the substrate and tip, respectively This form illustrates the approximately inverse polynomial overpotential dependence of the current in the superexchange mode. This dependence is generally much weaker than the exponential dependence for the electrochemical current, cf. below. Equation 7-3 is a useful rationale when the overpotential is varied over wide ranges and the intermediate electronic level is brought close to the Fermi level of the negatively biased electrode without actually being electronically populated, cf. the systems discussed below. This seems to be the behaviour displayed by the 6V6-system shown in Fig. 7-25.[64]

8.2 Sequential Electron Transfer or "Hopping" Mechanisms

Equations 7-1, 7-2 and 7-3 apply when the molecular LUMOs or HOMOs are strongly off-resonance with the Fermi levels of the electrodes, *i.e.* $\Delta/\Delta E \ll 1$. As the level(s) are brought close to resonance by the overpotential or bias voltage, new charge transfer channels, where (strong) electronic-vibrational interactions are important, take over and the molecular levels become temporarily electronically populated in the process. For this reason this electron transfer mode is often denoted as electron or hole "hopping". The theoretical approach to the transition from superexchange to hopping is only partially resolved,[71,72] but the following transparent formalism, addressed in greater detail in Chapter 8, illuminates the nature of sequential two-step electron transfer in the *in situ* STM or nanogap electrode configuration,[77,78] cf. Fig. 7-19.

Two-step electronic *in situ* STM processes rest on the notion of two sequential single-electron transfer steps, at the substrate-molecule and molecule tip interfaces. Analogous multiple-step processes would, similarly involve multiple "hopping" along a chain of molecular groups. The latter was addressed in Chapter 6 and will be disregarded here. The formalism of electron transfer at a single electrochemical interface is known from electron transfer theory,[71,72] giving for the rate constants

$$k^{o/r} \approx 8\kappa_{tip}\rho_{tip}k_B T \frac{\omega_{eff}}{2\pi}\exp\left[-\frac{\left(E_r - e\eta - \gamma eV_{bias}\right)^2}{4E_r k_B T}\right] \tag{7-4}$$

$$k^{r/o} \approx 8\kappa_{substr}\rho_{substr}k_B T\frac{\omega_{eff}}{2\pi}\exp\left\{-\frac{\left[E_r + e\eta - (1-\gamma)eV_{bias}\right]^2}{4E_r k_B T}\right\}$$

where $k^{r/o}$ represents electron transfer from substrate to the oxidized molecule and $k^{o/r}$ electron transfer from the reduced molecule to the tip (at positive bias voltage.). Equation 7-4 shows that the processes are now thermally activated, k_B being Boltzmann's constant and T the temperature. E_r is the environmental nuclear reorganization free energy, reflecting now the strong importance of the electronic-vibrational interaction. ω_{eff} is the effective vibrational frequency of all the nuclear modes. Both the substrate overpotential and the bias voltage also control the activation free energy, γ being the fraction of the bias voltage drop at the redox centre. ρ and κ are, finally the metallic electronic density and the electronic transmission coefficient for interfacial electron transfer, respectively, at the interface indicated by the subscript.

These forms are broadly known from electrochemical electron transfer theory. The steady-state tunneling current in the two-step process is [77,78]

$$i_{tunn} = 2en_{o/r}\frac{k^{o/r}k^{r/o}}{k^{o/r} + k^{r/o}} \tag{7-5}$$

where $n_{o/r}$ is the number of electrons transferred in a single oxidation-reduction cycle. $n_{o/r}$ assumes values from unity up to several orders of magnitude. This reflects subtleties in the STM process which are absent in electrochemical processes at a single interface. Combination of eqns. 7-4 and 7-5 gives for symmetric contacts with strong molecule-electrode interactions

$$i_{tunn} = \tfrac{1}{2}en_{o/r}\frac{\omega_{eff}}{2\pi}\exp\left(-\frac{E_r - eV_{bias}}{4k_B T}\right)\left\{\cosh\left[\frac{\left(\tfrac{1}{2}-\gamma\right)eV_{bias} - e\eta}{2k_B T}\right]\right\} \tag{7-6}$$

The tunneling current displays, interestingly, a maximum at the overpotential

$$\eta = \eta_{\max} = \left(\frac{1}{2} - \gamma\right)V_{bias} \tag{7-7}$$

i.e. at the equilibrium potential for a symmetric contact, $\gamma = \frac{1}{2}$. This "spectroscopic" feature is crucially associated with the two-step, multi-phonon nature of the tunneling process and an important mechanistic fingerprint. The data on iron protoporphyrin IX by Tao[2] follow interestingly in all respects these expectations. This system has therefore been a real probe system for single (bio)molecule electron transfer. The blue copper protein azurin discussed above and further in Chapter 8, is another system that qualifies for this notion. Extension of the hopping model to the double-junction with the metallic nanoparticle and redox active bridging molecule,[79] has recently been reported.[80] We refer to Chapter 8 for further discussion of this model.

Equations 7-3 and 7-6 represent the two limits of molecular conductivity in a simplified useful form directly useful in real data analysis. One limit is superexchange with electron tunneling through strongly off-resonance electronic levels. The other one is "hopping" with successive electronic population and depopulation. By their physical nature and the form of the equations, the two modes are conceptually and physically quite different. Other systems and formal aspects are discussed in Chapter 8.

9. BIOMOLECULAR ADSORPTION AT NANOPARTICLES

Although *in situ* STM has not been used to image biomolecular adsorption at metal nanoparticles, it is expected that models derived for related adsorbates at single crystal electrodes will help the design of ligands with biomolecular functionality for nanoparticles. *In situ* characterisation of model single crystal electrodes has particular relevance due to the aspiration of designing functionalised nanoparticles compatible with the aqueous biological environment, which may then be applied *in vitro* and *in vivo*. With this aim in mind, some recent advances in the design and application of capping ligands with a biomolecular functionality are briefly reviewed here.

There has been intense interest in nanoparticles in the scientific community in recent years due to their potential for incorporation into functional structures such as single electron transistors, nanoscale switching devices or photonic materials.[79,81–91] Nanoparticles exhibit novel optical and electronic properties due to their structure and size, which in the limit leads to electronic quantisation effects.[92] Of particular interest is the interaction of these properties with other nanoparticles and/or other entities to produce nanoscale devices, for example, a nanometre-scale electronic switch.[79]

To take advantage of the properties of metallic nanoparticles in it is essential to assemble a ligand shell around them, since nanoparticles otherwise have a propensity to aggregate, particularly in aqueous solutions. The ideal ligand shell protects the nanoparticles from complex experimental environments, which may range from electrochemical to biological, and it should provide the means to couple specifically at defined stoichiometry functional groups. Moreover, the ligand shell should be of a defined thickness such that the geometry of assemblies of nanoparticles can be predicted.

Since thiols form strong covalent bonds with nanoparticles, existing ligand shells often possess one or more thiols, e.g., mercaptoundecanoic acid (MUA). Simple ligands such as MUA can produce a thin self-assembled monolayer, which is attractive, since it provides a defined chemical environment and the thickness of the ligand shell is controlled by the length of the alkyl chains. In the best cases, e.g., MUA, these ligand shells provide a degree of stabilisation in aqueous solutions.[93,94] However, the range of conditions under which MUA stabilised particles are stable is fairly narrow, which limits their use. In particular, functionalisation of the MUA-stabilised nanoparticles by reaction of carboxylic acid groups can only be accomplished stoichiometrically with a subsequent purification step, yet such particles are not compatible with many standard chromatographic media.

An alternative strategy is to use a thiolated polymer, e.g., DNA or PEG. Whilst this yields nanoparticles that are stable in a wider range of aqueous environments than MUA-stabilised particles,[95] the organisation of the polymer around the nanoparticle may be unpredictable. Moreover, the polymer may self-organise to form small domains, e.g., hydrophobic domains found in carboxymethyl dextran, with unwanted binding characteristics. Stoichiometric functionalisation will again require purification, which may prove difficult, e.g., due to the polymer shielding groups from the chromatographic medium.

Recently, Levy *et al.* have developed a pentapeptide ligand, NH_2-cys-ala-leu-asn-asn-COOH, which stabilises gold and silver nanoparticles to an extent comparable to polymers.[96] The small size of the pentapeptide means that the thickness of the ligand shell is controlled by its 3-D structure and hence its packing density. Peptides also offer many advantages over other types of ligands. Firstly, their synthesis is trivial, since it is automated and there are a number of companies that provide peptide synthesis services. Secondly, peptides provide a vast combinatorial chemical space that includes the 20 natural L-amino acids found in proteins, which possess side chains that cover a range of chemical functionalities, their D-equivalents and synthetic amino acids. Thirdly, the core sequence of the pentapeptide may be extended without adversely affecting nanoparticle stability. Therefore, by using a defined mixture of core pentapeptide and pentapeptide possessing a biochemical tag or synthon, e.g., biotin, functionalised nanoparticles are easy to produce; stoichiometrically substituted nanoparticles may be purified by using the chromatographic properties of the tag/synthon.

The pentapeptide NH_2-cys-ala-leu-asn-asn-COOH provides exciting new opportunities in nanoparticle research.[96] Preliminary data using a peptide library suggest that there may exist a family of peptides, which stabilise nanoparticles. Current hypotheses regarding the stability such peptides impart on nanoparticles focus on the ability of peptides to pack closely through hydrogen bonding between the amides of the peptide bonds and interactions (including hydrophobic) of the amino acid side chains. The assembly of the peptides on the surface of nanoparticles of diameter 15 nm should be similar to that occurring at other metallic surfaces. Consequently, the structure adopted by these peptides can be probed using the same techniques employed to study other biological molecules adsorbed to surfaces. Moreover, adlayers of such peptides will impart new properties to the surface. Therefore, the study of self-assembled monolayers of short peptides should provide a rich experimental system for the future, with promising opportunities for the immobilization of other biomolecules at surfaces and nanoparticles, with likely technological exploitation, for instance in surface based bioassays.

The use of surfaces as supports in bioassays has been widely adopted, both for applications in fundamental research, e.g., enzyme-linked immunosorbent assays (ELISA), microarrays, high throughput screening and for standard tests used clinically and in some over-the-counter self diagnostics. Indeed, surface-based bioassays represent a large industry. The widespread use of this assay format relates to the convenience of surfaces in relation to reagent addition and exchange and the ability to increase sample throughput by using multiple "spots" on a surface. In most commonly used surface-based bioassays, the reagent adsorbed to the surface is a peptide, protein, DNA or a (relatively) small molecule. The adsorbed molecule, by virtue of its known biochemical properties acts to capture specifically its target from a complex mixture. Detection systems employing fluorescence or enzymes are used to determine the presence of the captured molecule. However, there is usually no control over the orientation of the reagents adsorbed to the surfaces. This is despite the fact that there will be one orientation which will present optimally the recognition site in the adsorbed molecule and so provide the highest signal-to-noise ratio. Indeed, considering the importance of the orientation of the molecule adsorbed to the surface and its packing with respect to capture of target molecules from solution, it is perhaps surprising how little is known about these subjects in the context of bioassays. In this context, probe microscopies and spectroscopic methods developed for studying adsorbed molecules at the electrochemical interface could be particularly valuable.

The lack of fundamental understanding of the conformation(s) of adsorbed biological molecules in surface-based bioassays reflects the fact that hitherto detailed characterisations have been limited to the adsorption of simple systems consisting of metabolites, e.g., nucleic acid bases, amino acids and a very limited number of larger entities, e.g., metalloproteins. It is perhaps appropriate now to consider the study of more complex systems, in which the order of the primary adlayer is used to specify the order of subsequent layers. For example, the organisation of the peptides described above is driven by interactions of the cysteine thiol and amino group with the gold substrate, by interactions between the amides of the peptide bonds and by interactions between amino acid side chains. By controlling such interactions through the core and terminal sequences of such peptide monolayers it is possible that such systems will provide a route to complex but highly ordered multilayered assemblies.

10. CONCLUSION AND OUTLOOK

The probe microscopies have enabled *in situ* imaging of a wide range of biomolecular adsorbates enabling biomolecular assembly and arrangement on surfaces to be viewed at high resolution. This chapter has detailed a number of such systems ranging from the adsorption of individual DNA and RNA bases, to oligonucleotides and metalloproteins. However, *in situ* probe microscopy is more than just a high resolution imaging tool. Probe microscopies have recently made real inroads into studying electron transfer patterns in single biomolecular adsorbates, for instance the metalloenzyme azurin detailed here and in Chapter 8. Recently, STM and conducting AFM have been developed for analyzing charge transfer across individual molecular wires. As outlined in this chapter, such methods have been used to measure the electrical properties of both single and double stranded DNA. Even at this early stage in their development these single molecule electrical methods are already proving helpful in resolving conduction mechanisms across individual DNA strands. These methods may offer future possibilities in single molecule sensing; one could for instance envisage detecting ligand-receptor interactions and even dynamics down to the single molecule level.

REFERENCES

1. Li, W.H.; Haiss, W.; Floate, S.; Nichols, R.J., *Langmuir*, **1999**, *15*, 4875–4883.
2. Tao, N.J., *Phys. Rev. Lett.*, *1996*, *76*, 4066–4069.
3. Wackerbarth, H.; Grubb, M.; Zhang, J.D.; Hansen, A.G.; Ulstrup, J., *Angew. Chem. Int. Ed.* **2004**, *43,* 198–203.
4. Bloomfield, V.A.; Crothers, D.M.; Tinoco, I., *Nucleic Acids: Structure Properties and Functions*, University Science Books: Sausalito, California, 1999.
5. Sinden, R.R., *DNA Structure and Function*, Academic Press: San Diego, 1994.
6. Sowerby, S.J.; Cohn, C.A.; Heckl, W.M.; Holm, N.G., *Proc. Nat. Acad. Sci. USA* **2001**, *98*, 820–822.
7. Sowerby, S.J.; Heckl, W.M.; Petersen, G.B., *J. Mol. Evol.* **1996**, *43,* 419–424.
8. *Adsorption of Molecules at Metal Electrodes*, Series: Frontiers in Electrochemistry, Ross, P.N., Lipkowski, J., Eds., Vol. 1, VCH Publ. Inc.: New York, 1992.
9. Allen, M.J.; Balooch, M.; Subbiah, S.; Tench, R.J.; Siekhaus, W.; Balhorn, R., *Scanning Microscopy* **1991**, *5*, 625–630.
10. Allen, M.J.; Balooch, M.; Subbiah, S.; Tench, R.J.; Balhorn, R.; Siekhaus, W., *Ultramicroscopy* **1992**, *42*, 1049–1053.
11. Heckl, W.M.; Smith, D.P.E.; Binnig, G.; Klagges, H.; Hansch, T.W.; Maddocks, J., *Proc. Nat. Acad. Sci. USA* **1991**, *88*, 8003–8005.
12. Edelwirth, M.; Freund, J.; Sowerby, S.J.; Heckl, W.M., *Surf. Sci.* **1998**, *417*, 201–209.
13. Sowerby, S.J.; Edelwirth, M.; Heckl, W.M., *Appl. Phys. A, Mat. Sci. Proc.* **1998**, *66*, S649–S653.
14. Sowerby, S.J.; Edelwirth, M.; Heckl, W.M., *J. Phys. Chem. B* **1998**, *102*, 5914–5922.
15. Sowerby, S.J.; Petersen, G.B., *J. Electroanal. Chem.* **1997**, *433*, 85–90.
16. Sowerby, S.J.; Stockwell, P.A.; Heckl, W.M.; Petersen, G.B., *Orig. Life Evol. Biosphere* **2000,** *30*, 81–99.
17. Sowerby, S.J.; Edelwirth, M.; Reiter, M.; Heckl, W.M., *Langmuir* **1998,** *14*, 5195–5202.
18. Sowerby, S.J.; Heckl, W.M., *Orig. Life Evol. Biosphere* **1998,** *28*, 283–310.
19. Freund, J.E.; Edelwirth, M.; Krobel, P.; Heckl, W.M., *Phys. Rev. B* **1997,** *55*, 5394–5397.
20. Srinivasan, R.; Gopalan, P., *J. Phys. Chem.* **1993,** *97*, 8770–8775.
21. Srinivasan, R.; Murphy, J.C.; Pattabiraman, N., *Ultramicroscopy* **1992,** *42*, 453–459.
22. Srinivasan, R.; Murphy, J.C.; Fainchtein, R.; Pattabiraman, N., *J. Electroanal. Chem.* **1991,** *312*, 293–300.
23. Tao, N.J.; Derose, J.A.; Lindsay, S.M., *J. Phys. Chem.* **1993,** *97*, 910–919.
24. Tao, N.J. In *Imaging of Surfaces and Interfaces, Frontiers of Electrochemistry*, Lipkowski, J.; Ross, P.N., Eds., Wiley-VCH, Inc.: New York 1999, Vol.5, Chap. 5, pp. 211–248.
25. Cavallini, M.; Aloisi, G.; Bracali, M.; Guidelli, R., *J. Electroanal. Chem.* **1998,** *444*, 75–81.
26. Holzle, M.H.; Wandlowski, T.; Kolb, D.M., *Surf. Sci.* **1995,** *335*, 281–290.
27. Roelfs, B.; Bunge, E.; Schroter, C.; Solomun, T.; Meyer, H.; Nichols, R.J.; Baumgartel, H., *J. Phys. Chem. B* **1997,** *101*, 754–765.
28. Dretschkow, T.; Dakkouri, A.S.; Wandlowski, T., *Langmuir* **1997,** *13*, 2843–2856.
29. Dretschkow, T.; Wandlowski, T., *Electrochim. Acta* **1998,** *43*, 2991–3006.
30. Wandlowski, T.; Lampner, D.; Lindsay, S.M., *J. Electroanal. Chem.* **1996,** *404*, 215–226.
31. Holzle, M.H.; Wandlowski, T.; Kolb, D.M., *J. Electroanal. Chem.* **1995,** *394*, 271–275.
32. Holzle, M.H.; Kolb, D.M.; Krznaric, D.; Cosovic, B., *Ber. Bunsen Ges. Phys. Chem. Chem. Phys.* **1996**, *100*, 1779–1790.

33. Wandlowski, T.; Holzle, M.H., *Langmuir* **1996,** *12*, 6604–6615.
34. Nakagawa, T.; Tanaka, H.; Kawai, T., *Surf. Sci.* **1997**, *370*, L144–L148.
35. Zubay, G.L., *Biochemistry*, 4th Ed., McGraw-Hill: New York 1998.
36. Haiss, W.; Roelfs, B.; Port, S.N.; Bunge, E.; Baumgartel, H.; Nichols, R.J., *J. Electroanal. Chem.* **1998,** *454*, 107–113.
37. Willner, I.; Katz, E.; Willner, B.; Blonder, R.; HelegShabtai, V.; Buckmann, A.F., *Biosens. Bioelectron.* **1997,** *12*, 337–356.
38. Tengvall, P.; Lestelius, M.; Liedberg, B.; Lundstrom, I., *Langmuir* **1992,** *8*, 1236–1238.
39. Dakkouri, A.S.; Kolb, D.M.; Edelstein-Shima, R.; Mandler, D., *Langmuir* **1996,** *12*, 2849–2852.
40. Zhang, J.D.; Chi, Q.J.; Nielsen, J.U.; Friis, E.P.; Andersen, J.E.T.; Ulstrup, J., *Langmuir* **2000,** *16*, 7229–7237.
41. Zhang, J.; Chi, Q.; Nielsen, J.U.; Hansen, A.G.; Andersen, J.E.T.; Wackerbarth, H.; Ulstrup, J., *Russ. J. Electrochem.* **2002,** *38*, 68–76.
42. Xu, Q.M.; Wan, L.J.; Wang, C.; Bai, C.L.; Wang, Z.Y.; Nozawa, T., *Langmuir* **2001,** *17*, 6203–6206.
43. Hansma, H.G.; Weisenhorn, A.L.; Gould, S.A.C.; Sinsheimer, R.L.; Gaub, H.E.; Stucky, G.D.; Zaremba, C.M.; Hansma, P.K., *J. Vac. Sci. Techn. B* **1991,** *9*, 1282–1284.
44. Dunlap, D.D.; Bustamante, C., *Nature* **1989,** *342*, 204–206.
45. Beebe, T.P.; Wilson, T.E.; Ogletree, D.F.; Katz, J.E.; Balhorn, R.; Salmeron, M.B.; Siekhaus, W.J., *Science* **1989,** *243*, 370–372.
46. Youngquist, M.G.; Driscoll, R.J.; Coley, T.R.; Goddard, W.A.; Baldeschwieler, J.D., *J. Vac. Sci. Techn. B* **1991,** *9*, 1304–1308.
47. Driscoll, R.J.; Youngquist, M.G.; Baldeschwieler, J.D., *Nature* **1990,** *346*, 294–296.
48. Firtel, M.; Beveridge, T.J., *Micron* **1995,** *26*, 347–362.
49. Hansma, H.G., *Ann. Rev. Phys. Chem.* **2001,** *52*, 71–92.
50. Jing, T.W.; Jeffrey, A.M.; Derose, J.A.; Lyubchenko, Y.L.; Shlyakhtenko, L.S.; Harrington, R.E.; Appella, E.; Larsen, J.; Vaught, A.; Rekesh, D.; Lu, F.X.; Lindsay, S.M., *Proc. Nat. Acad. Sci. USA* **1993,** *90*, 8934–8938.
51. Andersen, J.E.T.; Kornyshev, A.A.; Kuznetsov, A.M.; Madsen, L.L.; Moller, P.; Ulstrup, J., *Electrochim. Acta* **1997,** *42*, 819–831.
52. Chi, Q.J.; Zhang, J.D.; Nielsen, J.U.; Friis, E.P.; Chorkendorff, I.; Canters, G.W.; Andersen, J.E.T.; Ulstrup, J., *J. Am. Chem. Soc.* **2000,** *122*, 4047–4055.
53. Friis, E.P.; Andersen, J.E.; Kharkats, Y.I.; Kuznetsov, A.M.; Nichols, R.J.; Zhang, J.D.; Ulstrup, J., *Proc. Nat. Acad. Sci. USA* **1999,** *96,* 1379–84.
54. Friis, E.P.; Andersen, J.E.T.; Madsen, L.L.; Moller, P.; Nichols, R.J.; Olesen, K.G.; Ulstrup, J., *Electrochim. Acta* **1998,** *43*, 2889–2897.
55. Friis, E.P.; Andersen, J.E.T.; Madsen, L.L.; Moller, P.; Ulstrup, J., *J. Electroanal. Chem.* **1997,** *431*, 35–38.
56. Iversen, G.; Friis, E.P.; Kharkats, Y.I.; Kuznetsov, A.M.; Ulstrup, J., *J. Biol. Inorg. Chem.* **1998,** *3*, 229–235.
57. Zhang, J.D.; Kuznetsov, A.M.; Ulstrup, J., *J. Electroanal. Chem.* **2003,** *541*, 133–146.
58. Albrecht, T.; Li, W.; Haehnel, W.; Hildebrandt, P.; Ulstrup, J., *Bioelectrochemistry* **2006**, *69*, 193–200.
59. Davis, J.J.; Hill, H.A.O., *Chem. Commun.* **2002**, 393–401.
60. Facci, P.; Alliata, D.; Cannistraro, S., *Ultramicroscopy* **2001,** *89*, 291–298.
61. Cui, X.D.; Primak, A.; Zarate, X.; Tomfohr, J.; Sankey, O.F.; Moore, A.L.; Moore, T.A.; Gust, D.; Harris, G.; Lindsay, S.M., *Science* **2001,** *294*, 571–574.
62. Xu, B.Q.; Tao, N.J.J., *Science* **2003,** *301*, 1221–1223.

63. Haiss, W.; Nichols, R.J.; van Zalinge, H.; Higgins, S.J.; Bethell, D.; Schiffrin, D.J., *Phys. Chem. Chem. Phys.* **2004,** *6*, 4330–4337.
64. Haiss, W.; van Zalinge, H.; Higgins, S.J.; Bethell, D.; Hobenreich, H.; Schiffrin, D.J.; Nichols, R.J., *J. Am. Chem. Soc.* **2003,** *125*, 15294–15295.
65. Leatherman, G.; Durantini, E.N.; Gust, D.; Moore, T.A.; Moore, A.L.; Stone, S.; Zhou, Z.; Rez, P.; Liu, Y.Z.; Lindsay, S.M., *J. Phys. Chem. B* **1999,** *103*, 4006–4010.
66. Ramachandran, G.K.; Tomfohr, J.K.; Li, J.; Sankey, O.F.; Zarate, X.; Primak, A.; Terazono, Y.; Moore, T.A.; Moore, A.L.; Gust, D.; Nagahara, L.A.; Lindsay, S.M., *J. Phys. Chem. B* **2003,** *107*, 6162–6169.
67. Xiao, X.Y.; Xu, B.Q.; Tao, N.J., *Angew. Chem. Int. Ed. Engl.* **2004,** *43*, 6148–6152.
68. Xiao, X.Y.; Xu, B.Q.; Tao, N.J.; *J. Am. Chem. Soc.* **2004,** *126*, 5370–5371.
69. Xu, B.Q.; Zhang, P.M.; Li, X.L.; Tao, N.J., *Nano Lett.* **2004,** *4*, 1105–1108.
70. van Zalinge, H.; Schiffrin, D.J.; Bates, A.D.; Haiss, W.; Ulstrup, J.; Nichols, R.J., *ChemPhysChem* **2006**, *7*, 94–98.
71. Kuznetsov, A.M., *Charge Transfer in Physics, Chemistry and Biology*, Gordon & Breach: Reading. 1995.
72. Kuznetsov, A.M.; Ulstrup, J., *Electron Transfer in Chemistry and Biology. An Introduction to the Theory*, Wiley: Chichester. 1999.
73. Zhang, J.D.; Bilic, A.; Reimers, J.R.; Hush, N.S.; Ulstrup, J., *J. Phys. Chem. B* **2005,** *109*, 15355–15367.
74. Datta, S., *Electronic Transport in Mesoscopic Systems*, Cambridge University Press: Cambridge. 1995.
75. Joachim, C.; Ratner, M.A., *Nanotechnology* **2004**, *15*, 1065–1075.
76. Wang, W.Y.; Lee, T.; Kretzschmar, I.; Reed, M.A., *Nano Lett.* **2004**, *4*, 643–646.
77. Zhang, J.; Chi, Q.; Kuznetsov, A.M.; Hansen, A.G.; Wackerbarth, H.; Christensen, H.E.M.; Andersen, J.E.T.; Ulstrup, J., *J. Phys. Chem. B* **2002**, *106*, 1131–1152.
78. Kuznetsov, A.M.; Ulstrup, J., *J. Phys. Chem. A* **2000**, *104*, 11531–11540. Errata: *J. Phys. Chem. A* **2001**, *105*, 7494.
79. Gittins, D.I.; Bethell, D.; Schiffrin, D.J.; Nichols, R.J., *Nature* **2000**, *408*, 67–69.
80. Kornyshev, A.A.; Kuznetsov, A.M.; Ulstrup, J., *ChemPhysChem* **2005**, *6*, 583–586.
81. Alivisatos, A.P.; Johnsson, K.P.; Peng, X.G.; Wilson, T.E.; Loweth, C.J.; Bruchez, M.P.; Schultz, P.G., *Nature* **1996**, *382*, 609–611.
82. Hostetler, M.J.; Green, S.J.; Stokes, J.J.; Murray, R.W., *J. Am. Chem. Soc.* **1996**, *118*, 4212–4213.
83. Ingram, R.S.; Hostetler, M.J.; Murray, R.W., *J. Am. Chem. Soc.* **1997**, *119*, 9175–9178.
84. Kiely, C.J.; Fink, J.; Brust, M.; Bethell, D.; Schiffrin, D.J., *Nature* **1998**, *396*, 444–446.
85. Leff, D.V.; Ohara, P.C.; Heath, J.R.; Gelbart, W.M., *J. Phys. Chem.* **1995**, *99*, 7036–7041.
86. Mirkin, C.A.; Letsinger, R.L.; Mucic, R.C.; Storhoff, J.J., *Nature* **1996**, *382*, 607–609.
87. Whetten, R.L.; Khoury, J.T.; Alvarez, M.M.; Murthy, S.; Vezmar, I.; Wang, Z.L.; Stephens, P.W.; Cleveland, C.L.; Luedtke, W.D.; Landman, U., *Adv. Mat.* **1996**, *8*, 428–433.
88. Brust, M.; Walker, M.; Bethell, D.; Schiffrin, D.J.; Whyman, R., *Chem. Commun.* **1994**, 801–802.
89. Brust, M.; Bethell, D.; Schiffrin, D.J.; Kiely, C.J., *Adv. Mat.* **1995**, *7*, 795–797.
90. Brust, M.; Fink, J.; Bethell, D.; Schiffrin, D.J.; Kiely, C., *Chem. Commun.* **1995**, 1655–1656.
91. Schultz, D.A., *Curr. Opin. Biotechnol.* **2003**, *14*, 13–22.
92. Baum, T.; Bethell, D.; Brust, M.; Schiffrin, D.J., *Langmuir* **1999**, *15*, 866–871.
93. Gittins, D.I.; Caruso, F., *Angew. Chem. Int. Ed. Engl.* **2001**, *40*, 3001–3004.

94. Gittins, D.I.; Caruso, F., *ChemPhysChem* **2002**, *3*, 110–113.
95. Tshikhudo, T.R.; Wang, Z.; Brust, M., *Mat. Sci. Tech.* **2004**, *20*, 980–984.
96. Levy, R.; Thanh, N.T.K.; Doty, R.C.; Hussain, I.; Nichols, R.J.; Schiffrin, D.J.; Brust, M.; Fernig, D.G., *J. Am. Chem. Soc.* **2004**, *126*, 10076–10084.

Chapter 8

CHARGE TRANSFER AND INTERFACIAL BIOELECTROCHEMISTRY AT THE NANOSCALE AND SINGLE-MOLECULE LEVELS

JINGDONG ZHANG,[1] TIM ALBRECHT,[1] QIJIN CHI,[1] ALEXANDER M. KUZNETSOV,[2] AND JENS ULSTRUP.[1]
[1]Department of Chemistry and NanoDTU, Technical University of Denmark, DK-2800 Lyngby, Denmark; [2]The A.N. Frumkin Institute of Physical Chemistry and Electrochemistry of the Russian Academy of Sciences, Leninskij Prospect 31, 119071 Moscow, Russia.

1. INTRODUCTION

Interfacial electrochemistry of biological molecules and macromolecules such as redox metalloproteins and DNA-based molecules is presently moving towards new levels of structural and functional resolution. Chapters in this volume have illustrated strikingly, first that underlying fundamentals of electron and proton transfer in redox metalloprotein and DNA-based systems are understood but also that new concepts and theoretical challenges are disclosed currently. This applies particularly to interfacial charge transfer in the composite inhomogeneous and anisotropic interfacial solid-electrolyte environment. As also shown, the transition of interfacial pure and applied bioinorganic electrochemistry is emerging out of powerful interdisciplinary efforts. These draw from comprehensive biomolecular electrochemical studies which have mapped working interfacial environments for retaining biological charge transfer function of these highly sensitive macromolecular systems. They have also come to draw from other biotechnology in protein and DNA-molecular functional tuning by use of mutant proteins and DNA-base variability. A complementary line is the introduction of *de novo*

O. Hammerich and J. Ulstrup (eds.), Bioinorganic Electrochemistry, 249–302.

synthetic metalloproteins such as the 4-α-helix heme proteins, electrochemical properties of which have also been studied broadly. The combination of protein and DNA biotechnology with the electrochemical interface has come to offer perspectives towards sophisticated biosensing and bioelectrochemical communication for electrical and other signal transfer between target molecules and external electrochemical circuitry. Strategic surface preparation, functional linker molecules, and biological redox chains of metalloproteins and metalloenzymes are all parts of this.

Physical electrochemistry underwent a remarkable evolution from the late 1970's, almost to be likened by a renaissance of electrochemical science.[1] This was prompted by close interaction between electrochemistry and both solid state physics and surface science. The introduction of single-crystal, atomically planar electrode surfaces with well-defined surface structures[2–5] and sometimes quite simple preparation procedures of such surfaces[4,5] were a major breakthrough. This also laid the foundation for other important electrochemical surface techniques and theory. These included spectroscopy (UV/Vis,[6,7] IR,[8,9] Raman,[10,11] and X-ray photoelectron spectroscopy[12]), Quartz Crystal Microbalance,[13] and other physical techniques, as well as both statistical mechanical[14,15] and electronic structural theories and computations,[16–18] warranted by the new electrochemistry. Only slightly later the scanning probe techniques, scanning tunneling (STM)[19,20] and atomic force microscopy (AFM)[21,22] signalled a lift both of surface science and of interfacial electrochemistry to a new unprecedented level of structural resolution.[23] *Atomic* resolution of pure metal and semiconductor electrode surfaces, and at least *sub-molecular* resolution of electrochemical adsorbates could now be achieved, opening new worlds of microscopic structures and processes, and new approaches to electrochemical nanotechnology.

Interfacial electrochemistry of proteins and DNA-based molecules is now reaching a level where similar boundary-traversing efforts are visible. As shown by chapters above, recent efforts have pointed to the feasibility of introducing similar state-of-the-art physical electrochemistry into interfacial bioelectrochemistry of redox metalloproteins and DNA-based molecules. At the same time this has increased options for improved voltammetric sensitivity at well-defined electrode surfaces, and for structural mapping of the bioelectrochemical solid-liquid interface to the level of *single-molecule* resolution. It is remarkable that molecules as large and fragile as redox metalloproteins and metalloenzymes adsorbed on atomically planar electrode surfaces can be mapped to single-molecule resolution in their functional state by a subtle physical phenomenon, *the quantum mechanical tunnel effect.*

These latter observations have opened new approaches to bioelectrochemistry at the levels of well-defined monolayers towards nanoscale and single-molecule levels. These openings have also offered challenges for fundamental *theory* of electron tunneling through biological macromolecules, the role of the metal centres etc. Environments such as STM directly in aqueous media, essential for control of biomolecular function, and configurations such as limited-size molecular assemblies (ultimately a single molecule) warrant such efforts. These relate both to the finite-size nature of the systems (stochastic as opposed to statistical properties dominating), and to the single- or supramolecular charge transfer patterns in nanogap electrode systems and other novel environments.

2. LONG-RANGE INTERFACIAL CHEMICAL AND BIOLOGICAL ELECTRON TRANSFER – A COMPREHENSIVE THEORETICAL FRAME

Condensed matter molecular charge (electron and proton) transfer theory, now developed to high sophistication,[24–29] offers comprehensive guiding frames for evolving bioelectrochemistry of redox proteins and DNA-based molecules. The theory addresses, overarchingly *two* fundamental features. One is the aspect of electron transfer or electron tunneling between the donor and acceptor group through intermediate protein or other molecular "matter". The other one is the nuclear environmental effects from local modes, collective protein and DNA nuclear dynamics, and external solvent motion. The strong interaction between the electronic system (the transferring and reorganizing electrons) and the polar environment is thus a crucial feature. In molecular detail, both need attention to the inhomogeneous, anisotropic environment of the electrode-solution interfacial region.

The increasingly high spatial resolution of interfacial electrochemistry of biological macromolecules here offers new challenges. These refer to novel techniques such as single-crystal metal surfaces, surface spectroscopy, and particularly, the ultra-high resolution microscopies, STM and AFM with atomic or sub-molecular resolution even for biological macromolecules. They refer, further, to novel charge transfer *phenomena* – which require novel theoretical notions – for example disclosed by STM of redox molecules in aqueous solution, *i.e.* in truly biological environment. We overview below some features and accompanying conceptual and theoretical notions of these parts of evolving bioelectrochemical science.

2.1 Theoretical Notions of Interfacial Bioelectrochemical Electron Transfer

Electron transfer between molecular donor and acceptor centres mostly involve a single donor and acceptor centre with all other electronic levels widely separated. The most conspicuous *differences* in electrochemical electron transfer are that one of the "reactants", namely the electrode constitutes a virtual continuous manifold of electronic states. Each level, with a wave function macroscopically scaled to the whole electrode volume, contributes infinitesimally to the electrochemical current, but the combined action of a macroscopically large number of levels ascertain that finite currents emerge. In addition, the electrode/electrolyte interfacial region is exposed to widely different forces from the metal and solution. This endows molecular orientations, ionic charge distributions, dynamic solvation etc. with features different from the homogeneous, isotropic conditions for molecular charge transfer in homogeneous solution.

2.1.1 The interfacial electrochemical electron transfer process

Established views of the electrochemical electron transfer process as a multiphonon electronic transition composed of contributions from all the electronic levels of the metallic electrode carries over to chemical and bioelectrochemical nanoscale systems. Notions in focus are illuminated by the following cathodic electrochemical current density form, $j(\eta)$ at the overpotential η (with an analogous form for the anodic process)[24–26]

$$j(\eta) = \int d\varepsilon f(\varepsilon)\rho(\varepsilon) j(\varepsilon;\eta)\,;\;\; j(\varepsilon;\eta) = e\Gamma_{ox}^{(1-\alpha)}\Gamma_{red}^{\alpha} W(\varepsilon;\eta) \qquad (8\text{-}1)$$

$W(\varepsilon;\eta)$ is the rate constant and $j(\varepsilon;\eta)$ the (infinitesimal) current density from a given electronic level ε, $f(\varepsilon)$ the Fermi function, $\rho(\varepsilon)$ the electronic level density, Γ_{ox} and Γ_{red} the population of the oxidized and reduced state, respectively, of the redox (bio)molecule close to the electrode surface, either in a monolayer film, or of mobile reactants, and e the electronic charge. α is the observable electrochemical transfer coefficient[30]

$$\alpha = -k_B T d \ln j(\eta) / d(e\eta) \qquad (8\text{-}2)$$

$W(\varepsilon;\eta)$ holds all the information about electron tunneling, driving force or overpotential, and environmental nuclear reorganization. The following equations apply broadly

$$W(\varepsilon;\eta) = \kappa_{el}\frac{\omega_{eff}}{2\pi}\exp\left\{-\frac{\left[E_r + e\eta - (\varepsilon - \varepsilon_F)\right]^2}{4E_r k_B T}\right\} \tag{8-3a}$$

$$\kappa_{el} = \kappa_{el}(\varepsilon;\eta) = \left[T_{\varepsilon A}(\varepsilon;\eta)\right]^2 \sqrt{\frac{4\pi^3}{E_r k_B T \hbar^2 \omega_{eff}^2}} \tag{8-3b}$$

where E_r is the environmental nuclear reorganization free energy, ω_{eff} the effective nuclear vibrational frequency, ε_F the Fermi energy of the electrode, k_B Boltzmann's constant and T the temperature. $\kappa_{el}(\varepsilon;\eta)$ is the electronic transmission coefficient. The most important part of $\kappa_{el}(\varepsilon;\eta)$ is the electron exchange energy, $T_{\varepsilon A}(\varepsilon;\eta)$ which couples the molecular acceptor level (A) with the metallic electronic level ε. Energy levels around $\varepsilon \approx \varepsilon_F$ mostly dominate, and combination of eqns.(8-1) and (8-3) gives

$$j = j(\eta) = e\Gamma_{ox}^{(1-\alpha)}\Gamma_{red}^{\alpha}k^{o/r}\ ;\ k^{o/r} \approx \kappa_{eff}\frac{\omega_{eff}}{2\pi}\exp\left[-\frac{\left(E_r + e\eta\right)^2}{4E_r k_B T}\right] \tag{8-4a}$$

$$\kappa_{eff} = 4\pi\kappa_{el}\left(\varepsilon_F;\eta\right)\rho\left(\varepsilon_F\right)k_B T\text{, if } \kappa_{eff} \ll 1\text{, otherwise } \kappa_{eff} \to 1 \tag{8-4b}$$

Equations (8-1)-(8-4) represent a range of interfacial (bio)electrochemical electron transfer systems. Equations (8-1)-(8-4) emerge out of similar formalism for chemical and biological electron transfer in homogeneous solution such as illustrated in the chapter by Winkler and Gray. The analytical transparency of the formalism prompts some observations that illustrate the physical nature of interfacial (bio)electrochemical electronic transitions and at the same time some limitations:

(a) The current holds an electronic tunneling factor, $\kappa_{el}(\varepsilon;\eta)$ or $\kappa_{eff}(\eta)$, and a nuclear activation factor. Nuclear activation in all the vibrational, protein conformational, and external solvent polarization modes, along with driving force effects precedes the electronic transition, which occurs in the non-equilibrium configuration at the crossing between the reactants' and products' potential surfaces, Fig. 8-1.

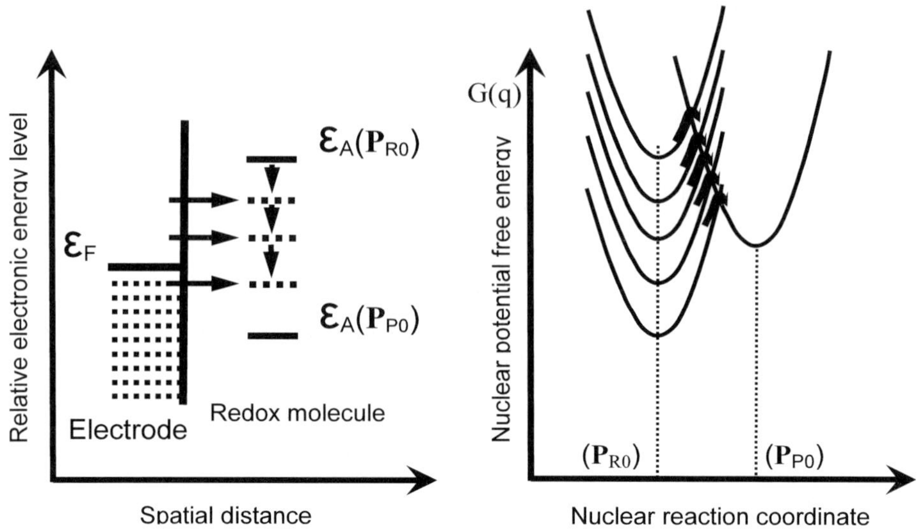

Figure 8-1. Left: Electronic energies of electrode and redox molecule at different solvent polarization **P**. $\mathbf{P}_{R0}$ and $\mathbf{P}_{P0}$ are the equilibrium values of **P** in the initial and final states, respectively. Right: Potential free energy surfaces for electrochemical (cathodic) electron transfer. From ref. 25 with permission.

(b) The *character* of the electronic factor differs significantly from electron transfer in homogeneous solution, being composed of an infinitesimally *small* single-level transmission coefficient, $\kappa_{el}(\varepsilon; \eta)$, and a macroscopically *large* number of contributing levels around the Fermi level, $4\pi\rho(\varepsilon_F)k_BT$. The "effective" transmission coefficient, $\kappa_{eff} = 4\pi\kappa_{el}(\varepsilon_F; \eta)\rho(\varepsilon_F)k_BT$ is therefore the electronically controlling factor. The process belongs to the diabatic limit of weak electronic coupling between the electrode and the redox centre when $\kappa_{eff} \ll 1$, while the adiabatic limit prevails when this coupling is strong, and $\kappa_{eff} \to 1$.

(c) In addition to the driving force, the activation energy holds the nuclear reorganization energy, E_r. The simple quadratic form in eqns.(8-3) and (8-4) rests on a linear response in all the local, conformational, and solvent modes to electric field changes associated with the charge transfer process. The conformational and solvent reorganization parts further depend on the temperature because the conformational and solvent equilibrium structures

themselves are changed when the temperature is varied. E_r is therefore a *free energy* quantity, with a temperature coefficient that can formally be identified as an activation entropy.

Molecular charge transfer theory has developed into much more powerful frames than implied by this simple formalism. Features that can be incorporated include[24,25]:

(1) General procedures for nuclear dynamics, to include anharmonic motion, frequency changes, mode mixing etc.
(2) Dynamic effects in electrochemical double layer and ionic atmosphere.
(3) Nuclear tunneling, important for proton and hydrogen atom transfer, cf. Section 2.2.
(4) Details of the electronic structure of the metal electrode surface.
(5) Interfacial electrochemical electron transfer via adsorbate states and electrocatalysis.
(6) Relaxation of the commonly applied Condon approximation in schemes of separating electronic and nuclear dynamics. This is important for long-range interfacial electron transfer.
(7) Theory of stochastic behaviour of the conformational and solvent dynamics, environmental relaxation, and multi-step processes via dynamically populated intermediate electronic states.

All these areas are covered in a broad literature, overviewed, for example in refs. 24 and 25. We do not here address all these elements of molecular charge transfer theory. Instead we discuss the two central factors in the interfacial (bio)electrochemical electron transfer process, first the nuclear reorganization (free) energy and then the electronic tunneling factor.

2.1.2 The environmental nuclear reorganization free energy

Strong interaction between localized electronic states and the nuclear environment is a crucial feature of most chemical and biological charge transfer processes. At the same time the inhomogeneous, anisotropic electrochemical interface and the complexity of the molecular structures of redox proteins and DNA-based molecules is a prodigious theoretical challenge. Comprehensive descriptions of solvation phenomena are, however, integrated constituents of contemporary molecular charge transfer theory. These rest on simple, "structure-less" dielectric continuum theory,[31–35] detailed extensions to include structural effects in the form of vibrational and spatial dispersion,[36,37] dielectric - say protein/aqueous solvent – interfaces,[38] and statistical mechanical views of the solutes and solvent.[39] Other reorganization phenomena, for example dynamic ionic strength features, can also be incorporated.[40,41] Structure-less dielectric continuum views including interfacial features apply to an extent to biological charge transfer but truly molecular approaches are effectively restricted to much simpler solute molecules.

The simple quadratic form of the activation (free) energy, however, rests on the very general assumption that whatever structural features characterize the environment, conformational, polarization, density, or other physical properties respond *linearly* to the field changes, *i.e.* changes in electric field, pressure, or other broadly defined forces. The quadratic form of the activation free energy, also commonly applied in interfacial electrochemical electron transfer of biological macromolecules is broadly valid independently of the *form* of solvation and reorganization free energies. We illustrate this briefly by comparing a few models of the environmental reorganization free energy and assessing subtler structural effects in the simplest continuum views.

The reorganization free energy is usually split in two parts. The local mode contribution is obtained in "standard" routines which require local potentials (say harmonic potentials) and vibrational frequencies in the reactants' and products' states. The collective modes associated with the proteins and the solvent, however, pose complications. One complication arises because classical electrostatics needs modification when the spatial extension of the electric field and charge distributions are comparable with the local structure extensions of the environment.[36] Other complications are associated with the presence of interfaces such as metal/solution, protein/solution, and metal/film/solution interfaces. These issues are only partly resolved, say by nonlocal dielectric theory and dielectric theory of anisotropic media.

The notion of "nonlocal" dielectric response implies, briefly, that the dielectric medium responds to the electric field, $\vec{D}(\vec{r})$ not only at the given point $\vec{r}$ but also at other points within a given spatial range Λ.[36] Λ is the solvent polarization correlation length and roughly a reflection of the local structural order in the medium, say one or two molecular diameters or so. Structural order and nonlocal dielectric features are important at distances approaching Λ. The dielectric polarization here no longer responds fully to the field due to the molecular structural limitations. The apparent dielectric constant is therefore smaller than in the bulk. This effect is thus rooted in the discrete structure of the medium at small distances and is clearly different from dielectric saturation which appears at strong electric fields where the polarization responds non-linearly to the field. Nonlocal effects are still within the linear regime. In the limit of very small distances, *i.e.* significantly smaller than Λ, the dielectric constant assumes the short-range "infrared" value $\tilde{\varepsilon} << \varepsilon_s$. $\tilde{\varepsilon} \approx 5$ for water , while $\varepsilon_s = 78$.

The following comparisons are illustrative. Starting with the form[31]

$$E_r^{\text{hom}} = \frac{1}{8\pi}\left(\frac{1}{\varepsilon_\infty} - \frac{1}{\varepsilon_s}\right)\int d\vec{r}\left(\vec{D}_R - \vec{D}_P\right)^2 \rightarrow e^2\left(\frac{1}{\varepsilon_\infty} - \frac{1}{\varepsilon_s}\right)\left(\frac{1}{a} - \frac{1}{R}\right) \tag{8-5}$$

where ε_∞ and ε_s are the optical and static dielectric constants, respectively, and $\vec{D}_R$ and $\vec{D}_P$ the electric inductions in the reactants' and products' state. The superscript "hom" refers to electron transfer in homogeneous solution. The second equation applies when the reactants are regarded as conducting spheres of radius a separated by the distance R. Eqn.(8-5) applies to a homogeneous structure-less dielectric medium when $R >> a$ so that repolarization effects in the space occupied by the reactant molecules can be disregarded. The simplest "nonlocal", *i.e.* solvent structural correction to the dielectric permittivity,[36]

$$\frac{1}{\varepsilon_{eff}(R)} = \frac{1}{\varepsilon_s} + \left(\frac{1}{\tilde{\varepsilon}} - \frac{1}{\varepsilon_s}\right)\exp\left(-\frac{R}{\Lambda}\right) \tag{8-6}$$

gives the following, almost equally simple form[42]

$$E_r^{\text{hom}} = e^2\left(\frac{1}{\varepsilon_\infty} - \frac{1}{\tilde{\varepsilon}}\right)\left(\frac{1}{a} - \frac{1}{R}\right) + e^2\left(\frac{1}{\tilde{\varepsilon}} - \frac{1}{\varepsilon_s}\right)\left\{\frac{1}{\Lambda} - \frac{1}{R}\left[1 - \exp\left(-\frac{R}{\Lambda}\right)\right]\right\} \tag{8-7}$$

In addition to ε_∞ and ε_s, the short-range, infrared dielectric constant, $\tilde{\varepsilon}$, has been introduced, important as the electron transfer distance approaches the local structural extension, represented by the solvent polarization correlation length, Λ. Eqn. (8-5) is recovered and $\varepsilon_{eff}(R) \rightarrow \varepsilon_s$ when $R \gg \Lambda$ and structural effects are unimportant, but E_r takes (much) smaller values when $R \rightarrow \Lambda$. As noted, this is, crudely speaking because the solvent molecules do not have space to reorient at small distances, and is not associated with dielectric saturation. Since $\varepsilon_\infty \approx 1.8$, $\varepsilon_s = 78$, and $\tilde{\varepsilon} \approx 5$ for water, this effect can be significant.

Physical effects and specific E_r-forms to illustrate analogous electrochemical electron transfer are much less transparent. The simple form analogous to eqn. (8-5) is[43]

$$E_r^{electr} = \frac{1}{2}e^2\left(\frac{1}{\varepsilon_\infty} - \frac{1}{\varepsilon_s}\right)\left(\frac{1}{a} - \frac{1}{2R_{electr}}\right) \quad (8\text{-}8)$$

where R_{electr} is the reactant-electrode distance. The ratio between E_r^{hom} and E_r^{electr} in this limit is thus simply $E_r^{\mathrm{hom}} / E_r^{electr} = 2$. This can be significantly modified by *two* extensions of the structure-less dielectric view. One is again the nonlocal dielectric features. The other one arises from electric field penetration *into* the metal, which then no longer acts merely as a hard wall and source or sink of electrons. Neither of these effects offers the same immediate formal transparency as eqn.(8-6) but do assume almost analytical, although cumbersome form. The effects have been analyzed in great detail, with *i.a.* the following observations:

- The field penetration effect is much smaller than structural and nonlocal solvent dielectric effects. Particularly, the increase of E_r with increasing electrode-molecule distance is retained, although attenuated. Attenuation can be appreciated as an image delocalization compared to the localized image charge contribution in eqn. (8-8).[37]
- Nonlocal solvent dielectric effects can, on the other hand, invoke rather drastic changes in the physical E_r-features.[37] This is reflected in significant differences in the absolute values of E_r, cf. below, but more importantly in the distance dependence of E_r. This dependence emerges opposite compared with the structure-less dielectric limit, *i.e.* E_r now increases with decreasing R close to the electrode surface, whereas E_r decreases with decreasing R in eqn.(8-8). The effect can become notable at distances of molecular contact with the electrode, *i.e.* a factor of two or so increase in E_r, and therefore the activation free energy. The effect also resembles expected E_r-behaviour for electrochemical electron transfer at electrode surfaces covered by a film of low dielectric constant, cf. below, and the physical origin is the same in both cases. The strong E_r-increase at small distances is rooted in a much stronger, *i.e.* much less dielectrically screened image charge on the metal side of the interface than for a structure-less dielectric, caused by the small "effective" solvent dielectric permittivity close to the electrode surface. These observations disclose the multifarious physical properties of the electrochemical interface but do not necessarily arouse needs to modify the broadly used quadratic activation free energy form.

– Even minor nonlocal effects, $\Lambda \approx$ 0.5-1 Å also drastically modify the reorganization free energy *ratio* for homogeneous and electrochemical electron transfer. Instead of $E_r^{\text{hom}} / E_r^{electr} = 2$, *i.e.* $E_r^{\text{hom}} > E_r^{electr}$ for structure-less media, much smaller values are obtained, even values *smaller* than unity, *i.e.* $E_r^{\text{hom}} < E_r^{electr}$ for small distances from the electrode surface and significant values of Λ.[37] Rate data for electrochemical electron exchange and self-exchange rate extracted from optical electron transfer for the ferrocene/ferrocinium couple in a range of polar solvents in fact display such patterns.[44]

Interfacial electrochemical electron transfer across thin dielectric films is, finally, a frequent environment in protein voltammetry (Fig. 8-2).

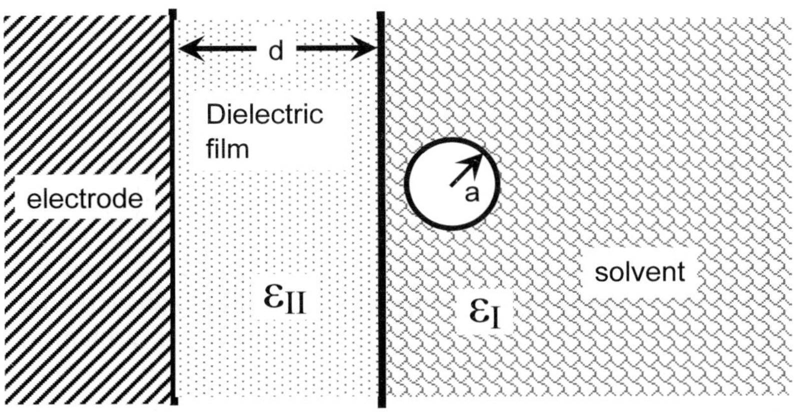

Figure 8-2. Schematic representation of a redox molecule of radius **a** in a solvent of dielectric constant ε_{I} approaching a planar electrode covered by a dielectric film of thickness **d** and dielectric constant $\varepsilon_{\mathrm{II}}$. This configuration is representative of electron tunneling across self-assembled monolayers, cf. Section 4.

The reorganization free energy for "simple ions" (*i.e.* conducting sphere-like) in a three-zone, solvent/film/metal dielectric environment has been addressed. The following extension of eqn.(8-5) must here be applied [45,46]

$$E_r^{film} = \frac{1}{8\pi}\left(\frac{1}{\varepsilon_\infty} - \frac{1}{\varepsilon_s^{film}}\right)\int_{film} d\vec{r}\left(\vec{D}_R - \vec{D}_P\right)^2 +$$

$$\frac{1}{8\pi}\left(\frac{1}{\varepsilon_\infty} - \frac{1}{\varepsilon_s^{solv}}\right)\int_{solv} d\vec{r}\left(\vec{D}_R - \vec{D}_P\right)^2 \qquad (8\text{-}9)$$

where ε_s^{film} and ε_s^{solv} are the static dielectric constants of the film and solvent, respectively. The optical dielectric constant is here taken as the same for the two regions. The induction integrals can be calculated by the method of image charges with respect to the metal/film and film/solvent interfaces. Different *conceptual* views give different values and distance behaviour of E_r.[45,46] We do not discuss these differences here. An outcome based on eqn.(8-9) of immediate relevance is, however, that the computed reorganization free energy increases with increasing film thickness. This has the same physical origin as in eqns.(8-5) and (8-8) but with formal contributions from infinite manifolds of image charges. The E_r-variation over accessible film thickness ranges, say alkanethiol monolayers of 5-20 Å thickness is, however, only from about 0.8 to about 0.9 eV, or a factor of 2-3 in the rate constant. This is much less than the variation of the tunneling factor, which entirely dominates the distance effect of interfacial electron transfer at film-covered electrodes.

These observations apply to protein film voltammetry as well but other observations are here important. A physical effect of some importance is that the protein/film interaction appears to depend on the film thickness. Some distance seems to be needed before this soft matter interaction relaxes into a stable pattern. This is seen for example by the biphasic distance dependence of the electrochemical rate constants of both cytochrome *c*[47,48] and azurin[49] at variable-length pure and functionalized alkanethiol monolayers. The other observation concerns the (dielectric) representation of the protein molecule itself which is poorly represented merely as a conducting metallic sphere. Such a representation would be restricted to the metallic redox centre while the protein part in these continuum views is rather represented as a separate region of low dielectric constant with boundaries constituted by the metallic centre on one side and the protein molecular surface on the other side, Fig. 8-3.[34,50] Further extension of eqn.(8-9) to forms such as

$$E_r^{prot-film} = \frac{1}{8\pi}\left(\frac{1}{\varepsilon_\infty} - \frac{1}{\varepsilon_s^{protein}}\right)\int_{protein} d\vec{r}\left(\vec{D}_R - \vec{D}_P\right)^2 +$$
$$\frac{1}{8\pi}\left(\frac{1}{\varepsilon_\infty} - \frac{1}{\varepsilon_s^{film}}\right)\int_{film} d\vec{r}\left(\vec{D}_R - \vec{D}_P\right)^2$$
$$+\frac{1}{8\pi}\left(\frac{1}{\varepsilon_\infty} - \frac{1}{\varepsilon_s^{solv}}\right)\int_{solv} d\vec{r}\left(\vec{D}_R - \vec{D}_P\right)^2 \qquad (8\text{-}10)$$

is thus needed. The first term represents the dielectric protein region, the second and third term the film and the solvent, respectively. Integrals such as these for a configuration such as shown in Fig. 8-3 can also be given analytical although cumbersome forms based on the method of image charges.

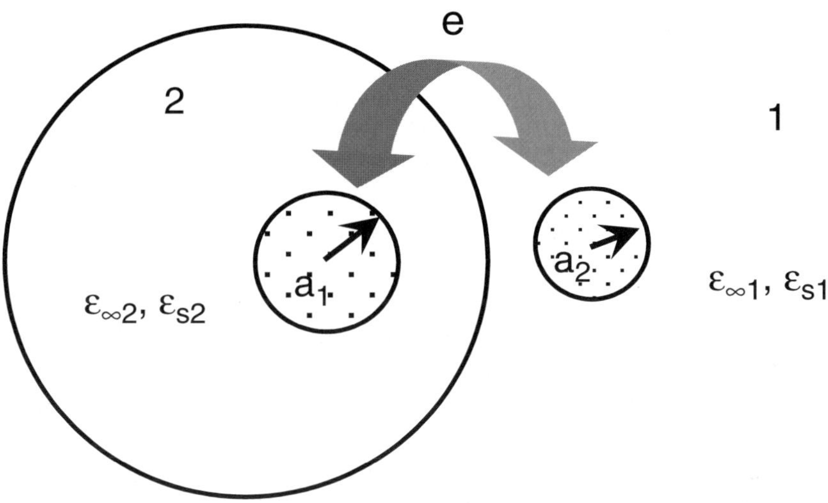

Figure 8-3. A model for electron transfer between a metalloprotein (left) and a small inorganic reaction partner (right) in aqueous solution. The shaded spherical regions of radius $\mathbf{a}_1$ and $\mathbf{a}_2$ are the electron transfer centres represented as conducting spheres. The optical and static dielectric constants are $\varepsilon_{\infty 1}$ and ε_{S1} in the aqueous region ("1") and $\varepsilon_{\infty 2}$ and ε_{S2} in the protein region ("2"). The reorganization free energy and inter-reactant work terms can be calculated exactly for this model.[25,33,38]

2.1.3 The electronic tunneling factor in long-range interfacial (bio)electrochemical electron transfer

Nuclear reorganization to a non-equilibrium configuration precedes electronic tunneling, and is followed by "percolation" of the electron along the electronically most facile route(s) in this configuration. The strong electronic-vibrational interaction thus leads to pre-organization prior to the electronic transition and nuclear relaxation subsequent to the transition. Inelastic tunneling effects "during" the tunneling step with excitation of local high-frequency nuclear modes are also feasible but disregarded here.

The tunneling factor in the weak-coupling limit of diabatic electron transfer poses challenges to state-of-the-art molecular dynamics and quantum chemical computations. Attractive analytical forms, however, illustrate the fundamental features. The tunneling process at the crossing of the potential surfaces, Fig. 8-1, can be viewed as "tunneling percolation" between the electrode and the molecular acceptor group through a "network" of intermediate groups, the energy levels of which identify the tunneling barrier.[51–55] The groups can be amino acid residues, single DNA-bases, or linker groups in molecular monolayers used to modify the electrode surfaces for gentle immobilization of proteins in well-defined surface orientations. In the form of eqn. (8-2) the electron exchange factor, $T_{\varepsilon A}$, can be given the approximate form,[51–53] (cf. references in ref. 25)

$$T_{\varepsilon A} = \sum_j \frac{\beta_{electr,1}\beta_{1,2}\ldots\beta_{j,A}}{\left(\varepsilon - \varepsilon_{1mol} - \Gamma_1\right)\left(\varepsilon - \varepsilon_{2mol} - \Gamma_2\right)\ldots\left(\varepsilon - \varepsilon_{jmol} - \Gamma_j\right)} \tag{8-11}$$

$\beta_{k,k+1}$ is the electron exchange factor for coupling between the k'th and (k+1)'th intermediate group, $\beta_{electr,1}$ couples the electrode with the nearest molecular group and $\beta_{j,A}$ the terminal bridge group with the acceptor group. The energy denominators are the energy gaps between the level ε in the electrode and the nearest off-resonance intermediate level, ε_k, and Γ_k ($\approx \beta_{k-1,k} + \beta_{k,k+1}$) is the exchange coupling for all couplings in which the k'th group is engaged.

Equation (8-11) can be simplified and the tunnel feature directly disclosed if: 1) only the shortest tunneling percolation path through nearest-neighbour interactions is important; and 2) all the intermediate group exchange interactions and energy gaps take the same values, β and $\Delta\varepsilon$, respectively, *i.e.* $\beta_{k,k+1} = \beta$ and $\varepsilon - \varepsilon_{kmol} - \Gamma_k = \Delta\varepsilon$. Eqn. (8-11) then reduces to

$$T_{\varepsilon A} = \frac{\beta_{electr,1}\beta_{NA}}{\beta}\left(\frac{\beta}{\Delta\varepsilon(\eta)}\right)^N = \frac{\beta_{electr,1}\beta_{NA}}{\beta}\exp\left[-\frac{1}{a}\ln\left(\frac{\Delta\varepsilon(\eta)}{\beta}\right)R\right] \tag{8-12}$$

where N is the number of intermediate groups. The explicit overpotential dependence of the energy gap, $\Delta(\eta)$, is shown. R is the distance between the electrode and the molecular acceptor *along the particular nearest-neighbour intermediate route*, and a the average structural extension of each group. Equation (8-12) discloses immediately the *directional* exponential distance decay associated with the tunnel effect, with faster distance decay the larger the energy gap and the weaker the electron exchange coupling between the groups.

Equations (8-11) and (8-12) can be inserted into eqns.(8-1)-(8-3) and the electrochemical current obtained. The following other transmission coefficient form is, however, also convenient. By introducing the energy "broadening" of the group linked to the electrode, *i.e.* the group "1", as[57,58]

$$\Delta_1 = \pi\left(\beta_{electr,1}\right)^2 \rho\left(\varepsilon_F\right) \tag{8-13}$$

combination of eqns.(8-1)-(8-3) and (8-12) gives

$$j(\eta) = 8e\Gamma_{ox}^{(1-\alpha)}\Gamma_{red}^{\alpha}\sqrt{\frac{\pi^3 k_B T}{E_r \hbar^2 \omega_{eff}^2}}\frac{\omega_{eff}}{2\pi}\left(\frac{\beta_{NA}}{\beta}\right)^2 \Delta_1 \exp\left[-\frac{2}{a}\ln\left(\frac{\Delta\varepsilon(\eta)}{\beta}\right)R\right]\exp\left[-\frac{(E_r + e\eta)^2}{4E_r k_B T}\right] \tag{8-14}$$

This form will be the conceptual basis for the analysis in Section 4.

Still another form of the electron tunneling factor can be recast in terms of the electrode-molecule electronic *density overlap*, *M(η)*, rather than electronic *wave function* overlap,[59]

$$j(\eta) \approx e\Gamma_{ox}^{(1-\alpha)}\Gamma_{red}^{\alpha}\mathrm{K}_{el}(\eta)\frac{\omega_{eff}}{2\pi}\exp\left[-\frac{\left(E_r + e\eta\right)^2}{4E_r k_B T}\right] \tag{8-15a}$$

$$\mathrm{K}_{el}(\eta) = \left(V_{electr,A}\right)^2\sqrt{\frac{4\pi^3}{E_r k_B T \hbar^2 \omega_{eff}^2}}M(\eta) \tag{8-15b}$$

$$M(\eta) = A\exp\left[-\gamma_{met}\left(a_{mol} - \bar{z}\right)\right] + B\exp\left[-\gamma_{mol}\left(a_{mol} - \bar{z}\right)\right] \tag{8-15c}$$

$V_{electr,A}$ is the physical electrode-molecule interaction (electrostatic forces, image charge effects etc.), *A* and *B* constants, and a_{mol} the distance of the molecular acceptor group from the electrode surface. $\bar{z}$ is the position of the metallic electron density front (say jellium front), γ_{met} and γ_{mol} the electronic density decay factors of the metal and molecule, respectively. γ_{mol} is often notably larger than γ_{met}. Eqn. (8-15c) then reduces to a single-exponential form dominated by the molecular properties. The merit of eqn. (8-15) is that it incorporates extensive density functional results of the metallic electronic interface and relates directly the electrochemical current to electrode charge, interfacial capacitance, and other interfacial properties. For example, the metallic electron density depends on the surface charge density, leading the (jellium) front $\overline{z}$, and therefore the density overlap, to expand and increase on negative electrode charging, and to contract and decrease on positive charging. This would lead to asymmetric current-overpotential relations, cf. eqns. (8-2), (8-3), (8-14), and (8-15) with a faster increase on the cathodic than on the anodic side. This effect could be detectable at "long" electron transfer distances and for metallic electrodes with intermediate electron density such as silver electrodes.[59]

2.2 Proton Transfer and Proton Conductivity

2.2.1 Electron transfer and proton or hydrogen atom transfer

Proton transfer and coupled electron/proton transfer are integrated parts of redox enzyme processes involving, for example conversion of dioxygen to hydrogen peroxide or water in membrane charge transport such as cytochrome *c* oxidase function, and in electrochemical electron transfer of protein films represented by the ferredoxins. Coupled electron/proton transfer extends all the way from separate electron and proton events to completely synchronous electron and proton transfer. In the former case the electron and proton transfer steps are still mutually influential, by the electric field changes accompanying the first of the step(s) in a sequence, for example reflected in charge-induced driving force effects. This constitutes one element of proton pumping. Another element is equivalent to hydrogen *atom* transfer. Being ubiquitously encountered in chemistry and biology, proton and hydrogen atom transfer have been in long-time experimental and theoretical focus. Recent reviews can be found in many chapters in the monograph in ref. 28 (see also refs. 60-65). We address here some concepts and theoretical notions of proton and hydrogen atom transfer central in systems discussed in this volume.

By involving transfer of a "light" particle between two much heavier molecular fragments, and with conspicuous quantum mechanical (tunneling) features already at room temperature, *proton* and *hydrogen atom* transfer display close analogies with *electron transfer*. The analogies extend to strong coupling to the environment and the vibrationally assisted tunneling character of both classes of processes. A particular perspective is that proton/hydrogen atom transfer is unique by offering a large kinetic deuterium or tritium isotope effect. This effect again offers subtle insight into the kinematics and quantum mechanical details of these broad system classes. Important *differences* between electron and proton/hydrogen atom transfer are associated with the synchronous bond breaking and formation in proton/hydrogen atom transfer, whereas the electron donor and acceptor in electron transfer processes preserve their integrity throughout the process. However, perhaps more importantly, while electron tunneling over long distances, *i.e.* tens of Ångströms is feasible, proton or hydrogen atom transfer is only feasible over much shorter distances, a fraction of an Ångström or so, even though the *equilibrium* proton/hydrogen atom transfer distances are significantly longer, say an Ångström or so.

This difference holds at least three implications. The first is that "gating" of the particle transfer is much more important than for electron transfer, and thermal activation along suitable donor-acceptor translational modes is an important contribution to the activation (free) energy.[61,62] It is the dominating contribution to the temperature coefficient of the kinetic isotope effects, caused by the shorter tunneling distance of the heavier isotope at the expense of the larger activation factor. A second implication is that the physical donor-acceptor interaction at the short distances involved is much stronger than in electron transfer. Proton transfer processes therefore frequently belong to the limit of so strongly adiabatic processes that "explicit" tunneling features (cf. below) are altogether concealed. Hydrogen atom transfer processes, with weaker donor-acceptor interactions and larger hydrogen atom transfer distances still display conspicuous quantum mechanical features, particularly large kinetic deuterium isotope effects with values up to orders of magnitude.[63,66,67]

A third implication, with direct bearing for example on trans-membrane proton transport directly follows from the first two implications. The notion of "long-range" proton transfer such as in trans-membrane proton transport or proton pumping is entirely different from the notion of long-range electron transfer. The latter is rooted in a clearly defined long-range single-electron transfer step over a distance which exceeds significantly the structural extension of the molecular donor and acceptor groups. In contrast, "long-range" proton transfer, for example along a membrane transport channel necessarily rests on a sequence of individual (very) short-range proton hops.[68] Donor and acceptor groups along the proton transport channel are amino or carboxylate groups lining the transport channel or excess water molecules in the channel. Likely sites have been identified, cf. Chapter 2, but at least two observations are here important. "Gating", *i.e.* conformational structural fluctuations of the lining proton transport groups is an essential and crucial feature of the overall process. This is entirely analogous to proton conductivity mechanisms in Nafion and other synthetic membranes.[69] The second observation is that water molecules as temporary proton transport sites formally leads to high-energy solvated H_3O^+ or OH^- intermediate states. This is itself not prohibitive for high-efficiency proton transport but does include intermediate steps of high activation (free) energy unless the pK_a-values of the water molecules and other proton transport groups are significantly different from the values in the bulk by strong electrostatic or other local interactions.

2.2.2 Proton transfer and proton conductivity

Large, *i.e.* > 10 values of the kinetic deuterium isotope effect, indicative of strong quantum mechanical tunneling features have long been known for proton transfer between poorly hydrogen bonded C-donor and –acceptor acids and bases.[70] Significantly larger values of the hydrogen/deuterium rate constant ratio, *i.e.* up to two or even several orders of magnitude have been reported for hydrogen atom transfer in frozen glasses at low temperatures[66] and in several enzyme processes at room temperature.[63,67] In either case the large effects are associated with "freezing" of the proton gating mode by the solid matrix or by the rigid protein framework. In most cases of proton transfer between O- and N-acids and –bases with strong hydrogen bonding along the proton transfer mode, the deuterium isotope effects are, however, small, *i.e.* weakly in excess of the stretching vibrational frequency ratio of $\sqrt{2}$.[28,68,70] Since the *equilibrium* structural proton transfer distance is still significant, say ≈ 0.8 Å or so, this is indicative of the strongly gated nature of the process. This appears at first to conceal the quantum mechanical nature of the transition, which is, however, maintained also in this limit, cf. below.

The discussion above offers a bio(electro)chemical scenario of proton transfer briefly summarized in Fig. 8-4 and the points below. Extensive analysis is given in refs. 62,63,65,68 and offers the following view:

- Proton transfer is coupled to the dynamics of *two* nuclear mode sets with widely different timescales, viz. the proton modes themselves and the environmental protein conformational and external solvent modes. The proton is trapped at the donor fragment at the equilibrium environmental conformational/solvent configuration in the reactants' state, but *fluctuations* in these nuclear coordinates induce a state of resonance between the proton vibrational levels in the reactants' and products' states. Environmental gating modes with their special status in proton transfer are parts of this process.
- Reorganization of the proton and electronic system parts proceed in this dynamic state of resonance and is followed by trapping of the proton in the products' state.
- The electronic system part follows smoothly ("adiabatically") the nuclear reorganization in the proton resonance state. The proton motion may, however, be obstructed by a prohibitive tunneling barrier. This limit is denoted as the "partially adiabatic" limit. Or, the proton may follow "adiabatically" the environmental nuclear

dynamics if the proton tunneling barrier is small. Attenuation of the proton tunneling barrier is part of the gating process as the barrier is narrower and shallower, the closer the gating mode dynamics has taken the donor and acceptor fragments.

– The partially adiabatic limit applies to *hydrogen atom transfer*, in redox enzymes or frozen glasses. The following rate constant forms are immediately palatable, cf. eqns. (8-1)-(8-3)[24,25,60–62]

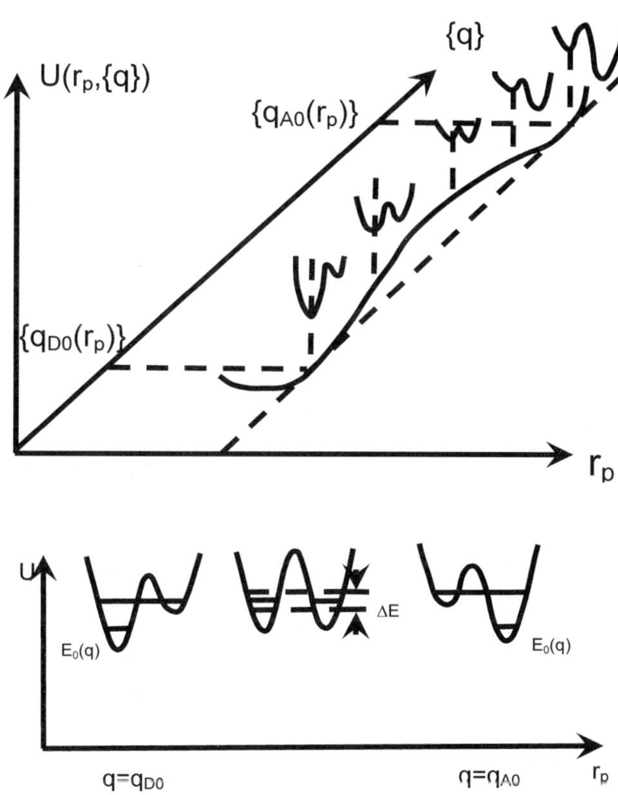

Figure 8-4. Two views of double-well potential free energy projections on the proton coordinate, r_p, in a proton transfer reaction. Different values of the environmental coordinate(s) are represented by q. The proton is trapped near the donor and near the acceptor, when q takes its equilibrium values in the initial, q_{i0}, and final state, q_{f0}, respectively. From ref. 62 with permission.

$$W_{RP} = Z_P^{-1} \sum_{v,w} \exp\left(-\frac{\varepsilon_{Rv}^P}{k_B T} \right) W_{RP}^{vw} \tag{8-16a}$$

$$W_{RP}^{vw} = \kappa_{vw}^P \frac{\omega_{eff}}{2\pi} \exp\left(-\frac{G_{vw}^{\neq}}{k_B T} \right); \quad Z_P = \sum_{v,w} \exp\left(-\frac{\varepsilon_{Rv}^P}{k_B T} \right) \tag{8-16b}$$

$$G_{vw}^P = \frac{\left[E_r + \Delta G^0 + \left(\varepsilon_{Pw}^P - \varepsilon_{Pw=0}^P \right) - \left(\varepsilon_{Rv}^P - \varepsilon_{Rv=0}^P \right) \right]^2}{4E_r} \tag{8-16c}$$

$$\kappa_{vw}^P = \left(\tfrac{1}{2} \Delta E_{vw}^P \right)^2 \sqrt{\frac{\pi^3}{E_r k_B T \hbar^2 \omega_{eff}^2}} \tag{8-16d}$$

$$\Delta\varepsilon_{vw}^P = \frac{1}{\pi} \hbar \Omega_P \exp\left\{ -\frac{1}{\hbar} \int_{left}^{right} \left\{ 2m_P \left[U_{vw}(r_P) - \varepsilon_{vw}^P \right]^{\frac{1}{2}} \right\} \right\} \tag{8-16e}$$

where m_P is the proton mass. The integration limits represent the extension of the proton tunneling barrier. We note:

(1) The same quadratic free energy relation of the activation free energy, eqn. (8-16c) as for electron transfer is recognized.

(2) Statistical averaging over all reactants', ε_{Rv}^P, and summation over products' proton vibrational states, ε_{Pw}^P are included. This is important for kinetic isotope effects but often reduces to complete dominance of the ground vibrational states $v = w = 0$.

(3) The transmission coefficient, κ_{vw}^P, is determined by the energy splitting of the proton vibrational levels at resonance, $\Delta\varepsilon_{vw}^P$. $\Delta\varepsilon_{vw}^P$ is in turn determined by the proton vibrational frequency, Ω_P, and the proton tunneling barrier, $U_{vw}(r_P)$ along the proton coordinate, r_P, at the non-equilibrium (fluctuational) environmental resonance configuration, cf. eqn. (8-16e).

The *character* of the proton transfer reaction changes entirely in the limit of strong proton donor-acceptor interactions. All immediate tunneling features are then concealed and the rate constant, eqn. (8-16b) reduces to

$$W_{RP}^{vw} = \frac{\omega_{eff}}{2\pi} \exp\left(-\frac{G_{vw}^{\neq}}{k_B T} \right) \tag{8-17}$$

Quantum mechanical features, however, remain by the quantum mechanical indices v and w. The tunneling features thus remain by the modified $G_{vw}^{\neq}$ -form

$$G_{vw}^{\neq} \rightarrow G_{vw}^{\neq} - \tfrac{1}{2} \Delta\varepsilon_{vw}^{P} \tag{8-18}$$

the proton tunneling features are now reflected quite differently, *i.e.* primarily by a lowering of the activation free energy rather than in the explicit appearance of a pre-exponential tunneling factor. The kinetic isotope effects are therefore also reflected differently, namely by a larger activation free energy, caused by a smaller splitting, $\Delta\varepsilon_{vw}^{P}$, for the heavier isotope.

By their highly efficient proton transport, proton conductors or proton pumping mechanisms such as in cytochrome *c* oxidase and other membrane-spanning protein complexes, or synthetic fuel cell-related membranes such as Nafion, invariably belong to the category of strongly interacting proton donor and acceptor groups. This is illustrated in generic form by key elements of the excess proton conductivity in aqueous solution, Fig. 8-5.[68,71,72] The dominating room temperature molecular species is likely to be the Zundel ion, $H_5O_2^+$ with the proton located symmetrically between two strongly solvating but highly labile water molecules. Proton conduction under the influence of an external electric field is effected by a subtle dynamic interplay between the strongly hydrogen bonding water molecules. Only very small structural shifts are needed to effect highly efficient proton translocation where by far most of the activation free energy is in the gating mode(s), with virtually no barrier for proton transfer remaining. Such views carry over to proton transport along flexible amino or carboxylate groups lining biological proton transport channels, or the proton transporting sulfonate groups in Nafion.

In conclusion, molecular charge transfer concepts and theory such as overviewed above and much more comprehensively in refs. 28,62,63,68 broadly frame proton and hydrogen atom transfer in biological systems and in interfacial bioelectrochemistry. Highly challenging incorporation of all the composite physical system properties, however, approach the boundaries of present state-of-the-art in condensed matter chemical rate theory.

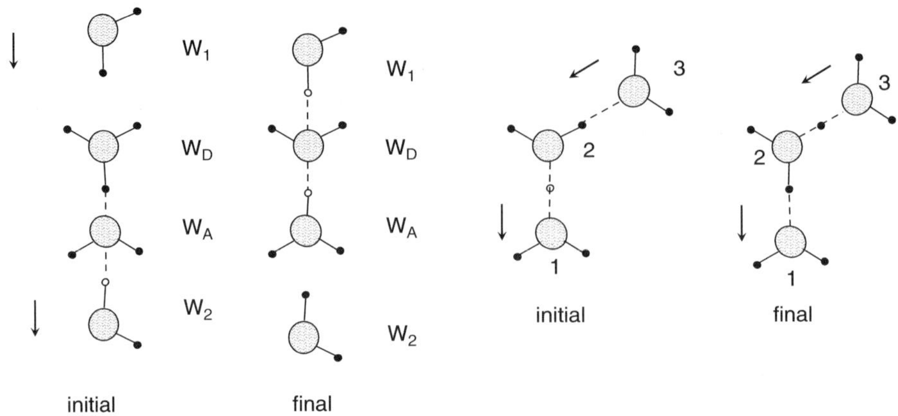

Figure 8-5. Top: "Generic" proton transfer mode in the Grotthuss mechanism.[68,71] The solvent molecule W_2 is initially hydrogen-bonded to the proton acceptor, A, while the solvent molecule W_1 near the proton donor D is less strongly bonded. Proton transfer is triggered by synchronous displacement and hydrogen bond formation and breaking. Bottom: Double-proton transfer in proton conduction via the Zundel complex $H_5O_2^+$. The proton is initially located symmetrically between the water molecules 1 and 2. In the final state the proton has been translocated to a symmetrical position between the water molecules 2 and 3. Mechanisms analogous to these can be expected to operate in proton conduction through biological and synthetic membranes.[69]

3. THEORETICAL NOTIONS IN BIOELECTROCHEMISTRY AT NANOSCALE AND SINGLE-MOLECULE LEVELS

3.1 Biomolecules in Nanoscale Electrochemical Environment

Previous chapters have discussed comprehensively protein and DNA-based function and bioelectrochemical systems. Single- and multi-functional monolayers of proteins, DNA-based molecules, and enzymes, and of monolayers of amino acids and DNA-bases as their building blocks testify towards bioelectrochemical control at the nanoscale and sometimes single-molecule levels. In these respects interfacial bioelectrochemistry is undergoing a process similar to what has been the case in physical electrochemistry of metallic nanostructures, molecular adsorbates, and other areas of physical electrochemistry.

One line in bioelectrochemistry is rooted in electrochemical techniques, spectroscopy, and other physical chemical techniques. Linear and cyclic voltammetry are central.[73–75] Other electrochemical techniques include impedance[77,78] and electroreflectance spectroscopy,[47,79] ultramicro-electrodes,[80] and chronoamperometry.[77,81] To this come: spectroscopic techniques such as infrared,[82] surface enhanced Raman and resonance Raman,[83,84] second harmonic generation,[85] surface Plasmon,[86,87] and X-ray photoelectron spectroscopy.[77,88,89] A second line has been to combine state-of-the-art physical electrochemistry with corresponding state-of-the-art microbiology and chemical synthesis. The former relates to the use of a wide range of designed mutant proteins,[90–93] the latter to chemical synthesis or *de novo* designed synthetic redox metalloproteins.[94–96]

Experimental approaches to bioelectrochemical systems include other techniques which introduce new environments for interfacial bioelectrochemical function. Introduction of single-crystal, atomically planar electrode surfaces[77,97] has opened a basis for the use of the scanning probe microscopies, STM and AFM, also for biological macromolecules. Importantly this extends to the electrochemical STM mode where electrochemical surfaces, adsorbate molecules,[23] and now also biological macromolecules can be mapped directly in their natural aqueous environment, with full electrochemical potential control (*in situ* STM and AFM).[77,98–103] Protein and DNA-association with metallic nanoparticles constitute other new combined techniques and microenvironment (ref. 104 and the Chapter by Willner and Willner), with extension to nanogap electrode configurations. Reactive (bio)molecules combined with *in situ* STM and nanoparticle configurations also disclose new interfacial electron transfer *phenomena*, which pose, therefore also new *theoretical* challenges. We first overview some of these and then discuss some systems which have been studied and analyzed within these new theoretical and conceptual frames.

3.2 Theoretical Frames and Interfacial Electron Transfer Phenomena

We focus here on the electronic conductivity of molecular monolayers or single molecules enclosed between a pair of metallic electrodes. We address specifically *in situ* STM of redox (bio)molecules but concepts and formalism to be discussed carry over to other metallic nanogap configurations. Importantly, in addition to the substrate and tip, a third electrode serves as reference electrode.[105–107] Such a scheme is shown in Fig. 8-6.

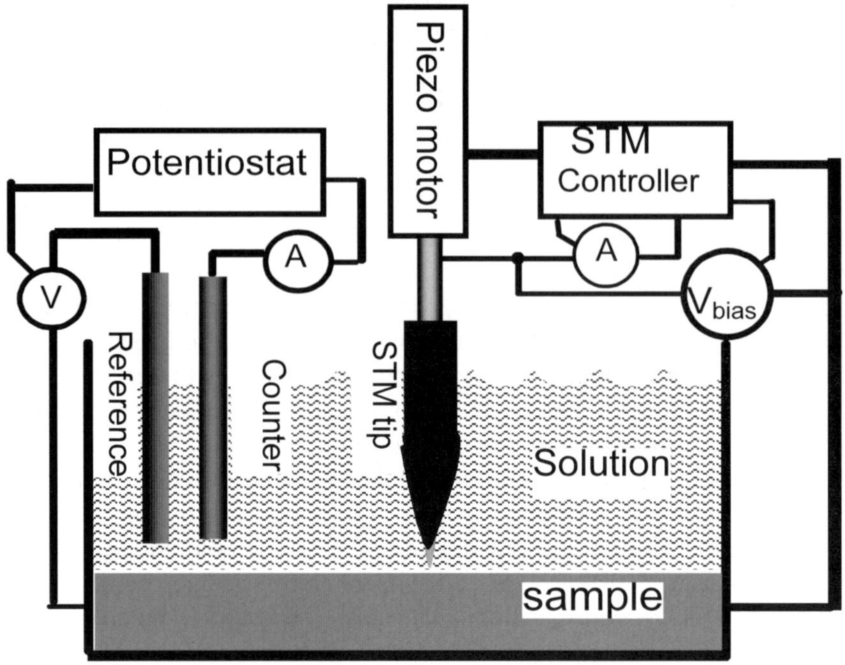

Figure 8-6. In situ STM configuration, schematic. In this four-electrode configuration the working electrode potential is controlled relative to the electrochemical reference electrode ("gate") and the (coated) tip potential relative to the working electrode potential. The counter electrode enables recording of electrochemical processes on the working electrode by the independently controlled tip electrode.

This ascertains electrochemical potential control of both substrate and tip. The reference electrode can also be regarded as a gate electrode in analogous three-electrode molecular transistor-like configurations. Importantly, the three-electrode configuration is the basis for *two* kinds of tunneling "spectroscopic" current-voltage relation, unique to the electrochemical configuration. One is the current-bias voltage relation as in STM in air or vacuum, but now with the additional notion that the substrate (over)potential is kept constant. The other one is the current-overpotential relation at constant bias voltage, *i.e.* the electrochemical substrate and tip potentials are varied in parallel relative to the common reference electrode. The two correlations offer mutually supportive insight in single-(redox) molecule electronics and can be extended to redox metalloproteins. Analogous dual-type correlations in other molecular three-level systems such as molecular Raman spectroscopy have been noted.[25,108]

3.2.1 Redox (bio)molecules in electrochemical STM and other nanogap configurations

Figure 8-7 shows a schematic view of a redox molecule enclosed in a nano- or molecular scale gap between *two* electrochemically controlled metallic electrodes.

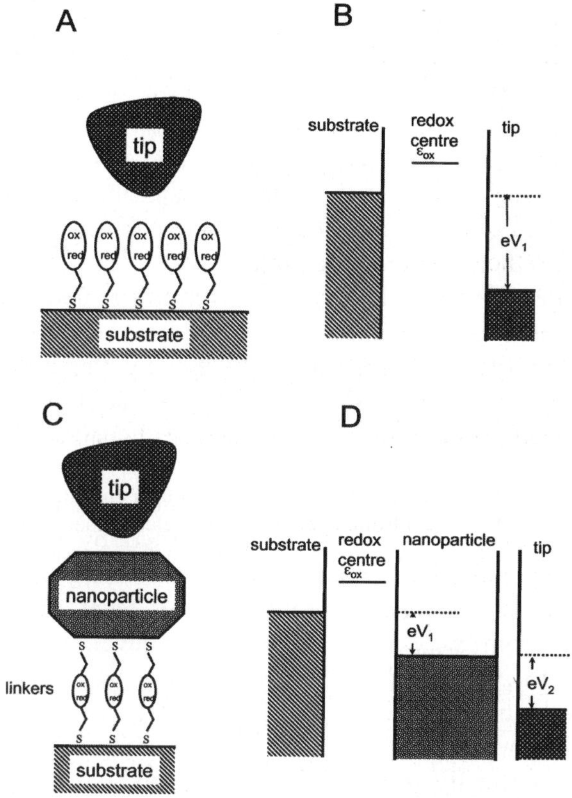

Figure 8-7. A: Schematic view of a redox molecule or metalloprotein in the gap between a substrate and a tip electrode in STM or between two nanogap electrodes. B: Electronic energy scheme for *in situ* STM of the redox molecule. C: Tunnel junction with a redox molecule and a metallic nanoparticle in a "hybrid" structure. D: Energy scheme corresponding to the "hybrid" structure with two potential drops.

We shall allude particularly to *in situ* STM of a redox (bio)molecule, but as noted, the configuration is representative broadly of redox molecules in electrochemically controlled nanogap electrode configurations and of single-molecule transistor-like configurations reported lately.[109–111] In contrast to the latter systems, which only apply in ultrahigh vacuum or at cryogenic temperatures, those represented by Fig. 8-7 and addressed below apply to room temperature and condensed matter environment. Figure 8-7 also shows an extension of the configuration in which a metallic nanoparticle is combined with a redox molecule and the combined supermolecule inserted in the gap. Configurations corresponding closely to both schemes in Fig. 8-7 have been reported[77,98–103,112–114] and characterized recently. The schemes prompt the following further observations:

(a) The scheme in Fig. 8-7, top is an extension of the scheme for electrochemical electron transfer at a single metallic electrode surface. Two electrode surfaces are now present, and their Fermi levels separated by the bias voltage, eV_{bias}, at given overpotential η, cf. Section 2. The overpotential is the substrate electrochemical potential, E_s, relative to the substrate equilibrium potential, E_s^0, *i.e.* $\eta = E_s - E_s^0$. The bias voltage is here viewed as the tip potential relative to the substrate potential, *i.e.* $V_{bias} = E_t - E_s$.

(b) The molecular redox level, say the oxidized, electronically "empty" level at electrochemical equilibrium is located well above the Fermi energy of the substrate electrode, *i.e.* by the reorganization free energy E_r. Figure 8-7 shows explicitly the configuration for *positive* bias voltage, so that the tip Fermi level is located *below* the substrate Fermi level.

(c) The energies can be shifted and controlled by three external factors. The *equilibrium* locations can be shifted by the electrochemical (over)potential. This raises the Fermi energy of the substrate electrode, $\varepsilon_{Fsubstr}$, relative to the redox level. If all energies are counted from $\varepsilon_{Fsubstr}$, the empty redox level, ε_{ox}, is thus lowered relative to $\varepsilon_{Fsubstr}$ with increasing negative η. The bias voltage is a second controlling factor. As the site of the redox centre is exposed to part of the bias voltage in the tunneling gap, the redox level is also shifted vertically relative to $\varepsilon_{Fsubstr}$ on *bias voltage* variation. Conformational and solvent polarization *fluctuations* of the same kind as in molecular charge transfer in homogeneous solution and in electrochemical electron transfer at a single interface are, finally, a crucial factor of the resulting (bio)molecular conduction process. In the *in situ* STM configuration, these novel interfacial bioelectrochemical phenomena are approaching the level of *single-molecule* resolution.

(d) The steady-state current through the nanoparticle configuration in Fig. 8-7, bottom offers similar dual type of current-bias voltage and current-overpotential relations. When the nanoparticle is large enough to behave as bulk metal (diameter ≥ 5 nm) the correlations will display "spectroscopic" features resembling those of the redox

group alone.[114,115] The correlations of very small particles are, however, expected to display significant electronic fine-structure caused by quantum size effects, *i.e.* discrete electronic structural effects of the small particle, which now is more alike a molecular structure. These points can be incorporated in a formalism,[116,117] elements of which we now overview.

3.2.2 New interfacial (bio)electrochemical electron transfer phenomena

The most transparent results in electrochemical *in situ* STM processes of redox molecules and biomolecules are obtained when the bias voltage is small, *i.e.*

$$\gamma \left| eV_{bias} \right| < E_r - e\eta \tag{8-19}$$

We consider this limit first. γ is the fraction of the bias voltage at the site of the redox centre. The bias voltage is thus in a sense a "probing energy tip", somewhat equivalent to the dispersion of probing electromagnetic radiation in optical absorption spectroscopy. In either case the "spectral resolution" is better, the narrower the probing tip. As the overpotential is raised, at fixed bias voltage the cathodic current first rises due to more favourable driving force, but then it drops as the overpotential is increased further since practically all active species are converted to reduced form with the energy level trapped below the Fermi levels. A different character of the tunneling process arises when the bias voltage is large, *i.e.* the opposite inequality of eqn. (8-19) applies. Either the reduced level, or at even larger bias voltage both the oxidized and reduced forms of the redox level are then trapped *between* the Fermi levels of the enclosing electrodes. Multi-electron transport (coherent or step-wise) then continues until the overpotential is raised still further and eqn. (8-19) again applies.)

An additional distinction between "weak" and "strong" electronic interactions between the molecular redox level and the electrodes is important. In the former limit the overall steady-state STM process can be viewed as two consecutive, environmentally relaxed interfacial *single-electron transfer steps*, each analogous to electrochemical electron transfer, Section 2, giving steady-state tunneling current forms such as

$$i_{tunn}^{weak} = e \frac{k^{o/r} k^{r/o}}{k^{o/r} + k^{r/o}} \tag{8-20a}$$

$$k^{o/r} \approx 8\kappa_{tip}\rho_{tip}k_B T \frac{\omega_{eff}}{2\pi}\exp\left[-\frac{\left(E_r - e\eta - e\gamma V_{bias}\right)^2}{4E_r k_B T}\right] \qquad \text{(8-20b)}$$

$$k^{r/o} \approx 8\kappa_{ssubstr}\rho_{substr}k_B T \frac{\omega_{eff}}{2\pi}\exp\left\{-\frac{\left[E_r + e\eta - \left(1-\gamma\right)eV_{bias}\right]^2}{4E_r k_B T}\right\}$$

(8-20c)

These rate constants forms are completely equivalent to the interfacial electrochemical rate constant forms in eqns. (8-2) and (8-3). $k^{o/r}$ is thus the rate constant for electron transfer from the reduced form of the molecule to the substrate and $k^{r/o}$ the rate constant for electron transfer from the tip to the oxidized form of the molecule.

Different forms and a different scenario emerge in the opposite limit of strong electronic interactions. Eqns. (8-20a)-(8-20c) are then replaced by

$$i_{tunn}^{strong} = 2en_{o/r}\frac{k^{o/r}k^{r/o}}{k^{o/r}+k^{r/o}} \qquad \text{(8-21a)}$$

$$n_{o/r} = \frac{\left|eV_{bias}\right|}{\Delta\varepsilon}; \qquad \Delta\varepsilon = \frac{1}{\kappa_{substr}\rho_{substr}} + \frac{1}{\kappa_{tip}\rho_{tip}} \ll k_B T \qquad \text{(8-21b)}$$

$$k^{o/r} \approx \frac{\omega_{eff}}{2\pi}\exp\left[-\frac{\left(E_r - e\eta - e\gamma V_{bias}\right)^2}{4E_r k_B T}\right] \qquad \text{(8-21c)}$$

$$k^{r/o} \approx \frac{\omega_{eff}}{2\pi}\exp\left\{-\frac{\left[E_r + e\eta - \left(1-\gamma\right)eV_{bias}\right]^2}{4E_r k_B T}\right\} \qquad \text{(8-21d)}$$

The difference from eqns. (8-20a)-(8-20c) is that the electronic transmission coefficients in the rate constants are now unimportant, but still appear in the quantity $n_{o/r}$ in eqn. (8-21a). The physical meaning here is, however, different from the meaning in eqns. (8-20a)-(8-20c). Due to the strong electronic molecule-electrode interactions, the first single-electron transfer event is followed by a large number (up to or exceeding a hundred or so) of subsequent events, whilst the occupied/reduced redox level relaxes through the energy window between the two Fermi levels. This is a novel electron transfer *phenomenon* associated specifically with the three-electrode nanoscale *in situ* STM configuration. It may also hold a clue to the frequent observation of large tunneling current densities (per molecule), which can be rooted partly in the large number of electrons transferred in a single molecular *in situ* STM process.

Eqns. (8-20) and (8-21) are directly suitable as frames for experimental data analysis. The following simple form derives directly from eqns. (8-20) and (8-21) for a symmetric electronic contact, *i.e.* when $\kappa_{substr}\rho_{substr} = \kappa_{tip}\rho_{tip}$,

$$i_{tunn}^{symm} = \frac{1}{2} e n_{o/r} \frac{\omega_{eff}}{2\pi} \exp\left(-\frac{E_r - eV_{bias}}{4k_B T} \right) \left\{ \cosh\left[\frac{\left(\frac{1}{2}-\gamma\right) eV_{bias} - e\eta}{2k_B T} \right] \right\}^{-1} \tag{8-22}$$

$n_{o/r} = 1$ or $\left| eV_{bias} \right| / \Delta\varepsilon$, cf. eqn. (8-21b) in the weak- and strong-coupling limit, respectively. This form discloses an immediate general implication of the formalism in eqns. (8-20)-(8-22), namely a pronounced ("spectroscopic") maximum in the tunneling current-overpotential relation (at given bias voltage). For the symmetric configuration, eqn. (8-22) the maximum appears at the overpotential

$$\eta = \eta_{\max} = \left(\tfrac{1}{2} - \gamma\right) V_{bias} \tag{8-23}$$

If the redox centre in the gap is exposed to half of the bias voltage drop, $\gamma = \frac{1}{2}$, then the maximum is at $\eta_{max} = 0$, *i.e.* at the equilibrium redox potential. This holds an important diagnostic dimension regarding the mechanism of the single-molecule electronic tunneling conduction mechanism. We shall return to specific data analysis based on this frame below.

Similar tunneling "spectroscopic" features are associated with the current-bias voltage relations, *i.e.* new electronic conduction channels open when the redox level is brought to cross into the energy region between the Fermi levels of the substrate and tip electrodes. The "spectroscopic" current-bias voltage peaks are now reflected in the *conductivity*, *i.e.* the tunneling current derivative with respect to the bias voltage, rather than directly in the tunneling current itself. Figures 8-9 and 8-17 below show experimental current-overpotential relations that follow closely this formalism and illuminate the effects of the external parameters, particularly the reorganization free energy and the potential distribution in the tunneling gap. The latter is here regarded as an external parameter but in principle liable to first-principles electronic structural computations.

The attractive formalism above applies to the limit of small bias voltage in the sense of eqn. (8-19). The notion of "small" or "large" thus itself depends on the overpotential. In the limit given by the opposite inequality of eqn. (8-19), according with the limit of "large" bias voltage the partially reduced level is trapped between the two Fermi levels rather than below both levels. The character of the tunneling process then changes. The current takes the form

$$i_{tunn}^{large-bias} = e\left(\frac{\hbar}{\Delta_{substr}} + \frac{\hbar}{\Delta_{tip}}\right) \tag{8-24}$$

where Δ_{substr} and Δ_{tip} are the broadenings of the molecular redox level caused by electronic interaction with the enclosing electrodes. In this state the redox level transmits coherently a large number of electrons, giving a high and activationless steady-state tunneling current. Δ_{substr} and Δ_{tip} are related to the electron exchange factors, $V_{electr,substr}$ and $V_{electr,tip}$ by

$$\Delta_{substr} = \pi\left(V_{electr,substr}\right)^2 \rho_{substr} \;;\quad \Delta_{tip} = \pi\left(V_{electr,tip}\right)^2 \rho_{tip} \tag{8-25}$$

cf. eqn. (8-13).

The following simple and useful interpolation formula represents the tunneling current at arbitrary bias voltage[115]

$$i_{tunn} = \frac{i_{tunn}^{small-bias}\, i_{tunn}^{large-bias}}{i_{tunn}^{small-bias} + i_{tunn}^{large-bias}} \tag{8-26}$$

where $i_{tunn}^{small-bias}$ is given by eqns. (8-20)-(8-22).

In situ STM molecular conductivity with the combined use of nanoparticle metallic (gold) particles together with molecular redox groups in the tunneling gap were illustrated in Fig. 8-7, bottom.[114,115] The substrate/redox group/Au-nanoparticle/tip junction constitutes a double, substrate/redox group/Au-nanoparticle and nanoparticle/tip contact. Under steady-state conditions the observable overall tunneling current, $i_{tunn}(V_{bias}, \eta)$ is given by each of the currents through the separate contacts, *i.e.*

$$i_{tunn}(V_{bias},\eta) = i_1(V_{bias,1},\eta) = i_2(V_{bias,2}) \ ; \quad V_{bias} = V_{bias,1} + V_{bias,2} \qquad (8\text{-}27)$$

where the subscripts "1" and "2" refer to the first, *i.e.* substrate/redox group/Au-nanoparticle and second, Au-nanoparticle/tip contact, respectively. Only the first current form depends explicitly on the substrate overpotential. Equation (8-27) offers again a computational scheme for both tunneling current/voltage relations, where $i_1(V_{bias,1}, \eta)$ follows the formalism above. Another specific form is needed for the Au-nanoparticle/tip contact. This could be[115]

$$i_2(V_{bias,2}) = e\frac{2\pi}{\hbar}V_{bias,2}\exp\left\{-\kappa_2 s\frac{2U_{02}}{3V_{bias,2}}\left[1-\left(1-\frac{V_{bias,2}}{U_{02}}\right)^{\frac{3}{2}}\right]\right\}$$

(8-28)

$$\kappa_2 = \frac{2}{\hbar}\sqrt{2m_2U_{02}}$$

appropriate for electron tunneling through a linear potential energy barrier. U_{02} is the barrier height at zero bias voltage, s the Au-nanoparticle/tip gap width, and m_e the electronic mass. Equations (8-27) and (8-28) together with eqns. (8-20)-(8-22) offer the following scheme for the double-tunneling contact shown in Fig.8-8 below, first at fixed overpotential and then at fixed bias voltage:

- Equation (8-27), together with the single-contact tunneling current forms, eqns. (8-20)-(8-22) provides two equation sets, at fixed overpotential:
- The "spectroscopic" tunneling current features are represented by

$$i_{tunn}(V_{bias},\eta) = i_1(V_{bias,1},\eta) \qquad (8\text{-}29)$$

where $V_{bias,1}$ is related to V_{bias} by eqn. (8-27).

- The current-bias voltage relations can be written in parametric form

$$i_{tunn} = i_1\left(V_{bias,1}, \eta\right)) \qquad \text{(8-30a)}$$

$$V_{bias} = V_{bias,1} + V_{bias,2}\left(V_{bias,1}, \eta\right) \qquad \text{(8-30b)}$$

$V_{bias,1}$ is thus the running variable. $V_{bias,2}$ as a function of $V_{bias,1}$ is computed from

$$i_1\left(V_{bias,1}, \eta\right) = i_2\left(V_{bias,2}, \eta\right) \qquad \text{(8-31)}$$

- Equation (8-30) also gives the tunneling current dependence on the overpotential at constant partial bias voltage $V_{bias,1}$. However, in order to keep $V_{bias,1}$ constant the total bias voltage, $V_{bias}(\eta)$ must be varied in a composite way determined by eqns. (8-30) and (8-31), at fixed $V_{bias,1}$.
- This scheme is discussed in some detail in ref. 115. A conclusion is that much of the conductivity behaviour of single-contact substrate/redox molecule/tip spectroscopic behaviour carries over to the double-contact tunneling junction. This is of interest in relation to reported data where Au-nanoparticles have been combined with redox molecules based on viologen moieties linked to enclosing gold substrate and tip electrodes by variable-length thiol-based linker groups.[114,119]

We proceed next to some recent data of direct relevance to what can be given the appellations of "nanoscale" and "single-molecule" electrochemistry and bioelectrochemistry.

4. INTERFACIAL ELECTROCHEMICAL CHARGE TRANSFER SYSTEMS TOWARDS THE NANOSCALE AND SINGLE-MOLECULE LEVELS OF RESOLUTION

The chapters by Winkler and Gray (Chapter 1), and by Wikström (Chapter 2) have testified to high detail and sophistication in the mapping of electron transfer and coupled electron and proton transfer of proteins and protein complexes in homogeneous solution and in protein-membrane systems. The chapters by Willner and Willner (Chapter 3), Butt and Armstrong (Chapter 4), Kelley and Hill (Chapter 5), and by Nichols *et al.* (Chapter 7) have pointed to other detail in the control and characterization of interfacial electron transfer and electrocatalysis in protein, enzyme, and DNA-based molecular films at monolayer levels of resolution. Other approaches along related lines are represented by the construction of interfacial biological electron transfer chains,[91] control of biological function by the use of ultramicroelectrodes[80] and the use of gold nanoparticles.[104] A, still somewhat limited, range of redox metallo*enzymes* at the level of protein or enzyme monolayers have, secondly, been addressed.[120] Target enzymes in immobilized surface configurations have included glucose oxidase,[121,122] fumarate oxidase[123] and succinate dehydrogenase,[124] nitrate[125] and nitrite reductases,[126,127] peroxidases,[128,129] DMSO reductase,[130] hydrogenases[131] and cytochrome *c* oxidase (ref. 132 and Chapter 3). These examples represent systems with complex interfacial and electrocatalytic functions, where novel detail could be expected by nanoscale and single-molecule approaches as achieved for transition metal complexes and the smaller blue copper, heme group, and iron-sulfur electron transfer metalloproteins, cf. below. In the following we discuss some case studies from our own group, that come close to these levels of interfacial bioelectrochemistry, and which warrant use of the theoretical frames in Section 3.

4.1 Electrochemical Nanoscale STM Spectroscopy of Transition Metal Complexes

Tao reported the first case of *in situ* STM spectroscopic features using Fe-protoporphyrin IX on highly oriented pyrolytic graphite as a target system.[133] Recent studies of the Os- and Co-complexes, Fig. 8-8 illustrate systematically elements of interfacial electrochemical electron transfer and *in situ* scanning tunneling spectroscopy at the nanoscale and single-molecule levels. The complexes adsorb in monolayers on Au(111)- and Pt(111)-electrodes.[134,135] Ordered domains in patches among largely disordered layers are formed under the conditions used. All the complexes display monolayer cyclic voltammetry with features suggestive of anion and other specific effects. The electrochemical kinetics of the Os-complexes is much faster than for the Co-complexes, with interfacial standard rate constants > 10^6 s^{-1} vs. (1-3)×10^3 s^{-1}. This follows patterns for electron exchange in homogeneous solution and can be ascribed to significant intramolecular reorganization energy and less favourable electronic coupling of the Co-complex. No significant structural changes accompany the electron transfer processes of the Os-complexes.

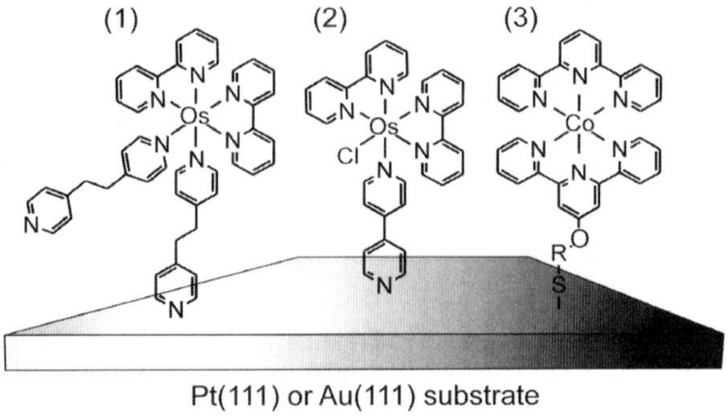

Figure 8-8. Transition metal complexes used in nanoscale and single-molecule electrochemical STM spectroscopy.

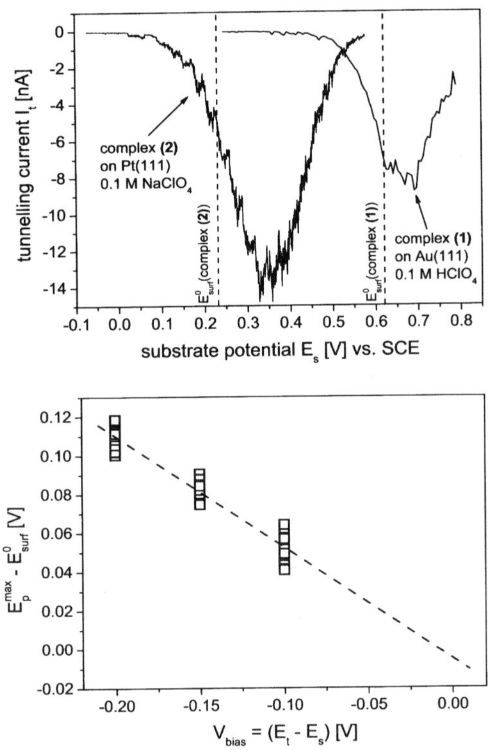

Figure 8-9. Top: Tunneling current/overpotential correlations of the two osmium complexes in Fig. 8-8. Constant bias voltage V_{bias} (= E_{tip}-$E_{substrate}$) = -0.15 V; 0.1 M $NaClO_4$ and 0.1 M $HClO_4$. Bottom: Effect of the bias voltage on the peak maximum (relative to the equilibrium potential of the surface-immobilized complexes, E^0_{surf}) of the tunneling current/overpotential relations, cf. eqns. (8-21) and (8-22). Data from refs. 134 and 135.

In situ STM tunneling spectroscopic data, *i.e.* tunneling current/overpotential correlations are shown in Fig. 8-9. These data follow, notably, clearly the pattern of sequential two-step interfacial electrochemical electron transfer. Both Os-complexes show a clear maximum which follows closely the equilibrium potentials, cf. eqns. (8-21)-(8-23). These are well separated, at 0.60 and 0.24 V (vs, SHE) for $[Os(bpy)_2(p2p)_2]^{2+/3+}$ and $[Os(bpy)_2(p0p)Cl]^{+/2+}$, respectively. The "on-off" current ratios are about 50 for the Os-complexes corresponding to about 1 nA current rise. Further, the peak potential follows a linear dependence on the bias voltage, Fig. 8-10, as expected from eqns. (8-21)-(8-23). The slope is -0.5 according with values of the potential distribution parameters ξ close to unity and γ close to zero. The Co-complex shows a much smaller peak current, *i.e.* 5-10 pA current rise instead of the 1 nA rise for the Os-complexes. This points to a significant role of the interfacial electrochemical electron transfer step between the substrate electrode and the redox centre as this step is three orders of magnitude faster for the Os-complexes than for the Co-complex, cf. above.

The transition metal complexes shown in Fig. 8-8 thus display a pattern which follows consistently the concepts of two-step electrochemical tunneling and the theoretical formalism above. Mapping and working principles of redox "switching" and "transistor"-like behaviour close to the single-molecule level of interfacial electrochemical electron transfer have thus been achieved. This can be compared with biological macromolecules addressed below.

4.2 Interfacial Electrochemistry of Redox Metalloproteins and Metalloenzymes Towards the Single-Molecule Level

4.2.1 Surface structures of ordered linker monolayers for protein voltammetry

With a few exceptions, redox metalloprotein voltammetry on gold surfaces is unstable or absent unless either the electrode is modified by chemisorbed monolayers of linker or promoter molecules, or the protein is modified by insertion of non-native amino acid residues. The linker molecules are usually thiol-containing molecules which adsorb strongly on the Au-surface.

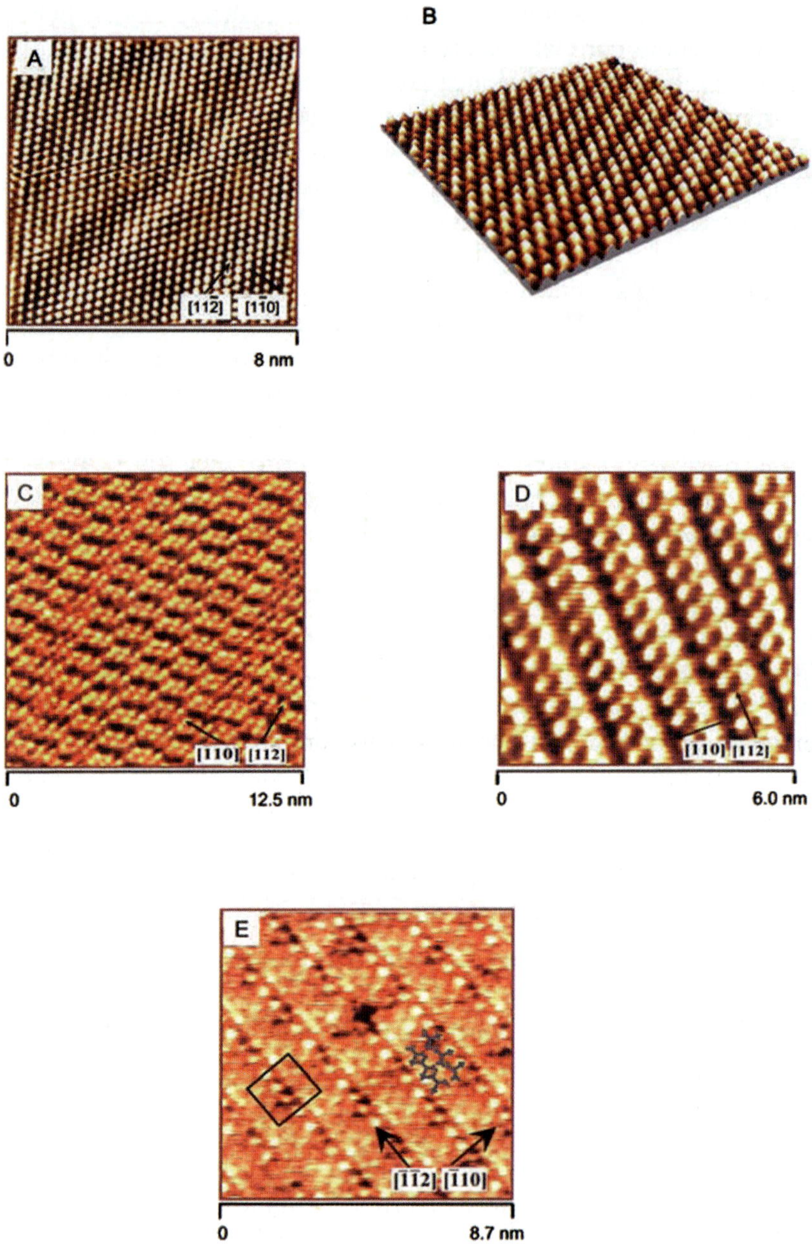

Figure 8-10. Overview of small linker molecules recently characterized to molecular resolution by *in situ* STM on Au(111)-electrode surfaces in aqueous buffer under potential control. A: Bare reconstructed Au(111)-surface. B: $(2\sqrt{3} \times \sqrt{3})R30^{o}$-4 butanethiol monolayer (hydrophobic).[136] C: $(2\sqrt{3} \times 6)R30^{o}$-6 monolayer of mercaptopropionic acid (negatively charged).[137] D: $(\sqrt{3} \times 4)R30^{o}$-2 monolayer of cysteamine (positively charged).[138] E: $(3\sqrt{3} \times 6)R30^{o}$-6 monolayer of cysteine/cystine (zwitterionic).[136]

The opposite end holds a functional group that interacts "gently" with the protein, ascertaining that the protein is immobilized in a well-defined orientation, retaining full electron transfer or enzyme functional integrity. Strategies for specific linker/protein interactions are broadly available but the interactions are often subtle. Closely related linker molecules can induce widely different protein voltammetric responses, and linker groups with no immediate expectable protein compatibility can arouse strong voltammetric signals.

The use of single-crystal electrodes offers both improved voltammetric resolution compared to polycrystalline electrodes, and enables surface structural characterization at the atomic level for pure and at the molecular level for the modified electrodes, directly in aqueous buffer solution. Figure 8-10 shows *in situ* STM images of a clean reconstructed Au(111)-surface and Au(111)-surfaces modified by a variety of thiol-based linker molecules. All of them form highly ordered monolayers on Au(111)-electrode surfaces in aqueous buffer solution. By their different terminal functional groups, the linker molecules have been efficient promoters in protein and enzyme voltammetry. Figure 8-10B shows a $(2\sqrt{3} \times \sqrt{3})R30^{o}$-4 monolayer of butanethiol with a hydrophobic surface.[136] Figure 8-10C shows a $(2\sqrt{3} \times 5)R30^{o}$-6 monolayer of mercaptopropionic acid[137] giving a negatively charged, hydrophilic surface, with a lattice structure composed of clusters of six mercaptopropionic acid molecules. Cysteamine with a terminal amine group gives a highly ordered positively charged $(\sqrt{3} \times 4)R30^{o}$-2 monolayer, Fig. 8-10D.[138] The unit cell contains two molecules giving two different *in situ* STM contrasts. This is supported by SERS studies of cysteamine monolayers on polycrystalline gold by Kudelski *et al.*, who assigned different C-S stretching frequencies to distinct surface conformations, *i.e. trans* and *gauche* conformers of cysteamine.[139] These have also been addressed comprehensively by a combination of molecular dynamics and density functional computations.[138] The two contrasts are assigned to two differently tilted molecular orientations where the strongest contrast arises from the most upright orientation. L-cysteine have shown an efficient promoter molecule for protein voltammetry but is also a natural amino acid and protein building block. The zwitterionic nature of the molecule gives highly ordered domains of a more specific $(3\sqrt{3} \times 6)R30^{o}$-6 structure, controlled by subtle electrostatic and hydrogen bond networks along the surface shown in Fig. 8-10E.[136] Each of the clusters contains *six* cysteine molecules. The surface structure is, notably, indistinguishable from that of the amino acid cystine, indicative that the disulfide group of the latter is cleaved on adsorption.

Most of the linker molecular monolayers are stable over broad potential ranges, limited by reductive and oxidative desorption and cleavage of the

Au-S bond. The adsorption process can be followed in real time through several intermediate phases, such as reported for cysteamine[138]. The *in situ* STM images shown in Fig. 8-10 thus offer an overall impression of the microenvironment for immobilized redox proteins "in voltammetric action".

4.2.2 Metalloprotein voltammetry at bare and modified Au(111)-electrodes

Electrochemical surfaces such as shown in Fig. 8-10, characterized and controlled to the single-molecule and atomically planar levels of structural resolution, represent microenvironments for comprehensive redox metalloprotein and metalloenzyme monolayer electrochemical function. Methodologies for protein electrochemistry have included linear, cyclic, fast-scan, and differential pulse voltammetry, electrochemical impedance and X-ray photoelectron spectroscopy, microcantilever sensor technology, and particularly STM, extending to *in situ* STM under full electrochemical potential control. Figure 8-11 shows an overview of some redox proteins studied in our group. The molecules include representatives of the three major classes of electron transfer metalloproteins, *i.e.* the blue copper,[49,77,100,101,140] heme group,[88] and iron-sulfur proteins[137] as well as the redox metallo*enzyme* copper nitrite reductase.[127,140,141] To this add *de novo* designed 4-α-helix proteins without[142] and with a heme group inserted,[143] see also the chapter by Willner and Willner (Chapter 3). Single-crystal, atomically planar bare and modified electrodes were used, and the proteins in their functional states were mapped to single-molecule resolution by *in situ* STM. In one case, *i.e. P. aeruginosa* azurin *in situ* tunneling spectroscopy with clear resonance features was achieved.[140] Other studies of azurin,[102,112,113,144] plastocyanin,[145] and yeast cyt *c*[103] have been reported elsewhere. *P. aeruginosa* azurin and a mutant azurin,[113] yeast (*Saccharomyces cerevisiae*) cyt *c*,[88,103] and a plastocyanin mutant with a disulfide group inserted[145] adsorb directly on Au-surfaces in functional electron transfer states via protein disulfide (azurin, mutant plastocyanin) or a thiol surface group (yeast cyt *c*). In most cases the functional integrity of the proteins is, however, retained much better after Au(111)-surface SAM modification, cf. Fig. (8-10) and below. Voltammetry, electrochemical impedance (azurin), XPS (azurin, yeast cyt *c*), and *in situ* STM (azurin, yeast cyt *c*, ferredoxin, Cu-nitrite reductase, *de novo* synthetic 4-α-helix proteins) have mapped the surface organization under conditions where the proteins are electrochemically functional. Microcantilever sensor technology has disclosed multi-exponential (or non-Langmuir) *kinetics* of yeast cyt *c* adsorption on gold-covered microcantilevers.[88] Intriguingly, yeast cyt *c* adsorbed on gold ultramicroelectrodes gives much more robust voltammetry than adsorption on atomically planar Au(111)-electrodes.[80] Cyt *c* voltammetry in the latter environment is dominated by a cyt *c* form in which the axial methionine ligand is replaced by a non-native ligand. Reasons for this difference can be that the protein surface thiol group is located slightly below the protein surface. Protein binding on an atomically planar surface would therefore be accompanied by conformational distortion in the protein surface region. This distortion could be less extensive on an ultramicroelectrode surface on which protruding atomic Au-surface structures could provide less protein-structure perturbing adsorption sites.

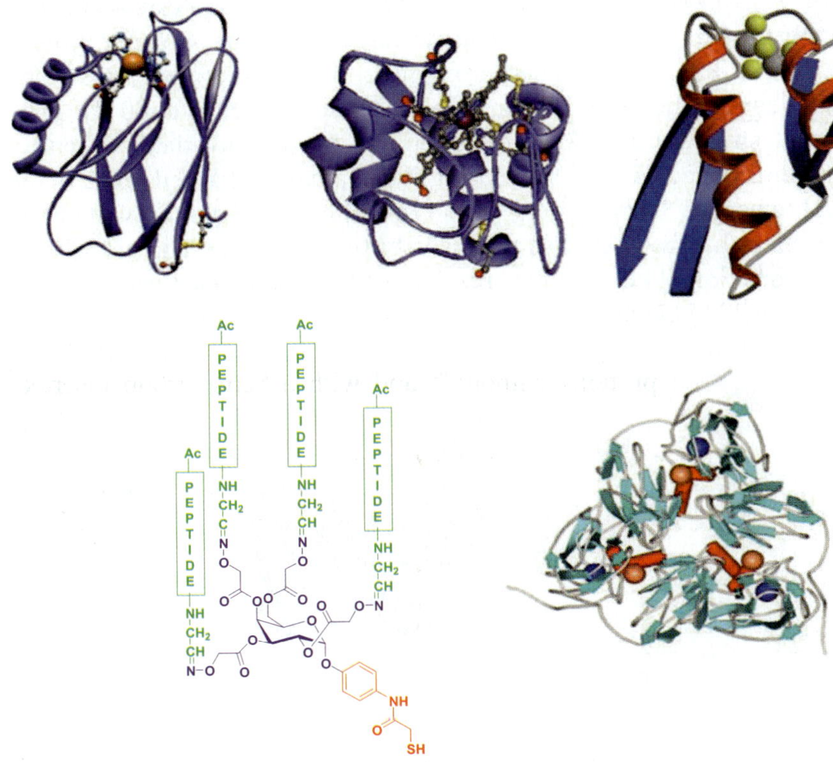

Figure 8-11. Overview of metalloproteins recently characterized by single-crystal-crystal protein film voltammetry and by *in situ* STM to single-molecule resolution. Top left: *P. aeruginosa* azurin.[49,77,100-102, 113, 140,143] Top, middle: *Saccharomyces cerevisiae* cyt *c*.[88,103] Top, right: *P. furiosus ferredoxin*.[137] Bottom, left: 4-α-helix-bundle synthetic protein.[142,143] Bottom right: *A. xylosoxidans* Cu-nitrite reductase.[127]

Robust voltammetry and *in situ* STM to molecular resolution have been achieved when the Au(111)-electrode surfaces are modified by linker molecules, Fig. 8-10, prior to protein adsorption. Comprehensive voltammetric data are available for horse heart cyt *c*[47,48] and *P. aeruginosa* azurin.[49,100,101,113,140,144] The latter protein, which we address in the next Section, has in a sense emerged as a "paradigm" for nanoscale bioelectrochemistry. We address first briefly two other proteins, viz. the electron transfer iron-sulfur protein *Pyrococcus furiosus* ferredoxin and the redox metalloenzyme *Achromobacter xylosoxidans* copper nitrite reductase.

Pyrococcus Furiosus Ferredoxin. *Pyrococcus furiosus* Ferredoxin (*Pf*Fd) is a small protein, with the molecular mass 7.5 kDa and a single [3Fe-4S] redox centre, Fig. 8-12. The thermophilic bacterial protein is stable in anaerobic environments and retains biological activity even at 80 °C. The X-ray crystal structure at 1.5 Å resolution has become available recently.[146] The protein surface is overall strongly negatively charged except for a positively charged "spot" around the [3Fe-4S] centre. Solution voltammetry of the protein has been studied extensively by Armstrong and associates.[147] Here we focus on *Pf*Fd monolayer voltammetry and surface organization at modified Au(111)-electrode surfaces.

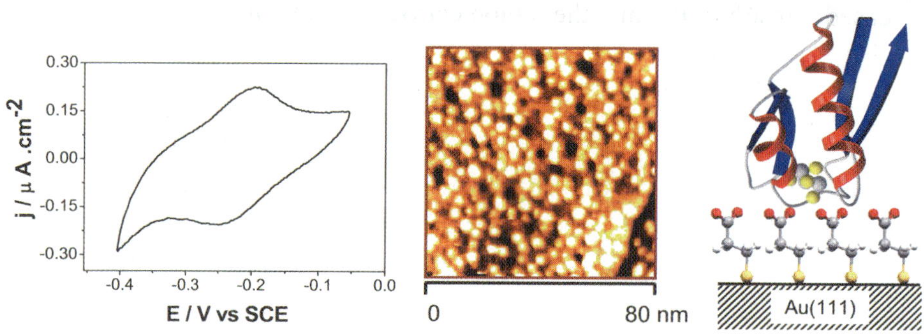

Figure 8-12. Left: Monolayer protein voltammetry of *P. furiosus* ferredoxin on an Au(111)-electrode modified by a mercaptopropionic acid (MPA) SAM, cf, Fig. 8-10C; 5 mM phosphate buffer, pH 7.9. Scan rate 5 mV s^{-1}. Middle: *In situ* STM image of *P. furiosus* ferredoxin molecules in electron transport action on the same MPA-modified Au(111)-electrode surface; Ar-atmosphere. Working electrode potential: -0.35 V (SCE), bias voltage: -0.35 V. Tunneling current 0.10 nA. Right: Schematic view of the *P. furiosus* ferredoxin molecule on the MPA-modified Au(111)-surface. From ref. 137 with permission.

Surface linking of fully active *Pf*Fd is achieved when *Pf*Fd is adsorbed on a highly ordered monolayer of mercaptopropionic acid, MPA, Fig. 8-10C.[137] The *Pf*Fd/MPA/Au(111)-electrode gives a single pair of redox peaks corresponding to the $[3Fe\text{-}4S]^{0/+}$ couple, Fig. 8-12. Suitable modification by binding to promoter molecules in solution induces a second pair of peaks corresponding to the $[3Fe\text{-}4S]^{-1/0}$ transition. Linking of the overall *negatively* charged protein to the *negatively* charged MPA/Au(111)-surface is intriguing and points to local positive charges around the surface [3Fe-4S]-cluster and perhaps local MPA-protonation as determining factors.

Figure 8-12 also shows an *in situ* STM image of the *Pf*Fd adlayer. Electron transferring protein molecules cover the whole surface, as seen by the high density of bright spots. The 3.0-3.5 nm diameter corresponds to the crystallographic dimensions of the *Pf*Fd molecule. The structural surface density accords with the coverage determined by voltammetry. This points to a standing issue in protein film voltammetry, namely the fraction of adsorbed protein molecules that retains voltammetric activity in the adsorbed state. This fraction appears to be about 60 % for *Pf*Fd/MPA/Au(111). In the cases of *A. xylosoxidans* copper nitrite reductase and *P. aeruginosa* azurin the fraction is close to 100 %, cf. below. A small part of the surface-confined *Pf*Fd molecules appears as dimers, perhaps reflecting the dimer structures that appear in the crystal lattice of *Pf*Fd. Fig. 8-12, right shows a schematic view of the *Pf*Fd molecule in action on the MPA/Au(111) surface.

Overall the single-crystal voltammetry, surface promoter sensitivity, and *in situ* STM of *Pf*Fd illuminate procedures and approaches to highly increased resolution in the bioelectrochemical mapping of redox metalloproteins, approaching the electrochemical resolution of small molecular adsorbate molecules.

Achromobacter Xylosoxidans Copper Nitrite Reductase. *A. Xylosoxidans* Copper Nitrite Reductase (CuNiR) is a trimeric metalloenzyme central in the global biological nitrogen cycle where it catalyzes the reduction of nitrite to nitrogen monoxide.[148] Each monomer (MW ≈ 36 kDa) contains a type I blue copper centre for electron inlet and a type II centre for catalytic NO_2^--reduction, Fig. 8-11. The type I centre resembles the type I Cu-centre in azurin but the hydrophobic surface area around the type I centre also holds negatively charged surface residues. The two centres are directly linked covalently via their ligand spheres, offering a facile route for intramolecular electron transfer between the two centres. A final rationale for the choice of CuNiR in efforts towards single-molecule metalloenzyme electrochemistry is that the substrate nitrite is a structurally small molecule, the topography of which will not be detected on the macromolecular enzyme background. What would be detected are electronic structural changes, including changes in the electrode-enzyme contact, induced by nitrite substrate binding to the immobilized enzyme.

Linking of CuNiR to modified Au(111)-electrode surfaces shows a composite pattern.[127,140,141] As an illustration, Fig. 8-13 shows the degree of electrocatalytic efficiency of CuNiR immobilized on Au(111)-electrode surfaces modified by a highly ordered monolayer of cysteamine, Fig. 8-10D. Figure 8-13 also shows an *in situ* STM image of the enzyme "in action" on the same surface as shown in Fig. 8-10D. These data show that strategic selection of linker molecules can provide both a suitable environment for high catalytic efficiency and *in situ* STM imaging of the enzyme in action. The rather low enzyme coverage (3-5 %) accords with the observed limiting catalytic currents and with reported intramolecular electron transfer rate constants of the enzyme in homogeneous solution. This suggests that the single-crystal surface ascertains that *all* the immobilized enzyme molecules have retained their catalytic efficiency in the adsorbed state.

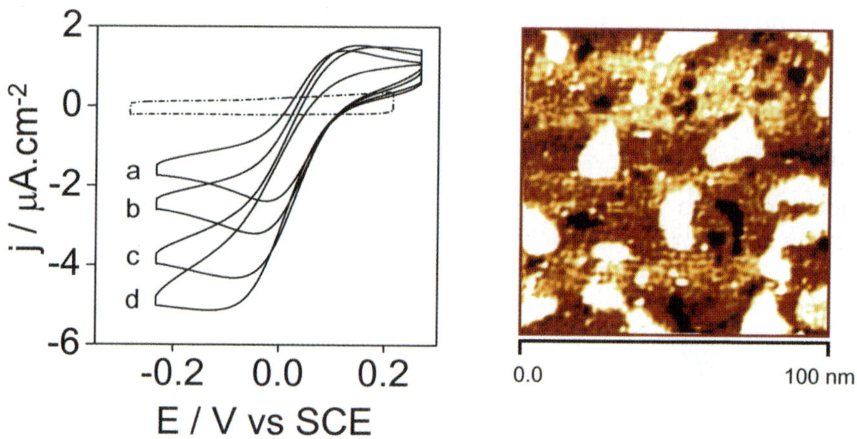

Figure 8-13. Left: Electrocatalysis of *A. xylosoxidans* Cu-nitrite reductase on cysteamine-modified Au(111)-electrode; 5 mM sodium acetate buffer, pH 6.0. Scan rate 10 mV s^{-1}. Dashed line: no KNO_2 present. Solid lines: Potassium nitrite concentrations (μM): a: 70, b: 110, c: 250, d: 800. Right: *In situ* STM of *A xylosoxidans* Cu-nitrite reductase. Same conditions. Potassium nitrite concentration 200 μM. Working electrode potential +0.38 V. Bias voltage -1.10 V. Tunneling current 0.1 nA. From ref. 127 with permission.

As for *Pf*Fd, the combination of voltammetry and *in situ* STM has thus addressed a standing issue in protein film voltammetry, viz. the fraction of immobilized protein molecules that retains full functionality on surface immobilization. Two other observations illuminate this. One is that the enzyme function is more subtle and sensitive to non-native environments than simpler electron transfer function. Retaining full CuNiR catalytic function testifies to the "strategic value" of the chosen cysteamine linker. The other observation is that the individual CuNiR molecules appear with different shapes, Fig. 8-13. Some show the triangular shape expected from the crystallographic structure. Others are round or oval, with little immediate structural correspondence. As, however, electronic conductivity rather than topographical shape is recorded, slightly different adsorption modes give different conductivity patterns. These stochastic effects are expected from quite fundamental principles. The fact that the apparently different orientations then contribute almost equally to the macroscopic electrocatalytic currents is notable. Tip interference could, however, also give such effects.

4.3 Interfacial Electrochemical Organization and Electron Transfer of the Blue Copper Protein Azurin – a Nanoscale Bioelectrochemical Paradigm

In separate ways horse cyt *c* and *P. aeruginosa* azurin have come to emerge as electrochemical "paradigms" for protein interfacial electrochemical electron transfer. Horse heart cyt *c* has been the key target molecule in the pioneering studies by Hill, Armstrong, Bowden, Niki, and Hildebrandt, both in the early stages of protein voltammetry and in later stages, where for example spectroscopic techniques were introduced in metalloprotein electrochemistry. *P aeruginosa* azurin has been a central target protein in recent efforts in protein electrochemistry, where physical electrochemical notions such as single-crystal electrode technology and *in situ* scanning probe microscopy have been in focus. The observations below illustrate the degree of interfacial bioelectrochemical resolution in protein film voltammetry achieved by combining well-defined interfacial electrochemical environments with suitably chosen target metalloproteins.

4.3.1 *P. aeruginosa* azurin on bare Au(111)-electrode surfaces

P. aeruginosa azurin is robust to surface manipulation, at the same time with "generic" natural (bio)chemical function, namely electron transfer. The protein can be immobilized in dense, although not ordered monolayers, directly on single-crystal Au(111)-surfaces. Adsorption is via the surface disulfide group structurally opposite to the Cu-centre, as clearly evidenced by interfacial capacitance and XPS data.[77] The latter display, interestingly, signals both from Au-S surface binding and from the S-ligands around the Cu-centre. The surface organization is illuminated by visually appealing *in situ* STM images, Fig. 8-14.

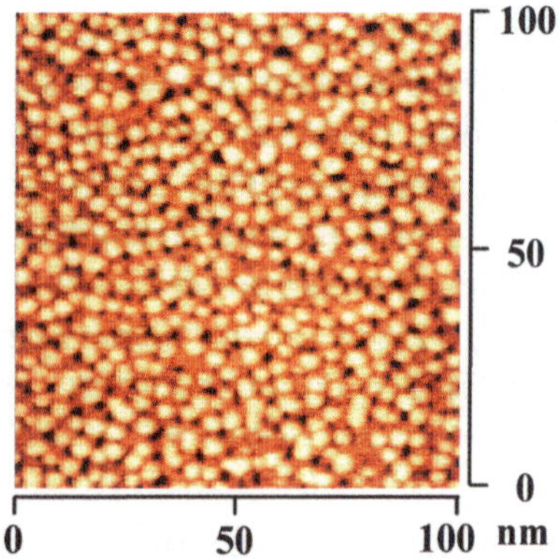

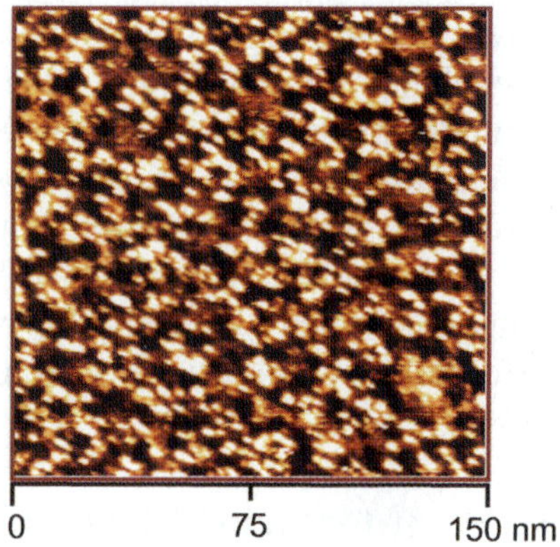

Figure 8-14. Left: *In situ* STM image of *P. aeruginosa* azurin on Au(111)-electrode surface. 20 mM ammonium acetate, pH 4.6. Working electrode potential -0.1 V (SCE), bias voltage 0.2 V. Right: *In situ* STM image of *P. aeruginosa* azurin on Au(111)-electrode surface modified by an undecanethiol monolayer. 20 mM ammonium acetate, pH 4.6. Working electrode potential +0.20 V (SCE), bias voltage -0.25 V. From ref. 140 with permission.

P. aeruginosa azurin retains voltammetric interfacial electron transfer function in the adsorbed state. This is notable as the protein structure is almost certain to be modified on adsorption via the surface disulfide bond. Disulfide bond breaking followed by Au-S bond formation is thus a likely adsorption scenario, but re-assembly of the protein structure via Au-S bond formation to the liberated thiol radicals is unlikely to recover the exact native structure.

Attempts to record *in situ* STM spectroscopy of *P. aeruginosa* azurin adsorbed directly on Au(111)-electrode surfaces have been reported.[118,144,145] While some indications of single-molecule redox features may have been discerned, these data are fraught with inadequate protein monolayer stability in the strongly confined STM gap, or by multifarious and less well-defined adsorption patterns. Several of these obstacles can be rectified by inserting the protein in a gentler, "biomimetic" electrochemical environment. This is strikingly illustrated by the completely different behaviour of *P. aeruginosa* azurin on Au(111)-electrode surfaces modified by highly ordered variable-length alkanethiol SAMs.[49,140] This particular environment offers a unique option for fine-tuning interfacial electrochemical redox metalloprotein performance towards optimal level of function.

4.3.2 *P. aeruginosa azurin* on variable-length alkanethiol modified Au(111)-electrode surfaces

Remarkably robust and "well-behaved" voltammetry is the first sign that Au(111)-electrode surfaces modified by variable-length alkanethiols constitute close to optimal environment for *P. aeruginosa* azurin adlayer interfacial electrochemistry.[49,140] Figure 8-15 discloses what appears as almost "ideal" voltammetry. A certain alkanethiol chain length, say above six or seven carbon atoms is needed for stable voltammetry. Chain lengths between six and ten carbon atoms give, notably, fast reversible monolayer voltammetry, while interfacial kinetic control, still with highly robust voltammetry takes over for longer chain lengths.

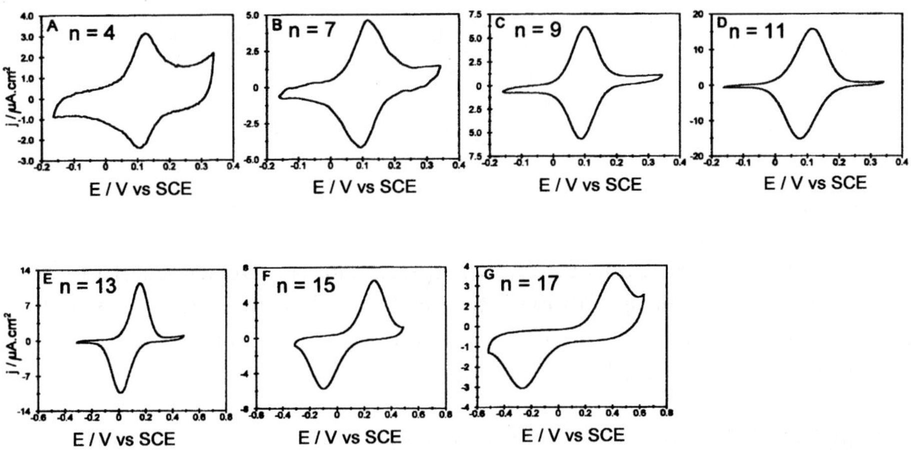

Figure 8-15. P. aeruginosa azurin on variable-length alkanethiol-modified Au(111)-electrodes. Cyclic voltammograms, 20 mM ammonium acetate, pH 4.6. The number of methylene groups (n) from A to G are: 4, 7, 9, 11, 13, 15, and 17. From ref. 49 with permission.

Such voltammetric data warrant kinetic analysis, theoretical framing, and consideration of *P. aeruginosa* azurin in approaches to nanoscale and molecular resolution by *in situ* STM of the protein in electron transfer action on the modified electrode surfaces. Figure 8-14, right shows an *in situ* STM image of the protein adlayer. The bright spots are individual azurin molecules "in action" Fig. 8-16B shows free energy relations from such data analysis. The quadratic relations can be associated with meaningful reorganization free energies, 0.3-0.5 eV, cf. Section 2.1 indicative of, expectably strong electronic coupling to the protein and solvent environment. Figure 8-16A shows a bimodal distance dependence of the interfacial electron exchange rate constant. No distance dependence is detectable below a chain length of about ten carbon atoms. Above this length exponential distance decay is observed. The decay factor is ≈ 1.0 per methylene group, which is strongly indicative that electron tunneling across the alkanethiol layer is now a controlling factor. Adsorption isotherms, and kinetic deuterium isotope effects represent other information available by such high-resolution voltammetry.[140,150]

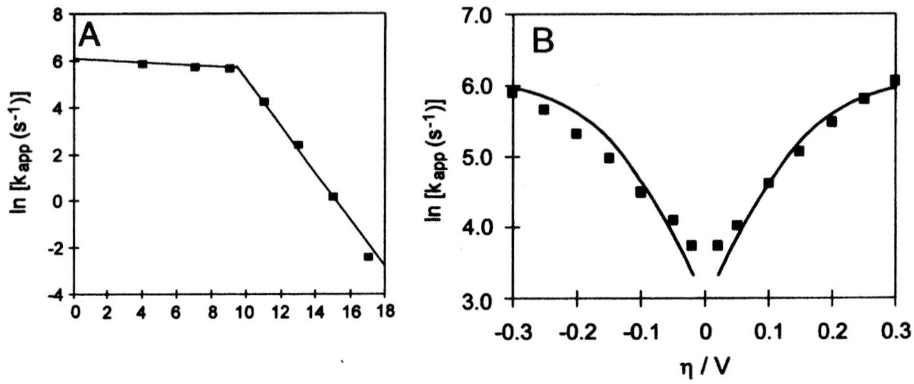

Figure 8-16. A: *P. aeruginosa* azurin on variable-length alkanethiol-modified Au(111)-electrodes. 20 mM ammonium acetate, pH 4.6. Dependence of exchange rate constant (s^{-1}) on the number of CH_2-units. B: *P. aeruginosa* azurin on $CH_3(CH_2)_{13}S$-modified Au(111)-electrode. 20 mM ammonium acetate. Dependence of apparent rate constant (s^{-1}) on overpotential. From ref. 49 with permission.

Comparison of the image in Fig. 8-14 with the voltammograms in Fig. 8-15 adds to the issue of the amount of adsorbed protein molecules that has retained interfacial electron transfer function. In the case of *P. aeruginosa* azurin on undecanethiol-modified Au(111)-electrodes virtually *all* adsorbed protein molecules retains full functionality. The variable-length, intermediate-size (eight-to-eleven carbon atoms) alkanethiols on Au(111)-electrode surfaces thus point to close to optimal function of immobilized *P. aeruginosa* azurin. Perhaps this is rooted in the "biomimetic" environment constituted by a delicate balance between the robust, but still structurally "soft", hydrophobic microenvironment of the intermediate-length Au(111)-alkanethiol layers. The gentle "pad" could thus be a particularly well suited environment for accommodating this hydrophobic functional protein. Similar "surface-padding" has been exploited in other recent studies of azurin mutants which have disentangled other issues of protein/surface linker interactions.[149]

P. aeruginosa azurin on alkanethiol-modified Au(111)-electrode surfaces is, finally, robust enough that single-molecule *in situ* scanning tunneling *spectroscopy* can be recorded. Resonance features of the same electronic origin as discussed in Section 3 and displayed by the Os-complexes discussed in Section 4.1 appear to emerge at the single-molecule level. Figure 8-17 shows a series of *in situ* STM images of a single azurin molecule on top of an Au(111)-electrode surface modified by a highly ordered undecanethiol SAM, recorded at different overpotentials around the equilibrium redox potential. The bias voltage is constant, *i.e.* substrate and tip potentials are varied in parallel as for the Os-complexes.

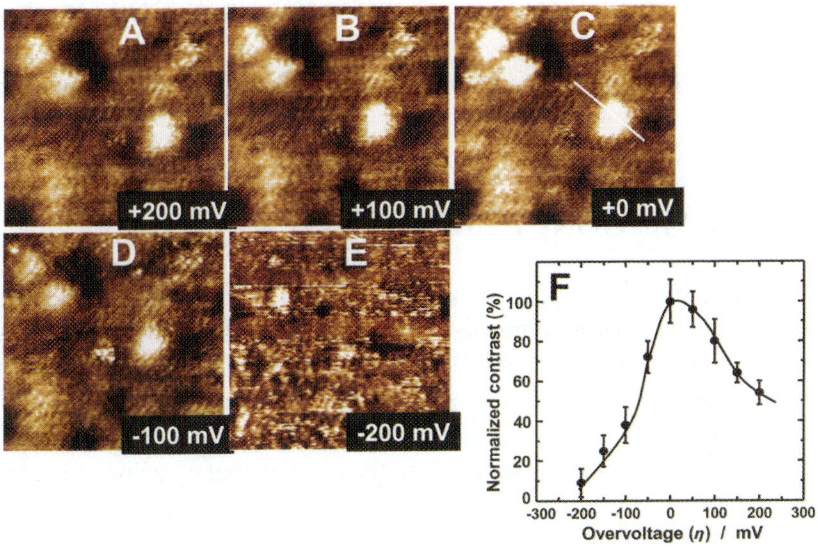

Figure 8-17. Left: Sequence of *in situ* STM images of redox-gated electron tunneling through a single *P. aeruginosa* azurin molecule on top of a highly ordered octanethiol (seen underneath the azurin molecule) on a Au(111)-electrode. 5 mM ammonium acetate buffer, pH 4.6. Bias voltage, V_{bias} = -0.2 V. Variable overpotential as shown. Right: Correlation between the normalized tunneling contrast and overpotential, showing a resonance close to the equilibrium potential. From ref. 150 with permission.

The high structural resolution is demonstrated by the single-molecule features not only of the azurin molecules but also of the ordered alkanethiol monolayer underneath. The tunneling spectra are recorded as normalized "apparent height" in the constant current mode rather than directly as current as for the Os-complexes. A robust resonance feature appears close to the equilibrium redox potential, with a maximum apparent height about nine times higher than the background. The bandshape adds other features such as a reorganization free energy around 0.3 eV, which is not far from the value obtained from the data in Fig. 8-15.

Overall the comprehensive data for electrochemical azurin surface dynamics have pointed towards meritorious properties of this particular redox metalloprotein as a target molecule for combining bioelectrochemistry with single-crystal electrodes and *in situ* scanning probe microscopies. The robust and well-defined protein film voltammetry, high-resolution *in situ* STM imaging, and feasibility of *in situ* scanning tunneling spectroscopy have, secondly, warranted the use of *theoretical* notions in the context of protein dynamics, to an extent comparable to that for transition metal complexes and other "small" molecules. This could apply generally for metalloprotein electrochemistry. Forthcoming efforts could be directed to redox metallo*enzyme* function at the same ultra-small nanoscale and single-molecule levels.

5. SOME CONCLUDING OBSERVATIONS

Single-crystal, atomically planar electrode surfaces have improved bioelectrochemical voltammetric resolution and paved the way for introducing the scanning probe microscopies, STM and AFM. These powerful technologies have increased the structural resolution of the (bio)electrochemical electrode surfaces to the molecular and sometimes sub-molecular levels. High-resolution images have been achieved for the intermediate-size "building blocks" of the biological macromolecules, i.e. DNA nucleobases and amino acids, both of which form highly ordered crystalline monolayers on Au(111)- electrode surfaces. Dynamic phenomena such as potential-induced phase transitions in the adlayers can also be followed. The degree of image detail in both the individual adsorbate molecules and patterns in their lateral organization holds clearly perspectives for the understanding of the interaction of "biological liquids" with solid surfaces.

Single-molecule resolution has been achieved for redox metalloproteins, larger redox metalloenzymes, and DNA-based molecules in surface-confined monolayers on metallic electrode surfaces, where the biological function in "natural" aqueous environment can be "controlled" by the electrochemical potential. Not only *structural* mapping of immobilized redox metalloproteins and DNA-based molecules can be achieved. Given adequate *theoretical* support, ET and redox enzyme *function* can also be addressed at this level of resolution. The Os-complexes and the redox metalloprotein, *P aeruginosa* azurin discussed above have particularly illuminated these perspectives that also offer surface mapping towards biological sensors and other bioelectrochemical systems with potential "device" function.

STM and *in situ* STM are theoretically demanding because the tunneling current through the molecules does not translate directly into molecular *shape* or topography. Long-range off-resonance molecular electronic conductivity is broadly understood in terms of electron exchange and energy gaps of the intermediate atomic or molecular orbitals, and energy broadenings of the orbitals closest to the substrate and tip. Electronic conductivity of molecules with low-lying redox states in nanogap electrode or *in situ* STM configurations displays a quite different pattern, namely sequential two- (or multiple-)step hopping through the redox state(s), induced both by external potential variation and environmental configurational fluctuations. Together with other experimental approaches, redox molecules and metalloproteins have been brought to display interfacial electrochemistry following these patterns at close to the single-molecule resolution. Theoretical notions rest on theory of interfacial electrochemical ET, but the novel environments of *in situ* STM has also disclosed new ET *phenomena.* "Switching" or "negative differential resistance" quite different from electrochemical ET at semi-infinite electrode surfaces is immediately conspicuous. "Coherent", multi-ET in a single *in situ* STM event when the redox level is strongly coupled to the electrodes is another "non-traditional" phenomenon. This chapter has discussed notions and theoretical formalism of these phenomena, and new nanoscale electrochemical and bioelectrochemical systems have been shown to accord with these views.

(Bio)molecular "electronics", enzyme electrochemistry, oligonucleotide monolayer organization, and DNA-based biological screening and hybridization, with levels of resolution towards the single molecule are exciting forthcoming perspectives. "Networks" of hybrid biomolecular structures in electrochemical nanogap electrode arrays, *in situ* STM, and coupled to nanoparticles and nanotubes, are novel targets. In a biotechnological perspective, *fundamental* bioelectrochemical innovation with theoretical support remains, however, a pre-requisite, where the accumulated mapping of redox metalloprotein and metalloenzyme structure and function in novel environments is crucial. Bioelectrochemistry and bioinorganic electrochemistry are thus presently in interesting stages of development.

ACKNOWLEDGEMENT

This work has been supported by the Danish Technical Science Research Council (Danish Research Council for Technology and Production, Contract no 26-00-0034 and the Russian Foundation for Basic Research (Grant No. 06-03-32193).

REFERENCES

1. *Non-traditional Approaches to the Study of the Solid-electrolyte Interface*, Furtak, T.E.; Kliewer, K.L.; Lynch, D.W., Eds., *Surf. Sci.*, **1980**, *101*.
2. *Interfacial Electrochemistry. Theory, Experiment and Applications*, Wieckowski, A., Ed., M. Dekker: New York, 1999.
3. Hubbard, A.T., *Acc. Chem. Res.* **1980**, *13*, 177-184.
4. Clavilier, J.; Faure, R.; Guinet, G.; Durand, R., *J. Electroanal. Chem.* **1980**, *107*, 205-209.
5. (a) Hamelin, A.; *J. Electroanal. Chem.* **1996**, *407*, 1-11. (b) Amelin, A.; Martins, A.M., *J. Electroanal. Chem.* **1996**, *407*, 13-21.
6. Furtak, T.E.; Lynch, D.W., *J. Electroanal. Chem.* **1977**, *79*, 1-17.
7. Kolb, D.M.; Kötz, R., *Surf. Sci.* **1977**, *64*, 96-108.
8. Bewick, A., *J. Electroanal. Chem.* **1983**, *150*, 481-493.
9. Hansen, W.N.; Osteryoung, R.A.; Kuwana, T., *J. Am. Chem. Soc.* **1966**, *88*, 1062-1063.
10. Furtak, T.E.; *J. Electroanal. Chem.* **1983**, *150*, 375-388.
11. Jeanmaire, D.L.; Van Duyne, R.P., *J. Electroanal. Chem.* **1977**, *84*, 1-20.
12. Nuzzo, R.G.; Allara, D.L., *J. Am. Chem. Soc.* **1983**, *105*, 4481- 4483.
13. Schumacher, R., *Angew. Chem. Int. Ed. Engl.* **1990**, *29*, 329-343.
14. Carnie, S.L.; Torrie, G.M., *Adv. Chem. Phys.* **1984**, *56*, 141-253.
15. Henderson, D.; Blum, L., *J. Electroanal. Chem.* **1982**, *132*, 1-13.
16. Badiali, J.P.; Rosinberg, M.R.; Goodisman, J., *J. Electroanal. Chem.* **1981**, *130*, 31-45.
17. Badiali, J.P., *Electrochim. Acta* **1986**, *31*, 149-154.
18. Kornyshev, A.A., In *The Chemical Physics of Solvation. Part C. Solvation Phenomena in Specific Physical, Chemical and Biological Systems*, Dogonadze, R.R.; Kálmán, E.; Kornyshev, A.A.; Ulstrup, J., Eds., Elsevier: Amsterdam, 1988, pp. 355-400.
19. Lustenberger, P.; Rohrer, H.; Christoph, R.; Siegenthaler, H., *J. Electroanal. Chem.* **1988**, *243*, 225- 235.
20. Wiechers, J.; Twomey, T.; Kolb, D.M.; Behm, R.J., *J. Electroanal. Chem.* **1988**, *248*, 451-460.
21. Engel, A.; Lyubchenko, Y.; Müller, D., *Trends Cell Biol.* **1999**, *9*, 77-80.
22. Gewirth, A.A.; Niece, B.K., *Chem. Rev.* **1997**, *97*, 1129-1162.
23. Kolb, D.M., *Angew. Chem. Int. Ed. Engl.* **2000**, *40*, 1162-1181.
24. Kuznetsov, A.M., *Charge Transfer in Physics, Chemistry and Biology*, Gordon & Breach: Reading, 1995.
25. Kuznetsov, A.M.; Ulstrup, J., *Electron Transfer in Chemistry and Biology: An Introduction to the Theory*, Wiley: Chichester, 1999.
26. Schmickler, W., *Interfacial Electrochemistry*, Oxford University Press: New York, 1996.

27. Marcus, R.A. and Sutin, N., *Biochim. Biophys. Acta* **1985**, *811*, 265-322.
28. *Isotope Effects in Chemistry and Biology*, Kohen, A., Limbach, H.-H., Eds., CRC Press: Boca Raton, 2005.
29. Dogonadze, R.R.; Chizmadzhev, Yu.I., *Dokl. Akad. Nauk SSSR, Ser. Fiz. Khim.* **1962**, *144*, 1077-1080; **1962**, *145*, 848-851.
30. Dogonadze, R.R., In *Reactions of Molecules at Electrodes*, Hush, N.S., Ed., Wiley: London, 1971, pp. 135-227.
31. Marcus, R.A., *J. Chem. Phys.* **1956**, *24*, 966-978.
32. German, E.D.; Kuznetsov, A.M., *Electrochim. Acta* **1981**, *26*, 1595-1608.
33. Kharkats, Yu.I.; Krishtalik, L.I., *J. Theor. Biol.* **1984**, *112*, 221-249.
34. Krishtalik, L.I.; Topolev, V.V., *Biochim. Biophys. Acta* **2003**, *1459*, 88-105.
35. Newton, M.D., *Adv. Chem. Phys.* **1999**, *106*, 303-375.
36. Kornyshev, A.A., In *The Chemical Physics of Solvation. Part A. Theory of Solvation*, Dogonadze, R.R.; Kálmán, E.; Kornyshev, A.A.; Ulstrup, J., Eds., Elsevier: Amsterdam, 1985, pp. 77-118.
37. Dzhavakhidze, P.G.; Kornyshev, A.A.; Krishtalik, L.I., *J. Electroanal. Chem.* **1987**, *228*, 329-346.
38. Iversen, G.; Kharkats, Yu.I.; Ulstrup, J., *Mol. Phys.* **1998**, *94*, 297-306.
39. Wolynes, P.G., *J. Chem. Phys.* **1987**, *86*, 5133-5136.
40. Dogonadze, R.R.; Kuznetsov, A.M., *J. Electroanal. Chem.* **1975**, *65*, 545-554.
41. Jensen, T.J., Gray, H.B.; Winkler, J.R.; Kuznetsov, A.M.; Ulstrup, J., *J. Phys. Chem. B* **2000**, *104*, 11556-11562.
42. Kornyshev, A.A.; Ulstrup, J., *Chem. Phys. Lett.* **1986**, *126*, 76-80.
43. Marcus, R.A., *Can. J. Chem.* **1959**, *37*, 155-163.
44. Phelps, D.K.; Kornyshev, A.A.; Weaver, M.J., *J. Phys. Chem.* **1990**, *94*, 1454-1463.
45. Kharkats, Yu.I.; Nielsen, H.; Ulstrup, J., *J. Electroanal. Chem.* **1984**, *169*, 47-57.
46. Liu, Y.P.; Newton, M.D., *J. Phys. Chem.* **1994**, *98*, 7162-7169.
47. Avila, A.; Gregory, B.W.; Niki, K.; Cotton, T.M., *J. Phys. Chem. B* **2000**, *104*, 2759-2766.
48. Murgida, D.H.; Hildebrandt, P., *J. Am. Chem. Soc.* **2001**, *123*, 4062-4068.
49. Chi, Q.; Zhang, J.; Andersen, J.E.T.; Ulstrup, J., *J. Phys. Chem. B* **2001**, *105*, 4669-4679.
50. Kharkats, Yu.I., Ulstrup, J., *Chem. Phys.* **1990**, *141*, 117-129.
51. Kuznetsov, A.M.; Ulstrup, J., *J. Chem. Phys.* **1981**, *75*, 2047-2055.
52. Larsson, S., *J. Am. Chem. Soc.* **1981**, *103*, 4034-4040.
53. Christensen, H.E.M.; Conrad, L.S.; Mikkelsen, K.V.; Nielsen, M.K.; Ulstrup, J., *Inorg. Chem.* **1990**, *29*, 2808-2816.
54. Gray, H.B.; Winkler, J.R., *Quart. Rev. Biophys.* **2003**, *36*, 341-372.
55. Skourtis, S.S.; Beratan, D.N., *Adv. Chem. Phys.* **1999**, *106*, 377-452.
56. McConnell, H., *J. Chem. Phys.* **1961**, *35*, 508-57.
57. Muscat, J.P.; Newns, D.M., *Progr. Surf. Sci.* **1978**, *9*, 1-43.
58. Kuznetsov, A.M., *J. Electroanal. Chem.* **1983**, *159*, 241-255.
59. Kornyshev, A.A.; Kuznetsov, A.M.; Nielsen, J.U.; Ulstrup, J., *Phys. Chem. Chem. Phys.* **2000**, *2*, 141-144.
60. Dogonadze, R.R.; Kuznetsov, A.M.; Levich, V.G., *Electrochim. Acta* **1968**, *13*, 1025-1043.
61. Levich, V.G.; Dogonadze, R.R.; German, E.D.; Kuznetsov, A.M.; Kharkats, Yu. I., *Electrochim. Acta* **1970**, *15*, 353-370.
62. Kuznetsov, A.M.; Ulstrup, J., *Can. J. Chem.* **1999**, *77*, 1085-1096.
63. Meyer, M.P.; Klinman, J.P., *Chem. Phys.* **1994**, *188*, 131-141.

64. Hammes-Schiffer, S.; Iordanova, N., *Biochim. Biophys. Acta, Bioener.*, **2004**, *1655*, 29-36.
65. Kiefer, P.M.; Hynes, J.T., *J. Phys. Chem. A* **2004**, *108*, 11793-11808; *J. Phys. Chem. A* **2004**, *108*, 11809-11818.
66. Siebrand, W.; Wildman, T.A., *Acc. Chem. Res.* **1986**, *19*, 238-243.
67. Kohen, A.; Klinman, J.P., *Acc. Chem. Res.* **1998**, *31*, 397-404.
68. Kornyshev, A.A.; Kuznetsov, A.M.; Spohr, E.; Ulstrup, J., *J. Phys. Chem. B* **2003**, *107*, 3351-3366.
69. Eikerling, M.; Kornyshev, A.A.; Kuznetsov, A.M.; Ulstrup, J., *J. Phys. Chem. B* **2001**, *105*, 3646-3662.
70. Bell, R.P., *The Proton in Chemistry*, 2nd Ed., Chapman and Hall: London, 1973.
71. Agmon, N., *Chem. Phys. Lett.* **1995**, *244*, 456-462.
72. Kuznetsov, A.M.; Ulstrup, J., *Russ. J. Electrochem.* **2004**, *40*, 1000-1009; *Russ. J. Electrochem.* **2004**, *40*, 1010-1018.
73. Hill, H.A.O., *Coord. Chem. Rev.* **1996**, *151*, 115-123.
74. Heering, H.A.; Hirst, J.; Armstrong, F.A., *J. Phys. Chem. B* **1998**, *102*, 6889-6902.
75. Armstrong, F.A., *J. Chem. Soc. Dalton Trans.* **2002**, 661-671.
76. Hirst, J., Armstrong, F.A., *Anal. Chem.* **1998**, *70*, 5062-5071.
77. Chi, Q.; Zhang, J.; Nielsen, J.U.; Friis, E.P.; Chorkendorff, I.; Canters, G.W.; Andersen, J.E.T.; Ulstrup, J., *J. Am. Chem. Soc.* **2000**, *122*, 4047-4055.
78. Bard, A.J.; Faulkner, L.R., *Electrochemical Methods: Fundamentals and Applications*, 2nd Ed., Wiley: New York, 2000.
79. Feng, Z.Q.; Imabayashi, S.; Kakiuchi, T.; Niki, K., *J. Electroanal. Chem.* **1995**, *394*, 149-154.
80. Heering, H.A.; Wiertz, F.G.M.; Dekker, C.; De Vries, S., *J. Am. Chem. Soc.* **2004**, *126*, 11103-11112.
81. Lamie, S.E.; Albracht, S.; Armstrong, F.A., *J. Am. Chem. Soc.* **2004**, *126*, 14899-14909.
82. Ataka, K.; Heberle, *J. Am. Chem. Soc.* **2003**, *125*, 4986-4987.
83. Murgida, D.H.; Hildebrandt, P., *Acc. Chem. Res.* **2004**, *37*, 854-861.
84. Seibert, M.; Picorel, R.; Kim, J.H.; Cotton, T.M., *Meth. Enzymol.* **1992**, *213*, 31-42.
85. Corn, R.M.; Higgins, D.A., *Chem. Rev.* **1994**, *94*, 107-125.
86. Knoll, W., *Ann. Rev. Phys. Chem.* **1998**, *49*, 569-638.
87. Brockman, J.M.; Nelson, B.P.; Corn, R.M., *Ann. Rev. Phys. Chem.* **2000**, *51*, 41-63.
88. Hansen, A.G.; Boisen, A.; Nielsen, J.U.; Wackerbarth, H.; Chorchendorff, I.; Andersen, J.E.T.; Zhang, J.; Ulstrup, J., *Langmuir* **2003**, *19*, 3419-3427.
89. Alessandrini, A.; Gerunda, M.; Facci, P.; Schnyder, B.; Kötz, R., *Surf. Sci.* **2003**, *542*, 64-71.
90. Faraday Disc. **2000**, *116*.
91. Gilardi, G.; Fantuzzi, A.; Sadeghi, S.J., *Curr. Opin. Struct. Biol.* **2001**, *11*, 491-499.
92. Chen, K.S.; Hirst, J.; Camba, R.; Bonagura, C.A.; Stout, C.D.; Burgess, B.K.; Armstrong, F.A., *Nature* **2000**, *405*, 814-817.
93. Davis, J.J.; Bruce, D.; Canters, G.W.; Crozier, J.; Hill, H.A.O., *Chem. Commun.* **2003**, 576-577.
94. Chen, X.X.; Discher, B.M.; Pilloud, D.L.; Gibner, B.R.; Moser, C.C.; Dutton, P.L., *J. Phys. Chem. B* **2002**, *106*, 617-624.
95. Willner, I.; Heleg-Shabtai, V.; Katz, E.; Rau, H.K.; Haehnel, W., *J. Am. Chem. Soc.* **1999**, *121*, 6455-6468.
96. Albrecht, T.; Li, W.W.; Ulstrup, J.; Haehnel, W.; Hildebrandt, P., *ChemPhysChem* **2005**, *6*, 961-970.

97. Chi, Q.; Zhang, J.; Friis, E.P.; Andersen, J.E.T.; Ulstrup, J., *Electrochem. Commun.* **1999**, *1*, 91-96.
98. Andersen, J.E.T.; Møller, P.; Pedersen, M.V.; Ulstrup, J., *Surf. Sci.* **1995**, *325*, 193-205.
99. Zhang, J.; Chi, Q.; Dong, S.; Wang, E., *Bioelectrochem. Bioenerg.* **1996**, *39*, 267-274.
100. Friis, E.P.; Andersen, J.E.T.; Madsen, L.L.; Møller, P.; Ulstrup, J., *J. Electroanal. Chem.* **1997**, *431*, 35-38.
101. Friis, E.P.; Andersen, J.E.T.; Kharkats, Yu.I.; Kuznetsov, A.M.; Nichols, R.J.; Zhang, J.; Ulstrup, J., *Proc. Nat. Acad. Sci. USA* **1999**, *96*, 1379-1384.
102. Facci, P.; Alliata, D.; Cannistrarro, S., *Ultramicroscopy* **2001**, *89*, 291-298.
103. Bonanni, B.; Alliata, D.; Bizzari, A.R.; Cannistraro, S., *ChemPhysChem* **2003**, *4*, 1183-1188.
104. Willner, I.; Katz, E., *Angew. Chem. Int. Ed. Engl.* **2000**, *39*, 1180-1218.
105. Kuznetsov, A.M.; Ulstrup, J., *Chem. Phys.* **1991**, *157*, 25-33.
106. Kuznetsov, A.M.; Sommer-Larsen, P., Ulstrup, J., *Surf. Sci.* **1992**, *275*, 52-64.
107. Widrig, C.; Schmickler, W., *J. Electroanal. Chem.* **1992,** *336*, 213-221.
108. Friis, E.P.; Andersen, J.E.T.; Madsen, L.L.; Møller, P.; Nichols, R.J.; Olesen, K.G.; Ulstrup, J., *Electrochim. Acta* **1998**, *43*, 2889-2897.
109. Liang, W.; Shores, M.P.; Bockrath, M.; Long, J.R.; Park, H., *Nature* **2002**, *417*, 725-727.
110. Park, J.; Pasupathy, A.N.; Goldsmith, J.I.; Chang, C.; Yaish, Y.; Petta, J.R.; Rinkoski, M.; Sethna, J.P.; Abruna, H.D.; McEuen, P.L.; Ralph, D.C., *Nature* **2002**, *417*, 722-725.
111. Kubatkin, S.; Danilov, A.; Hjort, M.; Cornil, J.; Brédas, J.-L.; Stuhr-Hansen, N.; Hedegård, P.; Bjørnholm, T., *Nature* **2003**, *425*, 698-701.
112. Davis, J.J.; Hill, H.A.O.; Bond, A.M., *Coord. Chem. Rev.* **2000,** *200*, 411-442.
113. Davis, J.J.; Hill, H.A.O., *Chem. Commun.* **2002**, 303-401.
114. Gittins, G.I.; Bethell, D.; Schiffrin, D.J.; Nichols, R.J., *Nature* **2000**, *408*, 67-69.
115. Kornyshev, A.A.; Kuznetsov, A.M.; Ulstrup, J., *ChemPhysChem* **2005**, *6*, 583-586.
116. Kuznetsov, A.M.; Ulstrup, J., *J. Phys. Chem. A* **2000**, *104*, 11531-11540. Errata: *J. Phys. Chem. A* **2001**, *105*, 7494.
117. Zhang, J.; Chi, Q.; Kuznetsov, A.M.; Hansen, A.G.; Wackerbarth, H.; Christensen, H.E.M.; Andersen, J.E.T.; Ulstrup, J., *J. Phys. Chem. B* **2001**, *105*, 1131-1152.
118. Zhang, J.; Kuznetsov, A.M.; Ulstrup, J., *J. Electroanal. Chem.* **2003**, *541*, 133-146.
119. Haiss, W.; Nichols, R.J.; Van Zalinge, H.; Higgins, S.J.; Bethell, D.; Schiffrin, D.J., *Phys. Chem. Chem. Phys.* **2004**, *6*, 4330-4337.
120. Armstrong, F.A., *Curr. Opin. Chem. Biol.* **2005**, *9*, 110-117.
121. Heller, A., *Phys. Chem. Chem. Phys.* **2004**, *6*, 209-216 and references cited therein.
122. Gorton, L.; Bremle, G.; Csöregi, E.; Jonsson, G.; Petterson, B., *Anal. Chim. Acta* **1991**, *249*, 43-54.
123. Heering, H.A.; Weiner, J.H.; Armstrong, F.A., *J. Am. Chem. Soc.* **1997**, *119*, 11628-11638.
124. Leger, C.; Heffron, K.; Pershad, H.R.; Maklashina, E.; Luna-Chavez, C.; Cecchini, G.; Ackrell, B.A.C.; Armstrong, F.A., *Biochemistry* **2001**, *40*, 11234-11245.
125. Anderson, L.J.; Richardson, D.J.; Butt, J.N., *Biochemistry* **2001**, *40*, 11294-11307.
126. Angove, H.C.; Cole, J.A.; Richardson, D.J.; Butt, J.N., *J. Biol. Chem.* **2002**, *277*, 23374-23381.
127. Zhang, J.; Welinder, A.C.; Christensen, H.E.M.; Ulstrup, J., *J. Phys.Chem. B* **2003**, *107*, 12480-12484.
128. Mondal, M.S.; Fuller, H.A.; Armstrong, F.A., *J. Am. Chem. Soc.* **1996**, *118*, 263-264.
129. Ferantopova, E.; Gorton, L., *Electroanalysis* **2003**, *15*, 484-491 and references cited therein.

130. Heffron, K.; Léger, C.; Rothery, R.A.; Weiner, J.H.; Armstrong, F.A., *Biochemistry* **2001**, *40*, 3117-3126.
131. Armstrong, F.A.; Albracht, P.J., *Phil. Trans. Roy. Soc. A. Math., Phys. Eng. Sci.* **2005**, *363*, 937-954.
132. Haas, A.S.; Pilloud, D.L.; Reddy, K.S.; Babcock, G.T.; Moser, C.C.; Blasie, J.K.; Dutton, P.L., *J. Phys. Chem. B* **2001**, *105*, 11351-11362.
133. Tao, N., *Phys. Rev. Lett.* **1996**, *76*, 4066-4069.
134. Albrecht, T.; Guckian, A.; Ulstrup, J.; Vos, J.G., *IEEE Trans. on Nanotechnology* **2005**, *4*, 430-434.
135. Albrecht, T.; Guckian, A.; Ulstrup, J.; Vos, J.G., *Nano Lett.* **2005**, *5*, 1451-1455.
136. Zhang, J.; Chi, Q.; Nielsen, J.U.; Friis, E.P.; Andersen, J.E.T., Ulstrup, J., *Langmuir* **2000**, *16*, 7229-7237.
137. Zhang, J.; Christensen, H.E.M.; Ooi, B.L.; Ulstrup, J., *Langmuir* **2004**, *20*, 10200-10207.
138. Zhang, J.; Bilíc, A.; Reimers, J.R.; Hush, N.S.; Ulstrup, J., *J. Phys. Chem. B.* **2005**, *109*, 15355-15367.
139. Kudelski, A., *J. Raman Spectr.* **2003**, *34*, 853-862.
140. Chi, Q.; Zhang, J.; Jensen, P.S.; Welinder, A.C.; Hansen, A.G.; Christensen, H.E.M.; Ulstrup, J., *Faraday Disc.* **2006**, *131*, 181-195.
141. Hansen, A.G.; Zhang, J.; Christensen, H.E.M.; Welinder, A.C.; Wackerbarth, H.; Ulstrup, J., *Israel J. Chem.* **2004**, *44*, 89-100.
142. Brask, J.; Wackerbarth, H.; Jensen, K.J.; Chorkendorff, I.; Zhang, J.; Ulstrup, J., *J. Am. Chem. Soc.* **2003**, *125*, 94-124.
143. Albrecht, T.; Li, W.; Haehnel, W.; Hildebrandt, P.; Ulstrup, J., *Bioelectrochemistry* **2006**, *69*, 193-200.
144. Alessandrini, A.; Gerunda, M.; Canters, G.W.; Verbeet, M.; Facci, P., *Chem. Phys. Lett.* **2003**, *376*, 625-630.
145. Andolfi, L.; Bonanni, B.; Canters, G.W.; Verbeet, M.P.; Cannistraro, S., *Surf. Sci.* **2003**, *530*, 181-194.
146. Nielsen, M.S.; Harris, P.; Ooi, B.L.; Christensen, H.E.M., *Biochemistry* **2004**, *43*, 5188-5194.
147. Fawcett, S.E.J.; Davis, D.; Breton, J.L.; Thomson, A.J., Armstrong, F.A., *Biochem. J.* **1998**, *335*, 357-368.
148. Adman, E.T.; Murphy, M.E.P., In *Handbook of metalloproteins*, Messerschmidt, A.; Huber, R.; Poulos, T.; Wieghardt, K., Eds., Wiley: Chichester, Vol.2, pp. 1381-1390.
149. Fujita, K.; Nakamura, N.; Ohno, H.; Niki, K.; Gray, H.B.; Richards, J.H., *J. Am. Chem. Soc.* **2004**, *126*, 13954-13961.
150. Chi, Q.; Farver, O.; Ulstrup, J., *Proc. Nat. Acad. Sci. USA* **2005**, *102*, 16203-16208.

Index